Advances in Grape and Wine Biotechnology

Edited by Antonio Morata and Iris Loira

Published in London, United Kingdom

IntechOpen

Supporting open minds since 2005

Advances in Grape and Wine Biotechnology
http://dx.doi.org/10.5772/intechopen.77885
Edited by Antonio Morata and Iris Loira

Contributors
Luís Filipe-Ribeiro, Fernanda Cosme, Fernando M. Nunes, Laurent Torregrosa, Charles Romieu, Anne Pellegrino, Markus Rienth, Antonio Morata, Iris Loira, Juan Manuel Del Fresno, Carlos Escott, María Antonia Bañuelos, Wendu Tesfaye, María Carmen González, Felipe Palomero, Jose Antonio Suárez Lepe, Pilar Baeza, Mar Vilanova, David Uriarte, Pedro Junquera, Jose Ramón Lissarrague, Ricardo Vejarano, Jackeline León-Vargas, Angie Gil-Calderón, Valeria Díaz-Silva, Francesca Venturi, Angela Zinnai, Chiara Sanmartin, Isabella Taglieri, Giuseppe Ferroni, Mike Frank Quartacci, Cristina Sgherri, Xiaoguo Ying, Gianpaolo Andrich, Gizella Győrffyné Dr. Jahnke, Zóra Annamária Nagy, Csaba Németh, János Májer, Gyöngyi Knolmajerné Szgeti, Dimitris Tsaltas, Dimitris Anagnostopoulos, Elena Kamilari, Ana Rosa Gutiérrez Viguera, Pilar Santamaría, Rosa López, Maria Del Patrocinio Garijo, Lucia González-Arenzana, Isabel López-Alfaro, Rocío Escribano-Viana, Vesna Maras, Jose Miguel Martinez Zapater, Javier Ibáñez, Pablo Carbonell-Bejerano, Carolina Royo, Nuria Mauri, Matthias Schmitt, Monika Christmann, Creina Stockley, Tamara Bucher, Kristine Deroover, María Arévalo-Villena, Ana Briones, Beatriz García-Béjar, Pilar Fernández-Pacheco, Pepe Heras

First published in London, United Kingdom, 2019 by IntechOpen
IntechOpen is the global imprint of INTECHOPEN LIMITED, registered in England and Wales, registration number: 11086078, The Shard, 25th floor, 32 London Bridge Street
London, SE19SG – United Kingdom
Printed in Croatia

British Library Cataloguing-in-Publication Data
A catalogue record for this book is available from the British Library

Additional hard and PDF copies can be obtained from orders@intechopen.com

Advances in Grape and Wine Biotechnology
Edited by Antonio Morata and Iris Loira
p. cm.
Print ISBN 978-1-78984-612-6
Online ISBN 978-1-78984-613-3
eBook (PDF) ISBN 978-1-78985-699-6

For EU product safety concerns:
IN TECH d.o.o., Prolaz Marije Krucifikse Kozulić 3, 51000 Rijeka, Croatia, info@intechopen.com or visit our website at intechopen.com.

Meet the editors

Antonia Morata is Professor of Food Science and Technology at the Universidad Politécnica de Madrid (UPM), Spain, specializing in wine technology. He is the coordinator of the Master in Food Engineering program at UPM. He is also a professor of enology and wine technology in the European Master of Viticulture and Enology, Euromaster Vinifera-Erasmus+. He is the Spanish delegate at the group of experts in wine microbiology of the International Organization of Vine and Wine (OIV). He has authored more than 60 research articles, two books, three edited volumes, two special issues, and thirteen book chapters.

Iris Loira is Assistant Professor of Food Science and Technology at the Universidad Politécnica de Madrid (UPM), Spain. She has authored twenty-four research articles, two edited books, and eight book chapters.

Contents

Preface

Written by experts from thirteen countries across three continents, *Advances in Grape and Wine Biotechnology* covers the technological and biotechnological management of vineyards and winemaking.

Part 1 of the volume focuses on vine biotechnology and plant technology. In Chapter 1, Prof. Torregrosa et al. describe the scientific use of a microvine as a plant model for vine biotechnology research under carefully controlled conditions, allowing for accelerated physiology and improving molecular biology and genetic studies. Chapter 2, by Prof. Martínez Zapater et al., delves into the somatic variations in vines at phenotypic and genomic sequencing levels and their potential applications in vine biotechnology. In Chapter 3, Dr. Jahnke discusses the use of microsatellite (SSR) markers in the identification of parents to see the evolution or pedigree of varieties. Chapter 4, by Dr. Maras, describes the characterization of Montenegrin varieties by ampelographic and genetic techniques. Finally, in Chapter 5, Profs. Baeza and Lissarrague explain the influence of hydric nutrition on vegetative behavior, plant production, and must and wine composition.

Part 2 focuses on wine biotechnology and yeast applications for wine production. Chapter 6, by Prof. Tsaltas, characterizes the dynamic differentiation of the microbiome from grapes to wine and the possibility of using it as a fingerprint tool to elucidate geographical origin. In Chapter 7, Prof. Gutiérrez et al. use molecular techniques to identify and differentiate yeast populations in several wineries from the Rioja wine region to see if the implantation of specific strains is common. In Chapter 8, Dr. Noble talks about the technique of using selective pressure factors in evolutionary engineering to improve the technological properties in selected yeasts. My team describes in Chapter 9 several biotechnological strategies to improve wine freshness through the use of selected non-*Saccharomyces* yeast species. Finally, in Chapter 10, Prof. Briones et al. suggest the interesting application of distillation by-products as a source of yeast for selection applications and the use of these strains in wine biotechnology. Chapter 11 by Prof. Vejarano describes the bioactive profile of wines and the nutraceutical properties of various wine molecules. In chapter 12, Dr. Venturi explains the main bottling, packaging, and closures possibilities and how they affect wine stability in the long term. Chapter 13 by Prof. Ribeiro focuses on the use of cork powder, a natural cork derivative, as a technological adjuvant to be used as adsorbent to improve sensory quality in wines contaminated by *Brettanomyces* and tainted with unacceptable levels of ethyl phenols. In Chapter 14, Profs. Schmitt and Christmann describe the physical technologies used to remove ethanol to obtain high-quality, low-alcohol wines. The final chapter by Dr. Bucher et al. is dedicated to the production of wines with low alcohol content through the application of viticultural, pre- and post- fermentative strategies to reduce or completely eliminate ethanol. Marketing and labelling procedures to promote these wines are also described.

We hope this book will help students, researchers, and winemakers to better understand the technological and biotechnological tools to improve wine quality.

Antonio Morata and Iris Loira
Universidad Politécnica Madrid (UPM),
Spain

Section 1

Vine Biotechnology

Chapter 1

The Microvine: A Versatile Plant Model to Boost Grapevine Studies in Physiology and Genetics

Anne Pellegrino, Charles Romieu, Markus Rienth and Laurent Torregrosa

Abstract

The microvine is a grapevine somatic variant. The Vvgai1 mutation results in a miniaturization of the vegetative organs of the plant keeping fruit size intact and a systematic conversion of tendrils into inflorescences. The physiological characterization of the vegetative and reproductive development of the microvine makes it possible to infer kinetic data from spatial phenotypes. This biological model allows experiments on vine and grape development in tightly controlled conditions, which greatly accelerate physiology, molecular biology, as well as genetic studies. After introducing the main biological properties of the microvine, main results from various research programs performed with the microvine model will be presented.

Keywords: research tools, microvine, grapevine model, physiology, genetics

1. Introduction

As a perennial fruit crop, the grapevine (*Vitis vinifera*) needs a long juvenile period before the reproductive cycle starts. Even vine cuttings from adult plants allow the production of fruits only from the second year. Moreover, during the adult phase, common cultivars produce reproductive organs only once per growing cycle (generally once per year) and per proleptic axis. These biological features, together with the large size of an adult vine, represent major drawbacks for precise physiological, ecophysiological, and omics experiments on the plant and fruit development under well-controlled conditions. Furthermore, those characteristics of normal vines slow down advances in genetics and breeding.

The microvine ML1 is a somatic variant obtained though somatic embryogenesis from Pinot Meunier cultivar. This phenotype results from a somatic mutation in the *Vvgai1* gene involved in gibberellin signaling. The mutation is originally present at the heterozygous state in the epidermal cells of Pinot Meunier, being responsible for its well-known hairy phenotype. However, the introduction of the mutation in all cell layers resulted in a miniaturization of all vegetative organs and in a conversion of tendrils into inflorescences, which leads to a continuous flowering and fruiting along vegetative axes.

The small size of the microvine renders this grapevine model very convenient for experiments in usual growth chambers, where a tight control of environmental factors (radiation, vapor pressure deficit (VPD), temperature, water and nutrient supplies) is possible, in contrast with experiments under vineyard conditions. Indeed, it is possible to grow the vines up to densities of 15–30 plants/m^2 and to limit their height to 1.2 m. Under such conditions, the most advanced fruits are mature 5–6 months after plantation of cuttings or seedlings, and the vegetative axis displays all developmental stages from young inflorescences (distal phytomers) to flowering, berry growth, and ripening (proximal phytomers). Under stable controlled conditions, the spatial gradients of vegetative and reproductive development of the microvine mimic well the temporal development of each phytomer, which allows to infer kinetic data from one-off spatial information along the proleptic axis.

In controlled conditions, microvine allows to experiment on berry development all year long, which greatly accelerates studies on physiology and molecular biology. Furthermore, by reducing the time lag between two generations and by increasing the precision of phenotyping, genetic approaches are much more efficient than the ones generally performed with macrovines. In the first section of the paper, we describe the genetic and molecular mechanisms underlying the phenotypes of the microvine and derived lines. Then, we review typical experimental designs that can be designed with the microvine. In the last section, we review recent project using this model to study grapevine development and fruit physiology and to identify quantitative trait loci (QTLs) of agronomic traits.

2. Biological origin of the microvine

2.1 Tissue chimerism and phenotypic consequences

The meristem of higher plants is organized in several cell layers. The outermost, which corresponds to epidermal cells, results from anticlinal divisions (i.e., following a plane of division perpendicular to the surface). This tissue which covers all the organs of the shoot system develops as a single cell layer [1]. Underneath, a multicellular zone, called L2 cell layer, is at the origin of all subepidermal tissues, following multidirectional divisions (i.e., primary structures but also lateral meristems, vascular cambium, phellogen, and their derivative tissues). No further, deeper cell layer (L3 cell layer), which forms in some species the core of shoot organs (pith), has been clearly identified in the grapevine yet [2].

In general, these cell lines that derive from initial cells located at the tip of the apical dome do not mix, unless there is an accident during cells multiplication. The organization in L1 and L2 cell layers is found in the various organs that derive from the shoot apical meristem (SAM) and in particular in the axillary meristems at the origin of caulinar organs. Because a somatic mutation is initially a single cellular event, it leads to the setting of chimeric tissues or organs, i.e., composed of cells of different genotypes and potentially displaying some phenotypic diversity [2]. When a somatic mutation appears laterally to a meristem, changes can only be distributed in the sector of the mutated organ. If the mutation occurs in an initial cell of a meristem, it can spread to all the tissues derived from the mutated cell. The resulting structure is a chimeric and periclinal genotype, i.e., including cell layers that are not all genetically identical. Periclinal chimeras can be stabilized by vegetative propagation, i.e., by cuttings or by grafting.

A somatic mutation can invade all the cell layers and spread uniformly to all derivative tissues, provided that the three following conditions are fulfilled: (i) the mutation is not lethal for the plant, (ii) the mutation appears in an initial cell within a meristem, and (iii) the mutation is established, by cell substitution in both L1 and L2 cell layers [2]. The probability of simultaneous occurrence of these three conditions being very low, most of the mutations therefore develop sectorially or periclinally and give rise to chimeric tissues and organs.

In the 1990s, thanks to the use of codominant genetic markers (microsatellites, RFLP), the existence of genetic chimerism has been demonstrated in several vine varieties. As such, Franks et al. [3] showed that Pinot Meunier can display up to three alleles for some loci, whereas a vine, having a diploid genome, can theoretically only show one allelic form per homozygous locus and two allelic forms for a heterozygous locus. Boss and Thomas [4] were able to de-chimerise Pinot Meunier by somatic embryogenesis. They characterized the resulting L1 and L2 genotypes and studied the associated phenotypes. This work showed that Pinot Meunier carries a mutation in *VvGAI1* gene in the L1 layer which confers the hairy phenotype to the variety (**Figure 1**).

Plants regenerated from L1 or L2 cells exhibited very different phenotypes. The plants obtained from the deepest cell layer (L2) no longer had a mutation at *VvGAI1* locus and presented phenotypic traits very close to Pinot Noir. Conversely, the plants derived from L1 cells that retained a mutated version of *Vvgai1* associated with a wild-type allele *VvGAI1* were dwarf and hairy and displayed a full conversion of all tendrils into inflorescences (**Figure 2**). This phenotype has been called microvine, due to the small size of the mutant.

Thus, the microvine has the *Vvgai1* mutation present in both cell layers that confers a very different phenotype from the Pinot Meunier from which it derives and which only bears the mutation in the L1 cell layer. Another interesting feature is related to the genetic status of the mutation in the microvine. Although it is present in both cell layers, the *VvGAI* locus is heterozygous, i.e., each cell is carrying a mutated allele *Vvgai1* is associated with a wild-type allele *VvGAI1*. Because *Vvgai1* is

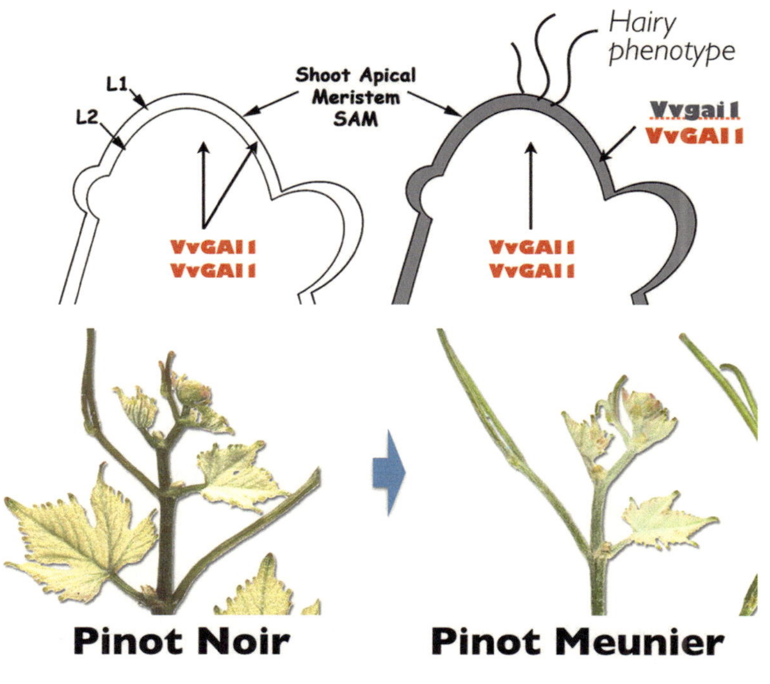

Figure 1.
Genetic structures of pinot noir and pinot Meunier and their respective apex phenotypes. Pinot Meunier is a somatic variant of pinot noir, which carries the mutation (Vvgai1) at heterozygous status. Localized in the epidermal cells (L1 cell layer), the mutation exacerbates the hairiness of vegetative organs of this variety (http://plantgrape.plantnet-project.org/en), without any other significant phenotypic change.

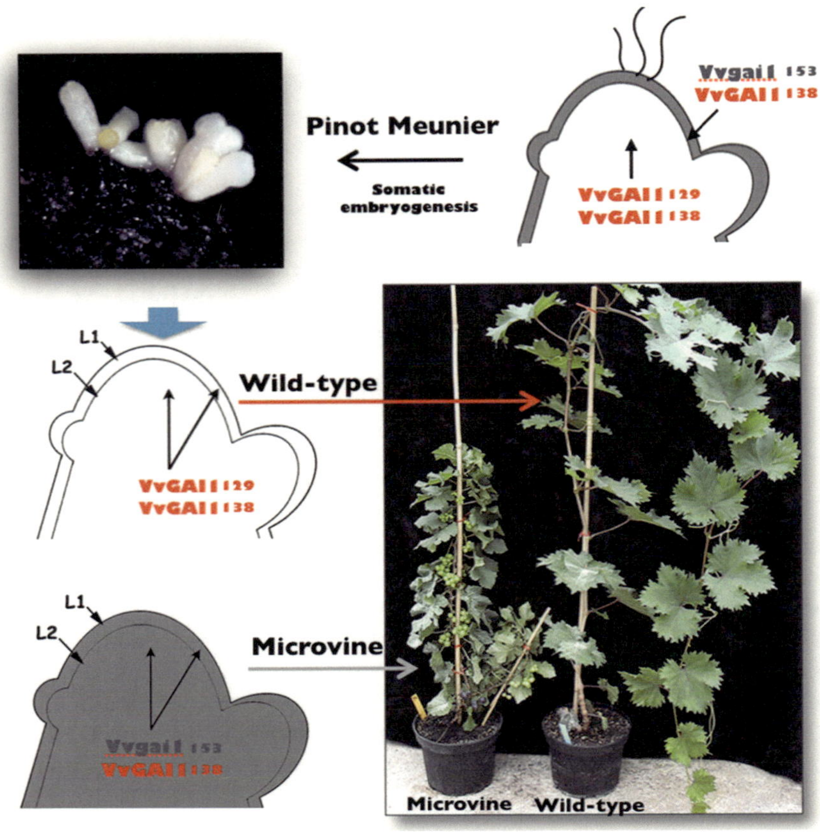

Figure 2.
By somatic embryogenesis from anthers of pinot Meunier, it is possible to obtain two types of plants. One, which no longer carries the mutation of VvGAI in the L1 and L2 cell layers, has a phenotype similar to pinot noir (large size, juvenility period, main production of clusters from proleptic axes, i.e., winter buds). The other, which carries the mutation of VvGAI in all the tissues, displays a miniaturized phenotype and extreme hairiness and produces inflorescences both in the winter buds and from the conversion of tendrils in inflorescences. In the figure, the numbers associated with VvGAI allele correspond to the nucleotide base length (bp) of the VVS2 microsatellite marker [4].

not a lethal mutation nor for the sporophyte or the gametophyte, this status can be rearranged by selfing in three genotypes:

i. Homozygous *VvGAI1/VvGAI1*, which corresponds to a vine without any mutation at the locus. The phenotype associated with this genetic status is non-dwarf, similar to classical macrovine varieties.

ii. Heterozygote *VvGAI1/Vvgai*, which corresponds to the same genotype and (dwarf) phenotype than the original microvine ML1.

iii. Homozygotous *Vvgai1/Vvgai1*, which corresponds to plants carrying both alleles in a mutated version. The phenotype associated with this status, called picovine, corresponds to an extreme dwarfism, with plants displaying very miniaturized shoot organs [4] (**Figure 3**).

Another interesting feature, linked to the heterozygous status *VvGAI/Vvgai1*, is the possibility to return to non-dwarf phenotype. Indeed, by crossing a microvine (*VvGAI1/Vvgai1*) with a classic grapevine variety, i.e., a macrovine (*VvGAI1/*

DOI: http://dx.doi.org/10.5772/intechopen.86166

Figure 3.
The three genotypes/phenotypes that can be obtained by selfing from the microvine (VvGAI1/Vvgai1): left, extremely miniaturized vines that carries the homozygous locus Vvgai1/Vvgai1, called picovines; middle, individuals with the same phenotype as the microvine, heterozygous for the mutation (VvGAI1/Vvgai1); and right, normal-sized plants that no longer carry mutated alleles, homozygous for the non-mutated form of the gene (VvGAI1/VvGAI1).

VvGAI1), it is possible to recover 50% of individuals with a microvine phenotype and 50% of individuals with the characteristics of a non-dwarf grapevine.

2.2 Molecular mechanisms associated with the mutation *Vvgai1*

The comparison of the allelic VvGAI forms present in Pinot Meunier and the microvine [4, 5] showed that the mutation corresponds to a modification of a single nucleotide in the DELLA motif of the protein, which is important for gibberellin signaling.

After transient transformation of epidermal onion cells, green fluorescent protein (GFP) fusions to *VvGAI1* and *Vvgai1* sequences responded differently to gibberellin applications. The GFP signal of the GAI1::GFP fusion disappears rapidly from the nucleus under the effect of gibberellins, which indicates its degradation following the hormonal stimulus. On the contrary, the gai1::GFP translational protein fusion remains insensitive to hormonal signaling, which indicates that the mutation in the DELLA motif abolishes the property of the protein to be degraded when triggered by gibberellins [5].

The GAI gene is known to be an important regulator of vegetative growth and reproductive development [6]. In grapevine, gibberellins, produced under shade, stimulate growth and inhibit the formation of inflorescences [7]. This effect is mediated by the nuclear protein GAI1, which, in its mutated form gai1, no longer transmits the hormonal signaling [5]. Thus, vegetative growth and the inhibition of the conversion of tendrils into inflorescences are no longer maintained which explains the dwarf phenotype and the continuous fructification along the stems. The characterization of the expression profiles of different isogenes of *VvGAI* revealed that *Vvgai1* is mainly expressed in vegetative organs such as buds and young leaves, while other forms are expressed in reproductive organs (unpublished data). For instance, *Vvgai2*, which does not have any mutation in the DELLA protein motif, is expressed in reproductive organs from flowering to ripening [5]. This explains why *Vvgai1* mutation does not interfere directly with berry developmental program which is similar to non-dwarf varieties.

3. Application of the use of the microvine

3.1 Vegetative development

Several experiments have been conducted outdoor and in controlled environments to characterize the vegetative development of the proleptic axis of the microvine [8]. Different day/night temperature treatments were applied (22/12, 25/15, 30/15, 30/20, 30/25°C), while VPD was maintained constant (about 1 kPa). These experiments showed that the vegetative organogenesis rhythm of the microvine is similar to that of non-dwarf vines. Indeed, its phyllochron (leaf emission rate) is around 24°C, similarly to other varieties of *V. vinifera* such as Grenache [10], and it fluctuates only slightly with temperature and radiation variations between experiments (photosynthetically active radiation (PAR) has been experimented from 19 to 25 mol.m^{-2}d^{-1}).

The duration of leaf and internode growth of the microvine is also similar to that of non-dwarf vines, lasting ca. 220°C (i.e., 20 days at 25/15°C) for leaves and ca. 150°C (i.e., 14 days under the same conditions) for internodes [9, 10]. The most significant phenotypic difference, induced by *Vvgai1*, is the size limitation of vegetative organs. The leaf area is reduced by half in the microvine compared to non-dwarf vines, and internodes are five times shorter. The dwarf phenotype is thus very valuable to conduct experiments under very well-controlled conditions in small growth chambers. Such property permits to study the impacts of single or combined abiotic factors (radiation, temperature, VPD, CO_2) on plant growth and development while minimizing uncontrolled biases arising from environmental fluctuation in field studies on perennial vines.

However, the shortening of the internodes increases leaf shading and promotes the development of fungal diseases as compared to non-dwarf vine. The control of powdery mildew (*Erysiphe necator*) on leaves and green berries or gray mold (*Botrytis*) on ripening fruits requires a strict phytosanitary management.

To improve the microclimate of the clusters, it is recommended to systematically remove the lateral branches to reduce the plant to a single proleptic axis and to systematically eliminate one leaf out of three, e.g., removing the leaves of all P0 phytomers which do not bear any inflorescence. Also, for the most fertile lines, it is necessary to control the number of ripening berries to avoid source/sink unbalance that could be prejudicial to the growth and the formation of new inflorescences as well as the accumulation of metabolites in the fruits. Because the microvine displays several levels of cluster at ripening stages, a good balance is achieved by limiting the number of ripening berries to 8–15 per cluster.

3.2 Reproductive development

The reproductive development of the microvine is divided into two distinct and simultaneously occurring patterns: (i) the fructification of proleptic shoots from preformed inflorescence primordia within winter buds and (ii) the continuous fruiting of proleptic and sylleptic axes resulting from the conversion of tendrils into inflorescences.

3.2.1 Fruiting from winter buds (two successive seasons)

In the grapevine, as for many other perennial fruit crops, fruit formation occurs during 2 consecutive years. The first step starts with the initiation and differentiation of inflorescence primordia in the winter buds prior to endo-dormancy until approximately the end of summer or beginning of autumn. During the subsequent cycle after the break of dormancy, approximately 2 weeks before budburst, the inflorescences resume their development and complete flower organogenesis and subsequently flowering in spring [6]. The level of differentiation of microvine winter buds (i.e., the number of preformed phytomers and inflorescence primordia) was analyzed during 80 days of growth under controlled environmental conditions (25/15°C day/night temperature, VPD 1 kPa, photoperiod 12 h). Two imaging methods were compared, the classic microscopy dissection and the noninvasive X-ray micro-tomography [11], with a resolution of 9 μm. These observations showed that winter buds of the microvine harbor a complex formed of primary, secondary, and tertiary buds of decreasing fertility, as non-dwarf vines [12]. The maximum fertility of the primary buds is two inflorescences in the microvine, whereas it can reach three or even four in some non-dwarf varieties. These inflorescences are inserted into phytomers n°4 to n°6 with an acropetal development as for macrovines [12, 13]. The lignification of the stem which develops from the vegetative axis base is concomitant with the slowdown of bud development and probably its entry into endo-dormancy, similarly as for non-dwarf vines [14].

3.2.2 Continuous flowering and fruiting (one single growing season)

The microvine has the particularity to develop inflorescences from tendrils along proleptic and sylleptic axes (**Figure 4**), which result in a continuous flowering and fruiting processes. A gradient of reproductive development stages is thus present simultaneously along the proleptic axis from the differentiation of inflorescences until maturity. This characteristic offers the opportunity to evaluate abiotic or biotic stress impacts on all reproductive stages of development along the proleptic axis simultaneously.

Top right, section of a winter bud analyzed by tomography. Bottom right, a longitudinal section of a winter bud exhibiting a lateral inflorescence primordium (IP) on the primary bud axis and a secondary preformed vegetative axis on the left side.

Figure 4.
Vegetative and reproductive development of the ML1 somaclone n°7, a microvine line regenerated from pinot Meunier cl. ENTAV 8 according to the method described by Torregrosa [15]. Top left, longitudinal section of an apex showing the preformation of 7–9 phytomers before emergence of caulinar organs. Upper middle, emergence of young inflorescences just below the apex. We note the very hairy appearance of the apex of the microvine ML1. On the middle, an 8-month-old ML1 microvine displaying all the sequences of the reproductive development from flowering to fruit ripening. Bottom left, a focus on the phytomers carrying bunches shifting from green to ripening stages and the concomitant lignification of the shoot (leaves have been removed for the clearness of the photograph). Top right, section of a winter bud analyzed by tomography. Bottom right, a longitudinal section of a winter bud exhibiting a lateral inflorescence primordium (IP) on the primary bud axis and a secondary preformed vegetative axis on the left side.

The synchronism between vegetative development and fruiting of the microvine also simplifies the study of their interactions compared to macrovines. The impact of contrasted source/sink balance on fruiting can be easily studied by manipulating shoot or fruit load (number of growing axes and/or number of leaves/inflorescence per axis). The continuous fruiting was found to be stable under standard environmental conditions (25/15°C day/night temperature, VPD 1 kPa, photoperiod 12 h) and when the leaf area to fruit fresh weight was less than 1 $m^2.kg^{-1}$. On the contrary, the capacity of flowering is strongly altered in the presence of abiotic or biotic stresses. High temperature (> 33°C), low radiation levels (PAR < 15 $mol.m^{-2}.j^{-1}$), or high VPD (>3 kPa) can induce inflorescences abortion and disrupt the continuity of the reproductive gradient along stem axes. The sensitivity of inflorescence development was found higher when the C reserves (starch) were reduced, in particular, in young plants. Thus, although it is possible to obtain fruiting organs from 5-month-old microvine cuttings, it is advisable to use 1-year-old or older plants that are much less susceptible to inflorescence abortion [16]. In experiments conducted in our lab, we obtained successive cycles of fruiting for at least 5 years without repotting.

The size of inflorescences of microvines is smaller (10–50 berries per cluster in average) than that of macrovines [17–19]. However, flowers and young fruits of the microvine do not display a very high abscission rate as observed in non-dwarf varieties. The development of flowers and berries is identical to non-dwarf vines. Flowering (50% of open flowers) occurs 320°C GDD (growing degree days) after the phytomer emission (i.e., 30 days at 25/15°C), which is comparable to the duration between budburst and flowering in the non-dwarf vines [18]. Ripening (onset of sugar loading) starts at ca. 500°C GDD (i.e., 47 days at 25/15°C) after flowering, and the physiological ripening (when metabolite loading stops) is reached at ca. 900°C GDD (i.e., 80 days at 25/15°C) after flowering or 30 days after the start of sugar loading. This behavior is similar in macrovines [18, 20]. Thus, berries of the

ML1 microvine reach a final individual size of 1.2 g, comparable to that of cv. Pinot meunier, from which this line derives. At physiological ripening, berries contain about 0.8 mmol berry-1 of soluble sugars in non-limiting water supply conditions which is similar to other varieties of *V. vinifera* (**Figure 5**).

3.3 Genetics and genomics

3.3.1 Genetic mapping and pre-breeding

The microvine provides different advantages over non-dwarf vines to speed up or facilitate genetics. Since the mutation is transmissible by hybridization and has a codominant effect, it is possible to cross microvines (*VvGAI1/Vvgai1*) or picovines (*Vvgai1/Vvgai1*) with non-dwarf genotypes, i.e., without the mutation (*VvGAI1/VvGAI1*), to create microvine segregating populations. In the first case, 50% of individuals will display the microvine phenotype, while using picovines as parent, 100% of the progeny exhibit a dwarf behavior.

The *VvGAI1* gene is located on chromosome n°1, while the QTL determining grapevine flower sex is located on chromosome n°2. That means both loci segregate independently, and it is therefore possible to use female microvines or picovines, which facilitates crosses by avoiding the time-consuming emasculation and reducing the risk of selfing [19]. On the other hand, when a female microvine (f/f) is crossed with a hermaphrodite genotype (H/f, the most common genotype in *V. vinifera* varieties), the population will be composed of 50% of female plants and 50% of hermaphroditic plants. For instance, by crossing between the

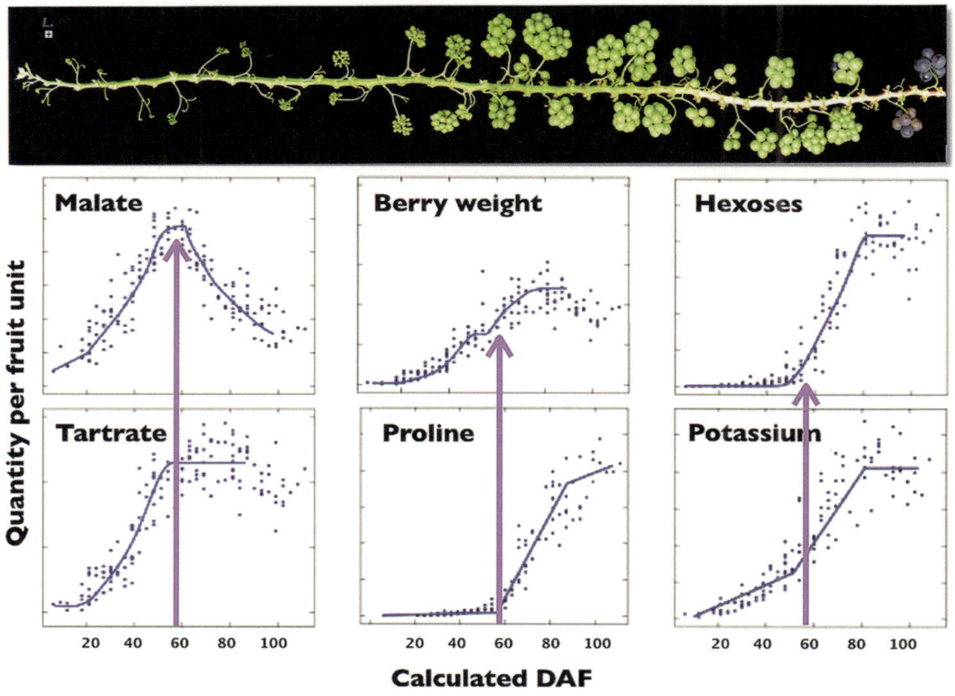

Figure 5.
Spatiotemporal distribution of the reproductive developmental stages from flowering to ripening. On the abscissa, the calendar time in DAF (days after flowering) was recalculated for each phytomer converting the corresponding plastochron index in thermal time and inferred in calendar time with the phyllochron. Kinetics of fresh fruit weight and the contents of major primary metabolites and potassium are presented in quantity per fruit unit.

PV00C001V0008 [19] and the fleshless berry mutant of the ugni blanc [21], a range of genotypes and phenotypes can be obtained [5].

This progeny is composed of 100% microvines (since the female parent has a *Vvgai1/Vvgai1* genotype) and a very small proportion of individuals with both hermaphrodite flowers and pigmented berries. Indeed, these two characters are present at the homozygous recessive state in one parent (f/f and n/n) and in the heterozygous dominant state in the other (H/f and N/n). It should be noted that since ugni blanc is heterozygous at the sex locus (H/f), while the picovine is f/f, selecting hermaphrodite individuals leads to a segregation distortion in the progeny of the genetic traits determined on the chromosome n°2.

As the microvine produces inflorescences as long as vegetative growth is maintained, it becomes possible to cross all year around without being hampered by seasonality. Under standard thermal and photoperiodic conditions (25/15°C day/night temperature, VPD 1 kPa, photoperiod 12 h), the microvine produces two to three new inflorescences per week, which enables to make hybridizations during long periods in repeating the crosses on the same plants. This also reduces the number of plants required for crosses and therefore experimental space while spreading the hybridization effort over selected and potentially long periods.

One to two months after a cross, it is possible to start harvesting seeds [22] to rescue zygotic embryos, which makes possible to establish a population maintained and amplifiable by micropropagation or microcuttings [23]. After a few micropropagation cycles, in vitro plants can be acclimatized to greenhouse conditions, and the first grapes are obtained within 12 months after the crosses. Thus, in less than a year, it is possible to start the study of the characteristics of the fruits and to proceed to new crossings to recover F2 populations. These speed up genetic mapping studies because it becomes possible to link a genotype and a phenotype in either F1 or F2 progenies in a few months instead of several years when using macrovines [23, 24].

Moreover, if a trait can be inherited through such crosses, it is possible to recover non-dwarf phenotypes (*GAI1/GAI1*) that can be directly proposed as breeding material. Indeed, 50% of the individuals from a cross between a microvine (*VvGAI1/Vvgai1*) and a macrovine (*VvGAI1/VvGAI1*) exhibit the same biological properties as conventional non-dwarf varieties. Thus, the microvine can be used both for the identification of QTLs of interest and also to combine or pyramid characters of interest in a pre-breeding perspective.

3.3.2 Functional genomics

The biological properties of the microvine are also of great interest for functional genomics [26]. Indeed, grapevine, as other perennial plants, is a difficult plant model to study the genes regulating the development of reproductive organs. The difficulty comes from its long juvenile period, its discontinuous fructification from winter buds, and the handling of large plants. The genetic transformation of classical varieties [28] requires several years to obtain adult plants and study the phenotypes linked to the ectopic expression of candidate genes.

With microvine, starting from embryogenic tissues compatible to *Agrobacterium tumefaciens*-mediated transformation (**Figure 6**), it is possible to recover transgenic fruiting plants in less than 1 year [19]. As for classical genetics, it is then easy to derive F2 lines to establish transgenic loci at homozygous state for further studies. In addition, the microvines have a very good aptitude for transformation by *Agrobacterium rhizogenes*, allowing to obtain transgenic organs stabilizable in axenic culture in a few weeks [25, 29, 30].

Figure 6.
From competent embryogenic tissues (top left), it is possible to regenerate transgenic plants in a few months and obtain reproductive organs in less than a year. This allows the study of the regulation of flower and fruit development within shorter delays than with the non-dwarf vines. On the right, a microvine line V9 overexpressing the gene VvHB was identified as a major regulator of the development of the flesh in grapevine fruit [27]. Using genetically modified microvines, it is possible to segregate the transgenes in different genotypic configurations or combine them with various other transgenic traits or not.

4. Temporal inference of spatial observations obtained on the proleptic axis

We have tested the possibility of converting spatial observations (along the proleptic axis) into temporal dynamics at a given stage of vegetative or reproductive development.

4.1 Temporal conversion of spatial profiles

Under controlled and stable environment (25/15°C day/night temperature, VPD 1 kPa, photoperiod 12 h), the development of the proleptic axis of the microvine is stable. The phyllochron is constant reaching ca. 24°C. The growing dynamics of leaves (surface) and berries (volume) from continuous fructification was found to be constant at a given level of phytomer, regardless of the date of bud break [20]. The growth durations of leaves and berries (herbaceous phase) are ca. 220°C after the emission of the phytomer and 500°C after flowering, respectively, as mentioned in Section 2.2. The development of these organs is also spatially stable: the dynamics of leaf area and berry volumes (herbaceous phase) for all levels of phytomer are superimposed when they are represented as a function of cumulative thermal time after the emission of the corresponding phytomer.

Based on these outcomes, the conversion of spatial dynamics of leaf and berry development along the stems into time profiles was tested (**Figure 7**). For this purpose, the positions of the phytomers along the axis were converted into cumulated thermal time after their emission by multiplying their plastochron index (or rank position from the apex) by the phyllochron. The temporal profiles of leaf area and berry volume (green growth phase) resulting from this spatiotemporal conversion are similar to the real temporal profiles obtained at a given level of phytomer [8, 20, 31]. This property makes it possible to reconstruct temporal dynamics of

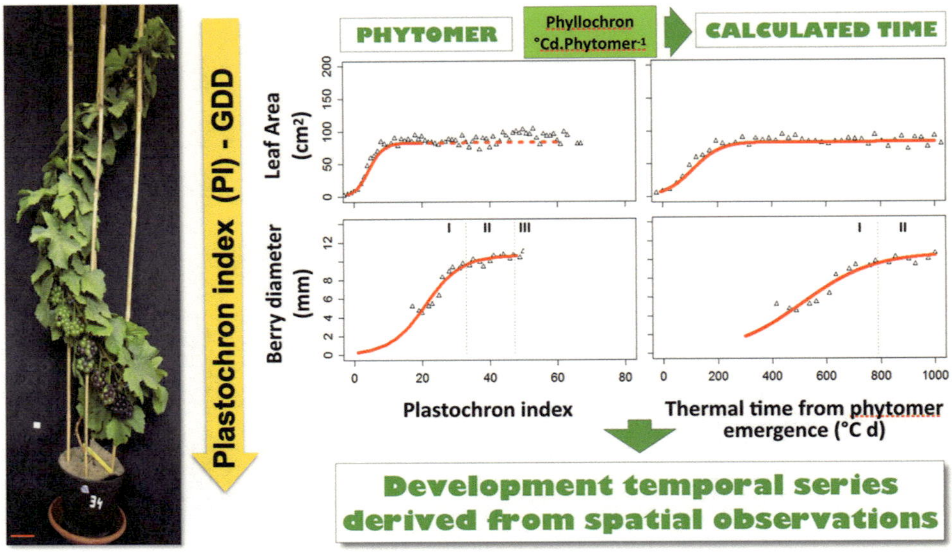

Figure 7.
Conversion of leaf and young berry growth data collected from the position along the microvine main shoot (plastochron index) into cumulated thermal time after phytomer emergence.

development from a single spatial observation of the axis at a given stage. The flow of biomass or metabolites within the organs and their responses to environmental constraints were then addressed using those calculated temporal profiles (Section 5.1).

4.2 Dynamics of inflorescence differentiation within winter buds

The spatiotemporal conversion approach presented above can also be used to characterize the evolution of winter bud development along the proleptic axis of the microvine [12]. Bud development was analyzed on microvines grown under standard environmental conditions (25/15°C day/night temperature, VPD 1 kPa, photoperiod 12 h), as explained in Section 3.2.1. The number of preformed phytomers initiated by primary axes within buds increases linearly as a function of the plastochron index (PI) of the proleptic axis in the non-lignified zone (PI < 25). The temporal dynamics of bud development were calculated from the spatial profiles using the proleptic axis PI x phyllochron. The primary axis of the bud displayed a maximum of six phytomers from IP 25 (lignified zone), i.e., 625°C or 57 days after its initiation (phyllochron of 24°C). A maximum of two inflorescence primordia was observed in this zone. The primordia of the first and second inflorescences, located between the preformed phytomers n°4 and n°6 of the primary axis, were initiated from IP 13 and 26 of the proleptic axis, respectively, corresponding to 325°C (or 30 days) and 650°C (or 60 days) after bud initiation. The timing of inflorescence primordium development in winter buds in non-dwarf vines [32] is similar to our observations on microvines. This pattern of winter bud development parameterized for the microvine can be used to evaluate, and potentially predict, the environmental stress impacts during the season 1 on the fructification potential of the season 2.

4.3 Dynamics of fruit development deriving from neo-formed inflorescence

The primary characterization of fruit development along a microvine axis showed that the microvine berry displays the two classical growth phases as observed for berries of macrovines [32, 33]. Microvine berry growth and metabolite accumulation were analyzed in details [34]. Ten microvines were grown under controlled conditions in a climatic room (30/22°C day/night temperature, photoperiod 14 h, VPD 1 kPa, PAR 400 mmol.m^{-2} s^{-1}). Sampling was performed when proximal fruits attained physiological maturity and when maximum berry volume was reached. Sampling of the present reproductive organs from fruit set to maturity was performed at the same time for each plant. Analysis of the main berry compounds (malic acid, tartaric acid, glucose, fructose, proline) has been carried out. To normalize the stages of development between plants, the spatiotemporal conversion described above was applied using the individual phyllochron of each plant.

The data presented in Rienth et al. [35, 36] shows that microvine fruit accumulates malic acid during the green growth stage for about 40 days after fruit set, until it ceases when the lag phase (herbaceous plateau), which separates the two growth phases, is reached. At the end of the herbaceous phase, at the 24 hours lasting véraison phase, the degradation of malic acid is triggered simultaneously with the accumulation of sugars and proline, which is often used as an indicator of ripening. These processes proceed throughout the second growth or ripening phase. With regard to tartaric acid, we found that it is also accumulated only during the first growth phase as for macrovines and that its amount remains quasi-constant during the ripening phase. The slight decreases in tartaric observed during ripening might be attributed either to enhanced tartaric precipitations as shown by Rosti et al. [37] or variations of microenvironment depending on bunche rank. At the end of green growth stage, the two major organic acids represent approximately 500 mEq, which is comparable to the acidity of the fruit of macrovines. The accumulation of sugars, triggered from the *veraison*, continues until the moment when the phloem unloading is slowed down (maximum volume of the fruit). From this point, the amount of sugars per berry remains constant, but the concentration increases due the loss of berry volume during over-ripening.

5. Examples of studies performed with the microvine

5.1 Impacts of temperature on carbon fluxes and fruiting

The impact of elevated temperature on growth and carbon distribution between vegetative and reproductive organs was investigated. Two contrasting thermal regimes with a difference of 8° C (30/20°C vs. 22/12°C day/night temperature) were imposed during a period of 450°C GDD. The VPD was 1 kPa and the PAR 19 mol. m^{-2}.d^{-1} for the two thermal regimes. The biomass, size, and carbon contents of the leaves, internodes, and berries were characterized from spatial observations at harvest and converted into temporal profiles according to the method described in Section 4. Only the organs that developed during heat treatments, i.e., vegetative phytomers younger than 450°C GDD at harvest and the reproductive phytomers, which were at pre-flowering stage at the beginning of experiments, were retained for analysis. Luchaire et al. [20, 36] have shown that high temperature accelerates the growth and the accumulation of biomass in vegetative organs (leaves and internodes) in thermal time, at the expense of the accumulation of sugars in internodes and the surface area to mass of the leaves (thinner leaves).

Under high temperature, the growth and accumulation of biomass of the fruit slowed down on a thermal time basis. Sugar loading of proximal phytomers (from the post-flowering stage to onset of heat treatment) was also delayed by ca. 400°C GDD at high temperatures. High temperatures increased inflorescence abortion rate (+ 20%) at pre-flowering stages, concomitantly with the beginning of sugar loading in the proximal clusters ripening [20, 36, 38]. These results suggest that high temperature decouples vegetative and reproductive development, increasing the total biomass of vegetative organs while reducing the accumulation of carbon reserves and hampering continuous fruiting.

5.2 Circadian variations of the grape transcriptome

Transcriptomic studies are difficult to run with macrovines grown outdoor because of the seasonality of fruiting and the day-to-day environment fluctuations. Thus, while transcriptomics is a very common approach today to understand the genetic mechanisms regulating grape development, no work has attempted to describe the circadian evolution of the grape transcriptome. The results published by Rienth et al. [39] were the first for a perennial fleshly fruit that addressed this topic. For this experiment microvines were grown in climatic growth chambers [40] under controlled environments (30/20°C day/night temperature, photoperiod 14 h, VPD 1kPA) for 3 months to encompass a complete reproductive cycle from flowering to ripening. When most proximal grapes reached physiological maturity, berry samples from two green and two ripening developmental stages were collected at different periods of the photo and nyctiperiod, and a whole genome transcriptomic analysis was carried out by Nimblegen® Vitis 12x microarrays.

All genes modulated during the day also showed some variation of expression at night, with 1843 genes that are only regulated at night. The detection of this very large number of specifically regulated genes during the night emphasized the importance of the regulatory mechanisms associated with the nocturnal fruit development. The comparison of differentially modulated transcripts between day and night at different stages showed that circadian regulation was very specific to the stage of development with only nine commonly deregulated genes between day and night at all stages. With respect to activated or deactivated functional categories, genes related to photosynthesis appear strongly repressed at night, in particular in young green berry, and several functional categories related to secondary metabolism (phenylalanine) and abiotic stress have shown strong overexpression at night at all developmental stages.

5.3 Effect of temperature on grape development

Until recently, the studies on the effect of temperature on grape development have only been performed using non-dwarf varieties, with the experimental limits associated with this model. Rienth et al. [41, 42] were the first to perform temperature experiments using microvines grown under tightly controlled environmental factors (photoperiod, light intensity, temperature, VPD, water, and mineral supply). This study was carried out with the ML1 microvine applying temperature gradients ranging from 12 to 35°C during 2 h to 4 weeks.

A first series of experiments focused on short-term stress effects (2 h, 35°C) of microvine fruits at different stages between green growth and ripening sampled during day and night. Nimblegen® Vitis 12x microarray assays revealed that a large number of genes (5653) respond to the increase in temperature, at all stages of development (**Figure 8**). Temperature effect was time and mainly development stage specific, with berries close to *veraison* being the most reactive to temperature elevation, especially for some categories such as anthocyanin synthesis which was specifically heat repressed at this stage. Furthermore, various genes of secondary metabolism (phenylalanine, anthocyanins) are repressed at the *veraison*, by high temperature with a larger number of genes regulated during the nocturnal phase.

Long-term thermal stresses (> 30 days) were also experimented using various temperature charts to several stages of grape development, taking into account

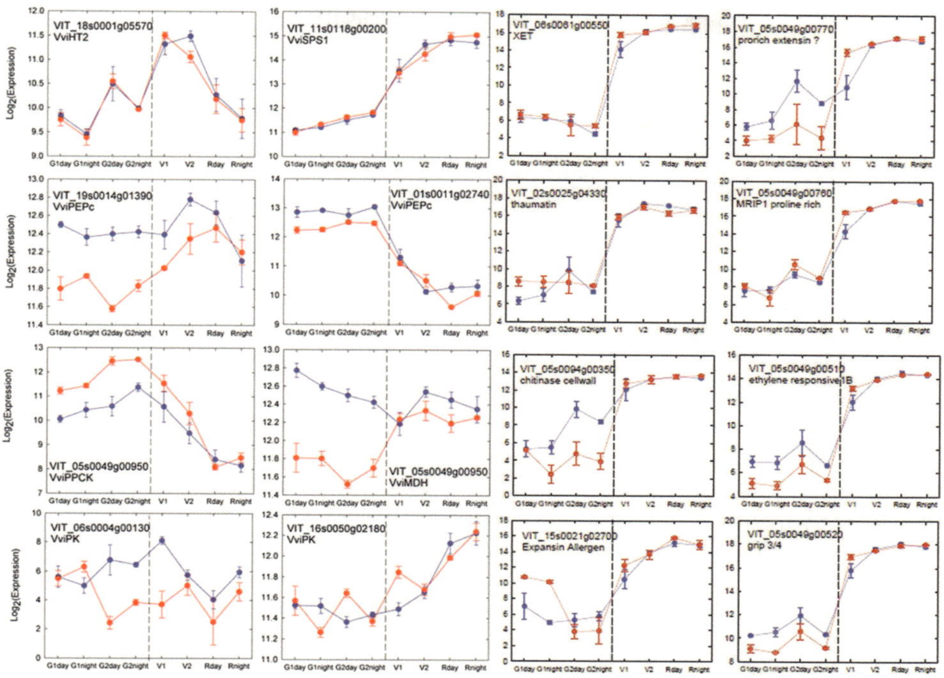

Figure 8.
Schema of the expression changes induced by temperature elevation for some genes of the central metabolism during the grapevine fruit development.

circadian variations of the transcriptome [41]. In these studies, we used high-throughput transcriptomic analysis through RNA-seq (Illumina technology). A total of 10,788 genes could be detected as a function of stage, temperature regime, and photoperiod. The importance of "heat shock"-type genes with highly variable expression patterns as a function of the duration of the stress, the circadian cycle, and the stage of development of the fruit has been highlighted. The rise in temperature led to an acceleration of fruit growth during the green growth phase. In fruit at the onset of ripening, the temperature increased the respiration of malic acid and delayed the accumulation of sugars and downregulating key genes of the flavonoid pathway. For the first time, a decoupling of sugar accumulation and malic acid respiration during ripening could be observed and related to the change in carbohydrate status of the plant as a function of temperature [9].

A number of genes known to display an induction at veraison and thereafter were confirmed in microvines displaying a remarkably stable expression pattern with respect to temperature (SPS1, sucrose phosphate synthase 1; XET, xyloglucanendotransglucosidase; thaumatin; MRIP, ripening-induced protein1-like precursor (proline-rich cell wall). However, other well-known ripening-induced proteins were induced in the cold in green stage (GRIP3/4, grape ripening-induced protein ¾, ethylene-responsive 1B, putative extensin proline-rich, cell wall chitinase). During the long-term low T° treatment, fruit transcriptomic analyses showed an overexpression of key enzymes linked to both glycolysis (PK, pyruvate kinase) and malic acid synthesis (PEPce, phosphoenolpyruvate carboxylase; MDH, malate dehydrogenase). Temperature variation also impacted posttranscriptional regulation mechanism such as the PPCK (phosphoenol pyruvate carboxylase kinase) which is overexpressed under heat. This gene expression pattern confirmed physiological observations of sugar-acid decoupling and suggests that under cool condition, where the plant energetic status is more comfortable due to lower vegetative growth and cellular respiration rate, malic acid respiration, as a supplemental energy source in the fruit, is not compulsory. In cool climate, the allocation of carbon to the fruit can support glycolysis, malate synthesis, and sugar accumulation into the vacuole. Conversely, under hot climate, cytoplasmic sugars could be limiting when the cell starts to accumulate sugar in the vacuole at the onset of ripening. Thus, the malate would be drained from the vacuole to supply energy through respiration and/or through H^+/sugar exchange at the tonoplast.

5.4 Identification of QTLs of development stable under fluctuating environments

Air temperature elevation combined with the shift of all phenological stages to warmer period is causing critical changes on vine yield and grape composition. Plant breeding has the potential to offer new cultivars with stable yield and quality under warmer conditions, but this requires the identification of stable genetic traits. The investigation about the stability of developmental QTLs with regard to abiotic factors is complicated with the non-dwarf varieties, because of its biological properties (long juvenile period, big size of the plants). Most of previous studies were carried out outdoors, in uncontrolled environmental conditions and with a relatively low experimental flow.

Houel et al. [25] reported the first experiment performed with microvines, to identify QTLs of vegetative and reproductive development, testing their stability under fluctuating environments. A F1 mapping population consisting of 129 microvines derived from the PV00C001V0008 x ugni blanc fleshless berry mutant was genotyped using an Illumina® 18 K SNPs chip (single-nucleotide polymorphism). Forty-three vegetative and reproductive traits were phenotyped over four vegetative cycles in the field, and a subset of 22 characters were measured over two climatic chamber culture cycles under two contrasting temperature regimes. Ten stable QTLs were identified for the development and composition of the berry and the leaf area on the parental genetic maps. A new major QTL accounting for up to 44% of variance of the berry weight was identified on the chromosome 7 in the ugni blanc parent. This QTL co-locates with QTLs of number of seeds per berry (accounting for up to 76% of the total variance), QTLs of fruit acidity before maturation (up to 35% of explained variance), and yield components such as the number of clusters and berries per cluster (up to 25% explained variance). In addition, a minor leaf surface QTL was found on the chromosome 4 in the same parent. This study which combined the use of microvine population to boost and facilitate the phenotyping with high-throughput genotyping technologies was innovative in grapevine genetics and also for perennial fruit crops. It allowed the identification of 10 stable QTLs, including the first QTLs of *V. vinifera* berry acidity detected in an intraspecific cross.

This progeny was also included in a study addressing the diversity for fruit volume, main sugar, and organic acid amounts in *V. vinifera* [43]. A panel of 33 genotypes, including 12 grapevine varieties and 21 microvine offspring, were characterized. Fruit phenotyping focused on two critical stages of fruit development: the end of green growth phase when organic acidity reaches a maximum and the physiological ripe stage when sugar unloading and water uptake stop. To determine the date of sampling for each critical stage, fruit texture and growth were carefully monitored. Analyses at both stages revealed large phenotypic variation for malic and tartaric acids as well as for sugars and berry size. At ripe stage, fruit fresh weight ranged from 1.04 to 5.25 g and sugar concentration from 751 to 1353 $mmol.L^{-1}$. The content in organic acids varied both in quantity (from 80 to 361 $meq.L^{-1}$) and in composition, with malic to tartaric acid ratio ranging from 0.13 to 3.62. At the inter-genotypic level, data showed no link between berry growth and osmoticum accumulation per fruit unit, suggesting that berry water uptake is not only dependent on fruit osmotic potential. The report showed that diversity for berry size, sugar accumulation, and malic to tartaric acid ratio could be exploited through crossbreeding.

These studies which (i) identified genotypes with contrasted fruit composition for compounds controlled by environmental factors and (ii) mapped QTLs of development, including for berry composition, provide interesting prospects to mitigate some adverse effects of climate warming on viticulture.

5.5 Identification of the genetic traits of aromatic character of cabernet sauvignon

Methoxypyrazines are a family of volatile compounds found in many fruits and vegetables and especially in grapes, providing herbaceous flavors (green capsicum aroma) to the wines of some varieties such as Cabernet Sauvignon or sauvignon blanc. While several methoxypyrazine biosynthetic pathways have been proposed, none of the metabolic steps have been genetically confirmed. Dunlevy et al. [24] used a F2 population derived from a F1 microvine obtained by crossing the Cabernet Sauvignon and a picovine. The Cabernet Sauvignon variety is capable of producing the molecule 3-isobutyl-2-methoxypyrazine (IBMP), the major compound associated with capsicum flavors, while the microvine that derives from Pinot Meunier produces very little amount of this compound. In F1 offspring, all individuals produced IBMP, suggesting a homozygote dominant genotypic status for this trait in Cabernet Sauvignon. The phenotyping of the F2 individuals identified 43 lines able to accumulate IBMP, while 21 individuals lacked this compound confirming the dominant homozygous genotype for Cabernet Sauvignon and the homozygous recessive genotype for picovine progenitor.

After genotyping and phenotyping, the entire F2 progeny, a 2.3 Mb locus determining IBMP accumulation in grape berries, was found on chromosome n°3. Of the 261 genes identified in the corresponding QTL, two candidate methyltransferase genes have been identified, *VvOMT3* and *VvOMT4*. Screening a collection of 91 grapevine genotypes differentially accumulating IBMP into the grapes indicated *VvOMT3* as the most likely candidate to explain the genetic determinism of the green capsicum trait in grapevine fruits. Moreover, the data suggested that the low level of methoxypyrazines found in most cultivated grape varieties resulted from human selection for mutations in methyltransferase. The markers identifying this locus are valuable tools for the selection of grape varieties that are aromatically typified by IBMP and recalling Cabernet wines.

5.6 Effect of application of exogenous stimulants of fruit metabolism

The microvine plant model which displays unique reproductive organ behavior offers new experimental options for grapevine fruit physiological studies, not only because of the size of the plants which facilitate experimental handling in greenhouse or growth cabinet but also because it is possible to study several developmental stages at once. Taking advantage of the biological properties of the microvine, two studies were recently performed to study the impact of exogenous compound application to the ML1 microvine grapes on the aroma accumulation during ripening. The first study was about the impact of vine-shoot aqueous extracts, which have been proposed as bio-stimulants to be sprayed to the canopy to modify wine aromatic profile. Sanchez-Gomez et al. [44] experimented the effect of vine-shoot extract foliar application on the composition of the grapes at 21 stages of development. The application was carried out from BBCH53 (detached inflorescences) to BBCH85 (berry softening) to reveal stage-specific responses of the accumulation of glycosylated aroma precursors at BBCH89 (ripe stage). Fifty grape sampling time points spreading to 86 days were established and normalized using the cumulative growing degree days parameter. The results confirmed that vine-shoot extract treatment had a positive impact on the accumulation of glycosylated compounds [45], especially aglycones such as alcohols, terpenes, and C13-norisoprenoids, with a higher impact when the treatment was applied at grape ripening stage.

The same approach was carried out to characterize the behavior of glycosylated aroma precursors in microvine fruits after foliar application of guaiacol. Previous outdoor experiments have showed that spraying guaiacol on vines could modify the contents of aroma compounds in grape and grape-derived wines. It was shown that such treatments could increase guaiacol glycoconjugates in leaves, shoots, and fruits of Monastrell variety, where there was a release of aglycone compounds during wine processing. However, the effect of such application and its timing on glycosylated aroma precursor pool remained unstudied. Sanchez-Gomez [46] studied the effect of guaiacol sprays when applied at several fruit developmental stages on glycosylated compound accumulation. The applications were carried out from phenological stage BBCH71 (fruit set) to BBCH85 (berry softening), to reveal stage-specific responses of the accumulation of glycosylated aroma precursors at BBCH89 (ripe stage). Data confirmed that guaiacol is an elicitor of the accumulation of glycosylated aromatic compounds in the microvine fruit, with a higher efficiency of application during ripening stages of the fruits. Geraniol, a terpene compound, exhibited the higher increase increment with up to 50-fold high concentration after guaiacol spraying than in the control.

6. Conclusions

The studies summarized here have shown that at a given phytomer level, the development of the vegetative and reproductive organs of the microvine exhibits comparable kinetics to those of non-dwarf vines grown outdoor. Given its original biological properties (small size, continuous fructification, possibility of inferring temporal observations from spatial data), this model can be used in fundamental studies on vine response to abiotic constraints or on fruit physiology under well-controlled environments. Thus, the microvine has already been used as a model in several scientific experiments on the effect of temperature on the vegetative and reproductive development, on changes in gene expression in grapes, and their plasticity under high temperature. This model has also shown its potential to accelerate conventional and reverse genetic approaches, including the identification of genetic determinants of developmental traits stable under fluctuating thermal conditions or major loci controlling the composition of the grapes. Studies are underway to use this model to study the impact of physical factors (drought, CO_2 concentration, temperature, etc.) on the development of the vine and the quality of the grapes but also to develop tools (markers of QTLs, pre-breeding lines pyramiding several agronomic traits of interest) for the selection of new varieties displaying original properties, i.e., traits of adaptation to climate changes.

Acknowledgements

These studies were supported by fundings from the following agencies or institutions: National Research Agency—Genopole (DURAVITIS project ANR-2010-GENM-004-01), Montpellier SupAgro, the departments EA (Environment-Agronomy) and BAP (Plant Biology and Improvement) of INRA, the Poupelain

Foundation, the European Eurasia 2 thesis mobility programs, EulaLink, and the Brazilian CNPq scientific cooperation program. Special thanks to Mark Thomas, Pat Corena, Don MacKenzy, and Ian Dry from CSIRO Agriculture (Adelaide) for mentoring and helping during some important steps of these experiments.

Author details

Anne Pellegrino[1], Charles Romieu[2], Markus Rienth[3] and Laurent Torregrosa[2]*

1 LEPSE, Montpellier University, CIRAD, INRA, Montpellier SupAgro, Montpellier, France

2 AGAP, Montpellier University, CIRAD, INRA, Montpellier SupAgro, Montpellier, France

3 University of Sciences and Art Western Switzerland, Changins, College of Viticulture and Oenology, Nyon, Switzerland

*Address all correspondence to: laurent.torregrosa@supagro.fr

References

[1] Torregrosa L. Génétique du développement et de la résistance chez la vigne. In: Carbonneau A, editor. Traité de la vigne: Physiologie, terroir, culture. 2nd ed. Paris, France. EAN13: 9782100726691: Dunod; 2015. pp. 237-262

[2] Torregrosa L, Fernandez L, Bouquet A, Boursiquot JM, Pelsy F, Martinez-Zapater JM. Origins and consequences of somatic variation in grapevine. In: Kole C, editor. Genetics, Genomics and Breeding of Grapes. Enfield, New Hampshire, USA: Science Publishers; 2010. pp. 68-92. DOI: 10.1201/b10948-4

[3] Franks T, Botta R, Thomas MR. Chimerism in grapevines: Implications for cultivar identity, ancestry and genetic improvement. Theoretical and Applied Genetics. 2002;**104**:192-199

[4] Boss PK, Thomas MR. Association of dwarfism and floral induction with a grape 'green revolution' mutation. Nature. 2000;**416**:847-850

[5] Torregrosa L, Rienth M, Luchaire N, Alcantara Novelli Dias F, Bigard A, Chatbanyong R, et al. The microvine, a biological model, very versatile and efficient to boost grapevine research in physiology and genetics. In: 39th OIV meeting. Bento Gonzalvez, Brazil; 24-28 October 2016

[6] Carmona MJ, Chaïb J, Martınez-Zapater JM, Thomas MR. A molecular genetic perspective of reproductive development in grapevine. Journal of Experimental Botany. 2008;**59**: 2579-2596. DOI: 10.1093/jxb/ern160

[7] Coombe BJ. Effects of growth retardants on *Vitis vinifera L*. Vitis. 1967;**6**:278-287

[8] Luchaire N, Rienth M, Nehe A, Torregrosa L, Romieu C, Muller B, et al. The Microvine: An ecophysiological model for grapevine. In: 18th Int. Symp. GiESCO, 7-11 July, Porto, Portugal. 2013

[9] Romieu C, Lecourieux D, Rienth M, Luchaire N, Chatbanyong R, Shahood R, et al. Temperature impacts on grapevine physiology and development at both plant and organ levels. In: Xth ISGPB, 13-15 june, Verona, Italy

[10] Lebon E, Pellegrino A, Tardieu F, Lecoeur J. Shoot development in grapevine (*Vitis vinifera*) is affected by the modular branching pattern of the stem and intra-and-inter shoot trophic competition. Annals of Botany. 2004;**93**:263-274

[11] Larabell CA, Nugent KA. Imaging cellular architecture with X-rays. Current Opinion in Structural Biology. 2010;**20**:623-631. DOI: 10.1016/j.sbi.2010.08.008

[12] Alcântara Novelli Dias F, Torregrosa L, Luchaire N, Houel C, Pellegrino A. The microvine, a model to study the effect of temperature on the grapevine latent bud. OenoOne (in revision)

[13] Carolus M. Description des stades du développement des primordia inflorescentiels durant l'organogenèse des bourgeons latents de la vigne (*Vitis vinifera* L. cv Merlot). Connaissance de la Vigne et du Vin. 1971;**2**:163-173

[14] Bernard AC. Contribution à l'étude de la biologie et des méristèmes des vitacées [Thèse d'Etat de l'Univ]. France: Montpellier; 1980. 216 p

[15] Torregrosa L. A simple and efficient method to obtain stable embryogenic cultures from anthers of *Vitis vinifera L.* Vitis; 1998;**37**:91-92

[16] Luchaire N, Torregrosa L, Gibon Y, Rienth M, Romieu C, Turc O, et al. Yield reduction under high temperature is paired with a low carbon status in microvine. In: Xth ISGPB meeting, Verona, Italy. 13-15 June 2016

[17] Pratt C. Reproductive anatomy in cultivated grapes—A review. American Journal of Enology and Viticulture. 1971;**22**:92-109

[18] Mullins MG, Bouquet A, Williams LE. Biology of the Grapevine. UK: Cambridge University Press; 1992. 252 p

[19] Chaib J, Torregrosa L, Mackenzie D, Corena P, Bouquet A, Thomas MR. The microvine—A model system for rapid forward and reverse genetics of grapevines. The Plant Journal. 2010;**61**:1083-1092

[20] Luchaire N, Rienth M, Romieu C, Nehe A, Chatbanyong R, Houel C, et al. Pellegrino A Microvine Fruitfulness Loss under Elevated Temperature Is Conditioned by the Initial Plant Vigor and Carbohydrate Pool. In: Exploring the Microvine developmental and physiological adaptation mechanisms to elevated temperature: characterization of the impact on carbon balance and its involvement in early reproductive development failure. PhD of Montpellier SupAgro. 2015. www.theses.fr/2015NSAM0011

[21] Fernandez L, Romieu C, Moing A, Bouquet A, Maucourt M, Thomas MR, et al. The grapevine fleshless berry mutation: A unique genotype to investigate differences between fleshy and non-fleshy fruit. Plant Physiology. 2006;**140**:537-547

[22] Chatbanyong R, Torregrosa L. A highly efficient embryo rescue protocol to recover a progeny from of the microvine. Vitis. 2015;**54**:41-46

[23] Bouquet A, Torregrosa L. Micropropagation of grapevine (*Vitis* spp.). In: Jain SM, Ishii K, editors. Micropropagation of Woody Trees and Fruits. Dordrecht, The Netherlands: Kluwer Academic Pub; 2003. pp. 3119-3352

[24] Dunlevy JD, Dennis EG, Soole KL, Perkins MV, Davies C, Bos PK. A methyltransferase essential for the methoxypyrazine derived flavour of wine. The Plant Journal. 2013;**75**: 606-617. DOI: 10.1111/tpj.12224

[25] Houel C, Chatbanyong R, Doligez A, Rienth M, Foria S, Luchaire N, et al. Identification of stable QTLs for vegetative and reproductive traits in the microvine (*Vitis vinifera* L.) using the 18K Infinium chip. BMC Plant Biology. 2015;**15**:205. DOI: DOI 10.1186/s12870-015-0588-0

[26] Vidal JR, Gomez C, Cutanda MC, Shrestha B, Bouquet A, Thomas MR, et al. Use of gene tranfer technology for functional studies in grapevine. Australian Journal of Grape and Wine Research. 2010;**16**:138-151

[27] Fernandez L, Torregrosa L, Terrier N, Sreekantan L, Grimplet J, Davies C, et al. Identification of genes associated with flesh morphogenesis in grapevine (*V. vinifera* L.) berry. Plant Molecular Biology. 2007;**63**:307-323

[28] Torregrosa L, Vialet S, Adivéze A, Iocco-Corena P, Thomas MR. Grapevine (*Vitis vinifera* L.). In: Wang K, editor. Agrobacterium Protocols. 3nd ed. New York, USA: Springer Science + Business Media; 2015. pp. 177-194

[29] Cutanda-Perez MC, Ageorges A, Gomez C, Vialet S, Romieu C, Torregrosa L. Ectopic expression of the *VlmybA1* in grapevine activates a narrow set of genes involved in anthocyanin synthesis and transport. Plant Molecular Biology. 2009;**69**:633-648

[30] Huang YF, Vialet S, Guiraud JL, Torregrosa L, Bertrand Y, Cheynier V, et al. A negative MYB regulator of proanthocyanidin accumulation, identified through expression quantitative locus mapping in grape berry. The New Phytologist. 2013;**203**:795-809. DOI: 10.1111/nph.12557

[31] Luchaire N, Rienth M, Romieu C, Torregrosa L, Gibon Y, Turc O, et al. Role of plant carbon status in yield components responses to elevated temperatures in microvine. In: 19th Int. Symp. GiESCO, 1-5 June, Gruissan, France. 2015

[32] Vasconcelos MC, Greven M, Winefield CS, Trought MCT, Raw V. The flowering process of Vitis vinifera: A review. American Journal of Enology and Viticulture. 2009;**60**:4

[33] Conde C, Silva P, Fontes N, Dias ACP, Tavares RM, Sousa MJ, et al. Biochemical changes throughout grape berry development and fruit and wine quality. Food. 2007;**1**:1-22

[34] Rienth M, Jaquerod A, Romieu C, Torregrosa L. La microvigne : un outil novateur pour la recherche et pour l'enseignement de la biologie de la vigne. Revue Suisse de Viticulture, Arboriculture, Horticulture. 2016;**48**:132-135

[35] Rienth M, Romieu C, Gregan R, Walsh C, Torregrosa L, Kelly M. Validation and application of an improved method for the rapid determination of proline in grape berries. Journal of Agricultural and Food Chemistry. 2014;**62**:3384-3389. DOI: 10.1021/jf404627

[36] Luchaire N, Mallent M, Bediee A, Muller B, Torregrosa L, Pellegrino A. The carbon balance of microvine plants is decreased under elevated temperature. In: EPSO Congress, 22-26 June, Dublin, Ireland. 2014

[37] Rösti J, Schumann M, Cleroux M, Lorenzini F, Zufferey V, Rienth M. Effect of drying on tartaric acid and malic acid in shiraz and merlot berries. Australian Journal of Grape and Wine Research. 2018;**24**:421-429. DOI: 10.1111/ajgw.12344

[38] Muller B, Luchaire N, Oury V, Tardieu F, Gibon Y, Torregrosa L, et al. Is failure of reproductive development under water or heat stress due to C shortage? In: SEB Symp., Brighton, 3-7 July, England. 2016

[39] Rienth M, Torregrosa L, Kelly M, Luchaire N, Grimplet J, Romieu C. The transcriptomic control of berry development is more important at night than at day. PLoS One. 2014;**9**:e88844

[40] Rienth R, Dauzat M, Pellegrino A, Lopez G, Torregrosa L, Romieu C. First observations of the microvine development under 100% LED (light emitting diodes) illumination. Vitis. 2012;**51**:167-173

[41] Rienth M, Torregrosa L, Luchaire N, Chatbanyong R, Lecourieux D, Kelly M, et al. Day and night heat stress trigger different transcriptomic responses in green and ripening grapevine (*Vitis vinifera*) fruit. BMC Plant Biology. 2014. DOI: 10.1186/1471-2229-14-108

[42] Rienth M, Torregrosa L, Gauthier S, Ardisson M, Brillouet JL, Romieu C. Temperature desynchronizes sugar and organic acid metabolism in ripening grapevine fruits and remodels its transcriptome. BMC Plant Biology. 2016;**16**:164. DOI: DOI 10.1186/s12870-016-0850-0

[43] Bigard A, Berhe DT, Maoddi E, Sire Y, Boursiquot JM, Ojeda H, et al. *Vitis vinifera* L. fruit diversity to breed varieties anticipating climate changes. Frontiers in Plant Science. 2018;**9**:455. DOI: 10.3389/fpls.2018.00455

[44] Sanchez-Gomez R, Torregrosa L, Zalacain A, Ojeda O, Bouckenooghe V, Schneider R, et al. The microvine, a plant model to study the effect of vine-shoot extract on the accumulation of glycosylated aroma precursors in grapes. Journal of the Science of Food and Agriculture. 2018;**98**:3031-3040. DOI: 10.1002/jsfa.8802

[45] Alem H, Rigou P, Schneider R, Ojeda H, Torregrosa L. Impact of agronomic practices on grape aroma composition: A review. Journal of the Science of Food and Agriculture. 2018;**98**:975-985. DOI: 10.1002/jsfa.9327

[46] Sánchez-Gómez R, Torregrosa L, Zalacain A, Ojeda H, Bouckenooghed V, Schneider R, et al. Behavior of glycosylated aroma precursors in microvine fruits after guaiacol foliar application. Scientia Horticulturae. 2019;**246**:3031-3040. DOI: 10.1016/j.scienta.2018.11.068

Chapter 2

Somatic Variation and Cultivar Innovation in Grapevine

Pablo Carbonell-Bejerano, Carolina Royo, Nuria Mauri, Javier Ibáñez and José Miguel Martínez Zapater

Abstract

Paradoxically, continuous vegetative multiplication of traditional grapevine cultivars aimed to maintain cultivar attributes in this highly heterozygous species ends in the accumulation of considerable somatic variation. This variation has long contributed to cultivar adaptation and evolution under changing environmental and cultivation conditions and has also been a source of novel traits. Understanding how this somatic variation originates provides tools for genetics-assisted tracking of selected variants and breeding. Potentially, the identification of the mutations causing the observed phenotypic variation can now help to direct genome editing approaches to improve the genotype of elite traditional cultivars. Molecular characterization of somatic variants can also generate basic information helping to understand gene biological function. In this chapter, we review the state of the art on somatic variation in grapevine at phenotypic and genome sequence levels, present possible strategies for the study of this variation, and describe a few examples in which the genetic and molecular basis or very relevant grapevine traits were successfully identified.

Keywords: somatic variation, genome sequence variation, somatic mutations, chimerism, Meunier phenotype, fruit color variation, seedlessness, Muscat aroma, forward genomics

1. Introduction

World viticulture is based on a wide diversity of cultivars, many of them autochthonous from their cultivated areas. In fact, almost 1500 grapevine wine cultivars are listed in the statistics published by the Wine Economics Research Centre at the University of Adelaide (Australia) every 10 years [1]. However, sixteen of those cultivars already occupy more than 50% of the world vineyard surface either because they belong to the few elite cultivars that are internationally recognized and grown in multiple wine regions across the world, or because they are widely grown in their regions of origin (**Table 1**). While this pattern responds to winemaking being a classical industry in which traditional cultivars are often preferred by producers and consumers, this also leads to the use of a limited genetic diversity, which represents a risk for the adaptation of viticulture and wine making to changing environments and market demands.

Global warming is changing climatic conditions in traditional winemaking regions [2, 3]. Along with the prolonged use of grapevine in monoculture and

Rank	Prime variety	Color	Origin	Area (hectares)	Share (%)	Cumulative share (%)
1	Cabernet Sauvignon	R	France	290,091	6.30	6.30
2	Merlot	R	France	267,169	5.81	12.11
3	Airén	W	Spain	252,364	5.48	17.60
4	Tempranillo	R	Spain	232,561	5.05	22.65
5	Chardonnay	W	France	198,793	4.32	26.97
6	Syrah	R	France	185,568	4.03	31.00
7	Garnacha Tinta	R	Spain	184,735	4.01	35.02
8	Sauvignon Blanc	W	France	110,138	2.39	37.41
9	Trebbiano Toscano	W	Italy	109,772	2.39	39,80
10	Pinot Noir	R	France	86,662	1.88	41.68
11	Mazuelo	R	Spain	80,178	1.74	43.42
12	Bobal	R	Spain	80,120	1.74	45.16
13	Sangiovese	R	Italy	77,709	1.69	46.85
14	Monastrell	R	Spain	69,850	1.52	48.37
15	Grasevina	W	Croatia	61,200	1.33	49.70
16	Rkatsiteli	W	Georgia	58,641	1.27	50.97

Selected from Wine Economics Research Centre [1].

Table 1.
Grapevine cultivars contributing to more than 50% of world vineyard surface.

globalization-related issues, global warming associates with the emergence of new pathogen and pest threats [4]. At the same time, consumers and new agriculture and food safety regulations are more and more demanding a reduction in the use of pesticides and fungicides in viticulture [5]. In this context, strategies to adapt viticulture in different regions to different models and markets are required to ensure the sustainability of the crop. Among the multiple possibilities that can be considered to this aim, strategies intending the genetic improvement and adaptation of elite and autochthonous varieties are very relevant to keep their intrinsic varietal values—these cultivars are traditionally related with wine quality and are indeed the basis of the most famous and expensive wines.

Grapevine varieties derive from the domestication of wild forms of the species *Vitis vinifera* [6]. Wild grapevines are dioecious, which obligates to outcrossing and results in highly heterozygous genotypes, a genetic feature that has been inherited by domesticated forms. This is the reason why vegetative propagation has been the preferred method to multiply selected grapevine varieties since ancient times, to keep the varietal attributes and shorten production lapses. In fact, most traditional cultivars in use nowadays derive from seeds that probably germinated several centuries ago and that have been vegetatively multiplied since that time to currently cover large vineyard surfaces as those shown in **Table 1**.

All species within the genus *Vitis* are cross-fertile and the identification of sources of genetic resistant for *Vitis vinifera* pathogens and pests mainly in other American or Asian species opened the possibility to improve grapevine varieties through classical breeding strategies. This approach has been successfully developed during the twentieth century and new resistant grapevine varieties have reached the markets with different success rates [7–10]. Furthermore, rising knowledge of the grapevine genome and the development of new genomics and molecular techniques in the last decade have triggered a renewed interest for breeding given, the pressure to reduce the use of pesticides in viticulture [10]. Genomics-assisted breeding represents an interesting and efficient strategy that has the potential to change the role of genetic materials in viticulture and wine making [11, 12]. Still, breeding ends in new grapevine genotypes that need to be registered as new varieties with new names, what generates bureaucratic problems delaying their commercial use and hindering their acceptance by the market.

Viticulture based on traditional varieties has relied on the phenotypic variation generated by spontaneous somatic mutations for the improvement and diversification of the crop. This variation has been traditionally selected by farmers along the history of viticulture to improve cultivars and adapt production to evolving conditions [6]. Later, along the twentieth century, this somatic variation became the basis of clonal selection. This strategy has the advantage that the derived clones keep the original cultivar name and are already mostly adapted to vineyard management practices and the wine making process as well as the market [13]. Varieties are considered to consist in groups of clones selected during vegetative propagation that share common features. When clones of the same variety have phenotypes different enough to be grown for the production of different wines, they can be considered as derived varieties [14] that could keep the name of the progenitor variety. For example, this is the case of Pinot Noir Blanc derived from Pinot or the recent Tempranillo Blanco derived from Tempranillo [15]. By the time being, the advent of new genomic and phenotypic techniques enables the identification of the origin and features of somatic mutations and the associated phenotypic variation, knowledge that can be exploited to efficiently improve the adaption of traditional cultivars to changing market and environmental demands. This strategy can be complementary to the development of new varieties by breeding and help understanding the genome diversity of traditional varieties and maintaining their production. In fact, part of the variation used in breeding programs is the result of somatic mutations selected along grapevine domestication as described in subsequent sections.

Along this chapter, we summarize what we have learned from the study of somatic variants in grapevine cultivars, their origin, their value, and the interest of their study. We also review several examples of very relevant grapevine traits that likely originated by somatic variation and that have been characterized at the molecular level and discuss how understanding the basis of this variation can now help to apply new technologies to the genetic improvement of grapevine cultivars.

2. Grapevine somatic variation

Despite vegetative propagation is used in grapevine to multiply plants that are identical to the original type, spontaneous phenotypic variation occasionally appears on some shoots (known as bud sports) as a result of somatic mutations [16]. From bud sports, the new variant phenotype can be established as a whole plant and, eventually, as a new variety, using the same propagation strategy. Bud sports can display any type of phenotypic variation at any organ, leaf, stem, bunch, berries, seeds, etc. Variation can affect reproductive traits that determine yield and quality such as fertility, cluster compactness, berry color, or flavor. Somatic variation can also affect vegetative traits including plant vigor, leaf morphology, or even disease susceptibility. There are some cultivars like Pinot Noir, Sultanina, or Italia [14, 16] for which a large number of somatic variants have been identified for multiple traits. The number of sports appearing in a given variety is expected to increase proportionally with its age and vineyard surface. In addition, the possibility that some genotypes are more prone to generate somatic variants has not been proven but cannot be discarded.

Spontaneous somatic variation results from the combination of mutations and cellular events. Initially, mutations take place in single meristematic cells associated with the DNA replication and cell division processes. Somatic mutations accumulate at a very low frequency. However, since current plants of traditional grapevine cultivars result from millions of mitotic divisions since the germination of the original seed, they accumulate a relatively large number of mutations in their genomes (see the next paragraph). For nuclear DNA, every somatic mutation can be considered to be heterozygous as they only affect one of the two existing genome copies per cell. These somatic mutations can range from single nucleotide substitutions to nucleotide insertions or deletions or even to large DNA sequence recombinations causing chromosomal reorganizations [16]. Other infrequent alterations include the change of ploidy level of the cell, reported in different varieties [17]. Somatic epimutations altering gene expression without affecting nucleotide sequence and causing new phenotypes have so far not been described in grapevine.

Most mutations do not have any effect on gene and cellular functions since only a small part of the genome sequence is involved in coding or regulatory functions [18]. Even mutations in coding sequences do not always generate amino acid changes in the encoded protein or if they do, still in many cases they behave as silent changes. Emergent somatic mutation will only affect one of the two copies of a given gene. This makes derived phenotypic effects to be mostly expected from dominant mutations either due to gains of function or haploinsufficiency. Independent recessive mutations causing loss of functional alleles in heterozygous loci carrying a null allele could also generate phenotypic effects although at low frequency. Importantly, deleterious mutations constraining essential cell functions will not accumulate because purifying somatic selection will prevent their propagation in the plant.

Cellular events associated with the stabilization of somatic mutations are conditioned by the tissue structure of plant meristems. The grapevine shoot apical meristem is organized in at least two cell layers, the outer L1 and the inner L2, from which all the cells of the plant derive [19]. These cell layers constitute almost closed compartments with very limited cell exchange between them. Cell division and differentiation in the L1 layer gives rise to all the epidermal cells of all the plant organs, while the L2 layer generates the cells that constitute all their internal tissues. The L2 cell layer is also responsible for gamete development within reproductive flower organs. Because mutations emerge spontaneously in either L1 or L2 layers, grapevine plants are genetic chimeras that carry slightly different genetic composition in L1- and L2-derived cell lines. In addition, vegetative multiplication from cuttings along centuries contributes to select and enrich part of the variation accumulating in the plant. At the same time, because of the lack of sexual reproduction, there is no purifying selection against mutations that could have deleterious effects on gametogenesis, fertilization, zygote formation, embryo development, seed germination, or juvenile growth.

To manifest a mutant phenotype in a given plant organ, the mutation has to propagate through cell division from the original mutant meristematic cell. Initially, mutant daughter cells occupy a meristem cell layer (either the L1 or L2) or sectors of it, which subsequently gives rise to mutant organs by additional cell divisions. Once the mutant genotype is propagated in the L1, the L2, or both cell layers of a shoot apical meristem, the mutation could be transmitted by bud propagation. Periclinal chimeras with somatic mutations fixed in only one meristem cell layer are quite stable in grapevine, and indeed, some varieties like Pinot Meunier (L1 mutant) [20] or Pinot Gris (L2 mutant) [21] are chimeras that are stably maintained through vegetative multiplication as we explain in sections below. If the mutant daughter cells colonize both meristem cell layers by migration of mutant cells to the wild type layer, bud multiplication from such mutant buds will fix the mutation in all tissues of derived plants. This is the case of white-berried variants derived from originally black-berried varieties such as Pinot Blanc [21]. Since plants do not have a separated germline, somatic mutations present in the L2 can be transmitted through sexual reproduction as far as they are not lethal in the haploid phase. Somatic mutations generating new interesting phenotypes, stabilized in grapevine plants as periclinal chimeras, or extended to all cell layers, have been selected as new clones of wine grape cultivars or as new derived cultivars [6, 14, 16].

3. Genome sequence variation within cultivars

Sequencing and de novo assembly and annotation of the first grapevine genomes [18, 22] provided a new body of knowledge and a new toolbox for the study of genome sequence diversity. Two different strategies were used for the first genome assemblies, a homozygous assembly based on PN40024, a partially inbred line derived from Pinot Noir, [18] or an assembly including both, consensus contigs of the two genome copies and independent contigs for each of the two haplotypes in more dissimilar genome regions of Pinot Noir (ENTAV 115) [22]. Both projects estimated a haploid genome size close to 500 Mb. More recently, long-read sequencing technologies such as PacBio are facilitating the release of haplotype-resolved assemblies, which are already available for the heterozygous grapevine cultivars Cabernet Sauvignon and Chardonnay [23, 24]. By the time being, the availability of reference genomes combined with the development of next-generation sequencing (NGS) technologies enable genome-wide analysis of the grapevine germplasm at affordable costs, which is extremely useful in genetic diversity studies as well as to search for mutations causing phenotypic variation [15, 24–26]. Although the use of these approaches to characterize somatic variation in grapevine is still scarce, an increasing number of publications are shedding light on the magnitude and type of variation that accumulates at the genome level within given cultivars.

Somatic SNV (single nucleotide variants) and small insertions/deletions (INDEL) mutations are often the result of errors in DNA replication taking place during mitotic cell division. While the frequency of INDEL may exceed that of single base substitutions due for instance to low resolution of polymerases at homopolymeric or short repeats, INDEL are more difficult to detect using high-throughput sequencing methods due to the same reason. The first attempt to detect somatic polymorphisms at a genome-wide scale in grapevine used 454 GS-FLX sequencing technology to compare three Pinot Noir clones to the sequences in the genome assemblies of the Pinot-related accessions PN40024 and ENTAV-115 [27]. In this study, mean rates of 1.6 SNV, 5.1 INDEL, and 35.2 mobile element movements per Mb were described among clones. Short-read sequencing technologies led by Illumina provide a framework to accurately detect SNV and are also useful to detect small INDEL. In this manner, genome resequencing of three clones corresponding to different morphotypes of the ancient Italian wine cultivar Nebbiolo identified between 16 and 26 clone-specific SNV per Mb of genome [28]. However, these numbers might be over-estimated considering that the validation success was 61% for a quality-trimmed sub-selection of SNV [28]. More recently, the re-sequencing of 15 clones of Chardonnay compared to a de novo genome draft assembly for this cultivar identified a much more reduced number of SNV using a stringent k-mer-based calling strategy variation [24]. The sum of SNV + INDEL ranged between 221 and 2 polymorphisms per clone (0.004–0.455 per Mb of genome), which corresponds to at least three orders of magnitude of lower rates than in the Nebbiolo study, despite that Chardonnay accessions corresponded to diverse geographical origins and phenotypes including seedlessness and berry color variation [24]. Concerning the putative impact of these polymorphisms, a total of 21 (0.07%) and 55 (3.4%) clone-specific variants were predicted as potentially altering protein function in Nebbiolo and Chardonnay, respectively, including one nonsynonymous substitution in the *VviDXS* gene as the possible origin of the Muscat flavor of one Chardonnay clone [24, 28]. Transcriptome re-sequencing (RNA-seq) can also be useful to identify polymorphisms in coding sequences. For example, an RNA-seq study comparing the seedless somatic variant Corinto Bianco to its seeded ancestor Pedro Ximenes identified 13 polymorphisms with 100% validation rate (12 SNV and one dinucleotide), all of them being heterozygous variants [29]. This is also important to be considered since, rather than resulting from direct base substitution mutations, some of the somatic SNV detected in sequencing studies might correspond to loss of heterozygosity (LOH) in hemizygous regions generated after somatic SV.

SV involves changes in the chromosome landscape. It includes inter- and intra-chromosomal translocations, deletions, and insertions (the last two types are generally considered as SV if >1 kb) including those caused by the movement of transposable elements (TE) [30, 31]. The rapidly growing number of genomic studies in multiple species is unveiling more complex forms of SV, collectively known as chromoanagenesis, and combines several of the previous features [32]. In addition to the activity of TE, SV often relies on mistakes in replicative processes or on DNA breakage during mitosis followed by illegitimate repair mechanisms [33–35]. Although SV is generally deleterious, it can accumulate along the multiplication of grapevine cultivars behaving as recessive heterozygous due to the absence of sexual reproduction [15]. Features such as changes in copy number and breakpoint joins have been used in genomic studies to detect SV between grapevine cultivars and somatic variants [15, 25, 36–38]. By far, the most recurrently described case of somatic SV in grapevine relates to hemizygous deletions of different sizes around the grape color locus on chromosome 2 that causes loss of berry color variants (see below). Smaller SV, translocations, and inversions have also been described in somatic variants differing in ripening time [25]. Genome-wide SV studies in a higher number of clones would be required to estimate the frequency of different types of SV independently of specific phenotypes or genome regions resulting from human selection.

TE are extremely frequent in plant genomes and correspond to sequences that have the ability to replicate and insert in different locations, either indirectly through an RNA intermediate (retrotransposons or class I) or directly by cut-and-paste mechanisms (transposons or class II) [39]. The transposition of these elements generates changes in genome size and can disrupt target loci upon insertion. In addition, TE can lead to SV and genome rearrangements due to noncanonical transposition events or to homologous recombination related with their repetitive nature [39]. Altogether, TE has a high potential to impact on organismal phenotypes. While all superfamilies of TE are represented in the grapevine genome, those in class I (e.g.: Non-LTR LINEs, LTR Ty1/copia, LTR Ty3/gypsy, and other LTR) are much more numerous (>100,000 copies in total) than class II superfamilies (hAT, PIF, Mutator, and CACTA) totaling about 3000 copies in the grapevine reference genome [18, 40]. Because ca 50% of the grapevine genomes involve mobile element-like/repetitive sequences [18, 38], it is reasonable that they could be a major driver of somatic variation emerging during the extensive vegetative multiplication of grapevine cultivars. In fact, emergent phenotypes in grapevine somatic variants have frequently been associated with the movement of TE altering gene expression [41–43], although, with the exception of color variants, their phenotypes have not been selected for production. While the use of molecular markers suggests that the TE genomic landscape can vary between grapevine clones [27, 44, 45], systematic studies are still required to determine the magnitude of somatic genome variation that accumulates associated to TE during the propagation of grapevine cultivars.

4. Nucleotide sequence variation underlying grapevine somatic variation

The availability of grapevine reference genomes and the advent of NGS technologies have paved the way for the identification of the nucleotide diversity underlying variation for relevant phenotypic traits in grapevine. Somatic variants are excellent tools for this goal, since they allow studying the mutation effect in a common genetic background when comparing somatic variants to the direct ancestor of the same cultivar. This facilitates the identification of the causal genes and gene variants. In fact, in the last years, the molecular and genetic basis of an increasing number of phenotypic traits has been elucidated using somatic variants as experimental systems.

We consider transcriptome RNA-seq comparisons as an excellent diagnosis tool for the screening of candidate genes because this technology has the potential to trace mutations that alter either gene expression or coding sequences. In our hands, the process starts with a careful phenotypic analysis comparing the progenitor normal plant and the somatic variants. Concurrently, we develop self-cross derived progenies of both genotypes for segregation analyses. The main objective of the phenotypic analysis is to understand the developmental origin of the emerged trait. In this manner, we can identify a target organ, tissue, and developmental stage in which the mutation is initially expressed and take samples of it from each variant to conduct a transcriptome comparison. The interpretation of gene biological function from the developmental and phenotypic variation can frequently be misleading since, as mentioned before, many of these mutations have dominant gain-of-function effects.

Under these premises, transcriptome comparison, both at gene expression and sequence levels, combined with the results of segregation analyses of mutant phenotype in self-cross populations of each variant can provide a preliminary identification of putative candidate genes. These candidates will have to be confirmed by directly comparing their sequences in normal and somatic variants of the same cultivar. Both in transcriptome and sequence analyses, it is important to consider the possible chimeric state of causal mutations in the somatic variants.

When the described approaches lead to the identification of sequence variation susceptible of generating the mutant phenotypic effect, it is still required to confirm that this sequence variation is the cause of the phenotype. When the responsible mutation is present in the L2 layer and can be transmitted through gametes, co-segregation of the mutant phenotype with the candidate sequence variants would support a causality relationship although it is not a definitive proof. Genetic transformation to restore normal or variant phenotypes can be a difficult and time-consuming alternative in grapevine. Other possibilities like allele-specific expression analyses or sequence characterization of a large number of variants or cultivars displaying the same phenotype have been used in different cases to proof that a candidate gene variant is in fact responsible for a relevant phenotypic effect [26, 42, 43, 46].

In the next section, we review several examples of studies taking advantage of somatic variants to understand the molecular genetics of four relevant grape traits.

4.1 Meunier phenotype

The Meunier phenotype accounts for the tomentose (hairy) phenotype of shoot tips and leaves in cultivar Pinot Meunier derived from Pinot noir. Plants derived by somatic embryogenesis from different L1 and L2 cell layers of Pinot Meunier showed different phenotypes, demonstrating that Pinot Meunier is a periclinal chimera carrying a mutant L1 line responsible for the Meunier phenotype. In addition, those plants regenerated from L1 somatic embryos displayed a new dwarf phenotype with short internodes [20]. Further characterization of those dwarf plants showed that they produced inflorescences and bunches in all nodes along the length of the shoots [47]. Grapevine nodes develop either inflorescences or tendrils that share a common ontogenetic origin from uncommitted primordia in grapevine [48]. Application of gibberellins (GAs) and GA biosynthesis inhibitors has been shown to modify tendril and inflorescence development in grapevine [49]. This phenotype suggested that the Pinot Meunier was associated with an altered response to gibberellins, what was confirmed by the high levels of active gibberellins detected in the dwarf plants paralleled by their insensitivity to the application of these hormones [47]. It is also similar to the phenotype of *gai* mutants of *Arabidopsis*, carrying mutations in GAI, a negative regulator of GA response [50]. In fact, dwarf plants derived from Pinot Meunier L1 were shown to carry a point mutation in a *GAI* homologous gene converting a leucine residue into a histidine within its conserved DELLA domain (**Figure 1**), the GA-sensitive domain unique to all members of this family of regulatory proteins [47]. The final proof confirming the role of this mutation in the origin of the dwarf phenotype came from the genetic analyses performed on self- and out-crosses of the mutant dwarf plants regenerated from the L1 of Pinot Meunier. The results showed that the mutated allele behaved in a semi-dominant manner, with homozygous mutant plants displaying a more extreme dwarf phenotype [47].

Similar hairy phenotypes have also been found in other cultivars given the names of some derived varieties like Garnacha “peluda” (hairy in Spanish), a name that refers to the tomentose phenotype. However, whether this phenotype has the same genetic and molecular basis as the Meunier phenotype has not been investigated. The Meunier phenotype constitutes a great example of how a relevant agricultural trait can be generated by a mutation in chimeric state in a somatic variant, a feature that is lost when the mutation is present in all plant cell layers. Because bibliographic references to Pinot Meunier or Schwarzriesling in Germany date back at least to the seventeenth century [51], this case itself proves the stability of periclinal chimeras in grapevine.

Understanding the molecular basis of this phenotype opens the possibility to recreate with genome editing tools such as CRISPR/Cas9 [52] the causative single point mutation or other point mutations known to have similar effects on the DELLA domain of the GAI regulatory proteins. However, the replication of the Meunier phenotype in the same or other cultivar backgrounds will be difficult because it will require the mutation to be stable only in the L1 cell layer, something that could require more sophisticated cell culture and plant regeneration techniques. This exemplifies the specificity of phenotypes resulting from chimeric states.

To end, it is important to mention that the capacity of these dwarf plants to flower rapidly from the initiation of the first tendril makes them useful model systems for genetic studies in grapevine [53]. In this case, genome editing to recreate mutations in the DELLA gene and regenerate whole mutant plants to obtain dwarf models in cultivars other than Pinot would be more feasible.

4.2 Muscat flavor

The Muscat flavor in grapevine describes an intense floral aroma present in the berries of some specific cultivars and their derived wines. It is linked to the high accumulation of monoterpenoids such as linalool, geraniol, nerol, citronellol, and α-terpineol, all having a low olfactory perception threshold [54]. This aroma has been strongly appreciated since ancient times and a family of closely related Muscat varieties was spread from the Eastern Mediterranean area by Greeks and Romans and can still be found with different names in many locations of the world [55]. The Muscat flavor has also been found in somatic variants of cultivars like Savagnin, Chardonnay, or Chasselas [46, 56].

Genetic analyses of Muscat aroma in grapevine have been performed in biparental progenies involving Muscat varieties [57, 58] and in self-cross derived populations of Muscat Ottonel and Gewurztraminer (a Muscat flavor somatic variant of Savagnin) [59]. Muscat aroma segregated as a dominant trait and at least one common major QTL responsible for Muscat aroma was detected in all progenies, located in linkage group 5. A positional candidate gene, 1-deoxy-D-xylulose-5-phosphate synthase (*VviDXS*), was proposed to account for the terpenoid overproduction phenotype [58, 59]. This gene encodes the first enzyme of the plastidial methylerythritol phosphate (MEP) pathway, which functions upstream in monoterpene and diterpene biosynthesis. Several investigations have shown that this enzymatic reaction is a biosynthetic step of the pathway that limits terpenoid biosynthesis in plants [60].

Based on those hypotheses, Emanuelli et al. [46] re-sequenced the *VviDXS* grapevine gene in a collection of 148 grape varieties, including Muscat-aromatic as well as other aromatic and neutral accessions. Among the SNP significantly associated with the presence of Muscat aroma, they identified the putative causal SNP responsible for the Muscat phenotype. This SNP is present in all Muscat varieties and generates a predicted nonneutral substitution of a lysine by an asparagine in residue 284 of *VviDXS*. Interestingly, Muscat-like aromatic somatic variants also displayed unique nonsynonymous mutations in close positions of the same

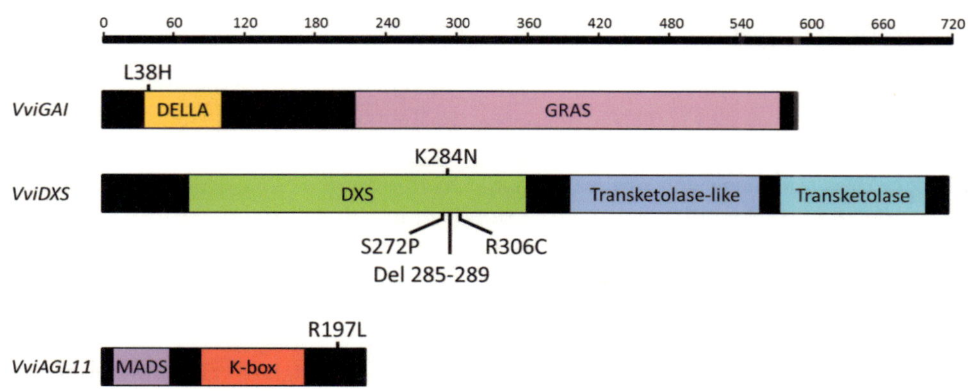

Figure 1.
VviGAI, VviDXS, and VviAGL11 proteins and mutations responsible for Meunier, Muscat, and Seedlessness traits. Protein domains are represented according to Pfam database. L38H mutation in the DELLA domain results in the lack of GA-response in pinot Meunier. K284N mutation is present in all Muscat varieties. Three additional independent mutations were identified in the same DXS domain in Muscat-like aroma somatic variants: S272P in chardonnay Musqué, deletion (Del) of five amino acids 285–289 in Chasselas Musqué, and R306C in Gewürztraminer. R197L mutation in AGL11 located in the C-terminus of the protein alters development and lignification of the seed coat.

domain of DXS protein (**Figure 1**). A serine substitution by a proline in position 272 in Chardonnay Muscat, an arginine substitution by cysteine in position 306 in Gewurztraminer, and a deletion involving five amino acids in position 285–289 on Chasselas Musqué [46]. Altogether, the correlation of independent nonlethal spontaneous Muscat mutations in this conserved DXS domain with the presence of Muscat aroma in all studied cases suggests its relevance in protein function. In fact, the Muscat amino acid substitution influences the enzyme kinetics by increasing its catalytic efficiency and it is also able to dramatically increase monoterpene levels in transgenic tobacco plants [60].

The closely related genetic relationships among Muscat varieties could be interpreted as resulting from the original selection of a somatic variant in which this characteristic aroma emerged and was propagated vegetatively. Occasional hybridization of this variant with other cultivars grown in ancient times as well as more recent directed hybridizations would have generated the currently available plethora of Muscat varieties (**Figure 2**). The identification of independent non-neutral amino acid substitutions or amino acid deletions in the same protein region clearly identifies *VviDXS* as a target gene to improve Muscat aroma through breeding (**Figure 1**). Specific nonneutral amino acid substitutions are not easily obtained from mutagenesis programs. However, the current catalog of known sequence variants in *VviDXS* provides several specific amino acid changes that could be recreated through genome sequence editing to introduce the Muscat flavor trait in any desired cultivar.

4.3 Berry color

Berry color is a very relevant trait determining consumer preferences in table grapes as well as the type of wines that can be elaborated from wine grape cultivars. In this way, red and *rosé* wines are made from black-berried cultivars, while white-berried cultivars are used for making white wines. In grapevine, berry color results from the biosynthesis and vacuolar accumulation of anthocyanins in berry skin cells during the ripening process from *veraison* stage. Variation for berry color is determined by a major locus on linkage group 2 [61, 62]. This berry color locus co-localizes with a cluster of tandemly repeated *VviMybA* genes [63]. Among them, *VviMybA1* and *VviMybA2* are expressed in the berry skin of black-berried cultivars from *veraison* [64]. The function of these transcription factors is required to trigger the expression of target genes such us UDP-glucose:flavonoid 3-O-glucosyltransferase (*UFGT*), encoding the limiting enzyme activity for the anthocyanin biosynthetic pathway [64, 65]. Original wild grapevines producing black berries and berry color diversity could have emerged as a result of somatic variation and be selected as a domestication trait in cultivated out-crossed forms [66]. Black-berried cultivars carry at least one functional copy of both *VviMybA1* and *VviMybA2* linked in a functional allele of the color locus. White-berried cultivars do not synthesize anthocyanins in the berry skin and they lack functional copies of these *MYBA* genes at the color locus. Most white-berried cultivars are homozygous for the canonical null allele of the locus in which *Gret1* retrotransposon insertion in the promoter of *VviMybA1* along with a small INDEL causing a frame-shift in *VviMybA2*, respectively, causes loss-of-function in the two genes [41, 64]. Most of the diversity in berry color observed among grapevine varieties has been related to nucleotide sequence variation in this locus [67].

In addition, intracultivar variation for berry color, useful to select new derived varieties, has also been associated to variation in the berry color locus. In this way, spontaneous red-berried variants identified in white-berried table grape cultivars like Italia or in wine cultivars were shown to derive from recombination, reverting the insertion of the *Gret1* retrotransposable element present in the promoter of *VviMYBA1* in white alleles, which at least partially recovers the expression of the gene [41, 68]. In other cases, red-berried variants emerged as new functional *MYBA* genes resulting from the recombination of nonfunctional homologous genes within the color locus [69]. On the other hand, black-berried cultivars heterozygous for the null allele occasionally display grape color variants with either red/gray or white berries depending on whether only the L2 or both L1 and L2 meristem cell layers (**Figure 3**), respectively, carry mutations at the color locus [21, 70–73]. Molecular characterization of red/gray and white berry somatic variants of Cabernet Sauvignon and Pinot Noir cultivars through Southern blots showed that the lack of berry skin anthocyanins was associated with deletion of the functional allele of the color locus [70, 74]. Later, the loss of heterozygosity along the color locus has been used to size the extent of deletions [21, 72, 73]. This heterozygosity loss has been directly related to spontaneous deletions involving the functional color locus allele and resulting in hemizygosity at the grape color locus, leaving only the null allele [15].

Altogether, these results demonstrate that intracultivar color variation appearing in either white or colored cultivars is mostly associated to structural variation at the color locus on linkage group 2, in combination with cellular events generating different chimeric situations and color patterns. Gain of color variants generally correspond to recombination events within the locus that generate gain-of-function mutations and dominant phenotypes. Loss of color variation seems to be restricted to black-berried cultivars heterozygous for a functional allele at the color locus and is associated with different deletions or complex chromosomal rearrangements eliminating this single functional copy [15]. Based on this information, bud irradiation with physical mutagens increasing the frequency of recombination and deletion could be a strategy to generate new color variants in grapevine.

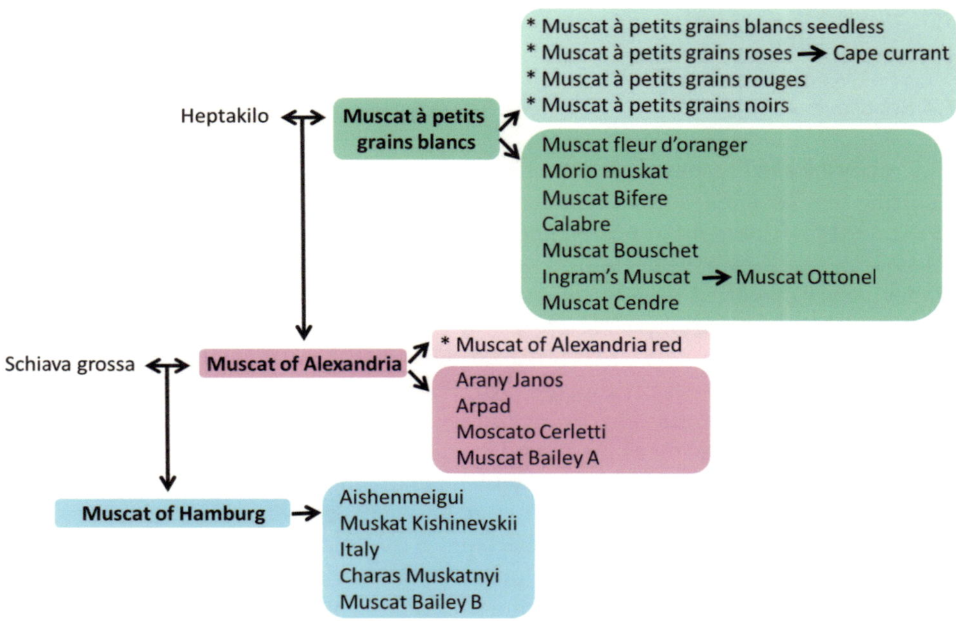

Figure 2.
Genetic relationships among Muscat varieties. Muscat à petits grains blancs is the progenitor of ancient variety Muscat of Alexandria and the putative ancestor of all the Muscat varieties. From them, additional Muscat varieties are derived by spontaneous or directed hybridizations (see [55] and Vitis *International Variety Catalog (http://www.vivc.de)) as well as through somatic variation (*).*

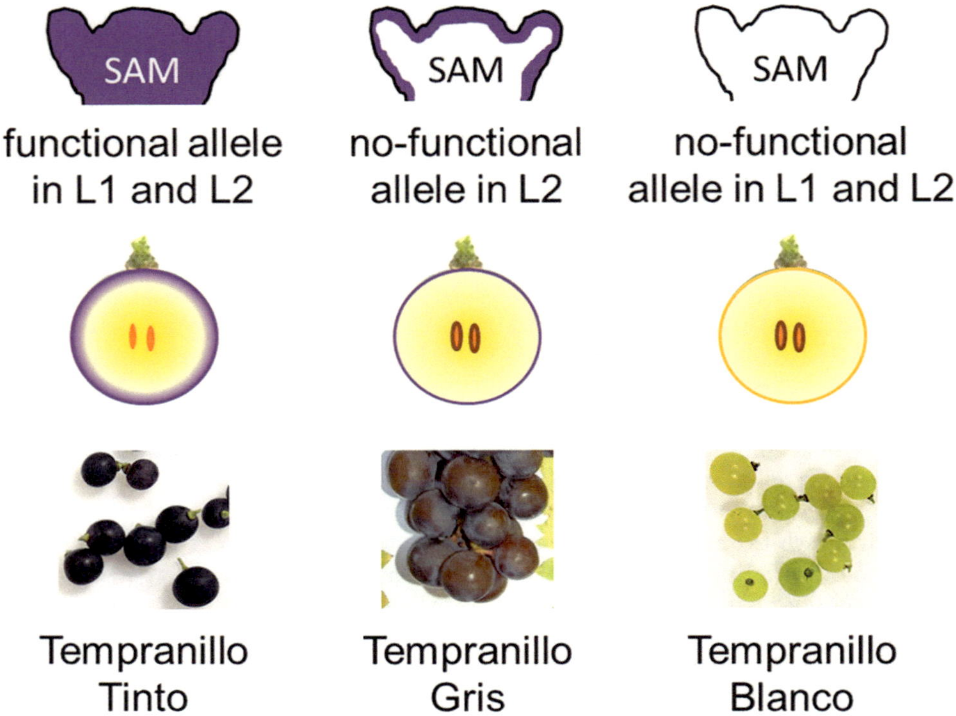

Figure 3.
Proposed genetic composition of shoot apical meristem (SAM) and berry color in Tempranillo somatic variants. L1 (outer) and L2 (inner) layers are represented in SAM, purple color indicates that cells in the layer carry a functional allele at the color locus and white color indicates the lack of functional color alleles in the cells. One functional allele in both meristematic layers is enough to develop black berries, while periclinal chimera with a mutant L2 cell layer in the SAM gives rise to gray color berries, the lack of functional alleles in both meristem layers yields white berries.

4.4 Seedlessness

Grapevine seedlessness is one of the best examples of cultivar innovation resulting from original somatic mutation in table grapes. Somatic variants defective in seed development appeared spontaneously along the history of grapevine cultivation and they have been propagated vegetatively [6]. Seedless variants in grapevine are classified into two major classes: (i) parthenocarpy, when fruits are set and develop without fertilization resulting in small berries free of seeds [75] and (ii) stenospermocarpy, when fertilization and embryo formation is not altered but later seed development is aborted [76]. Parthenocarpic varieties have been widely used for the production of Corinto seedless raisins, but, as their sterility makes sexual transmission of the causal mutation impossible, the use of this trait remains limited to those genotypes in which parthenocarpy appeared spontaneously. Recent work in Corinto Bianco, a parthenocarpic variant derived from Pedro Ximenes cultivar [75], has pointed out to meiotic alterations precluding the development of viable gametes as the origin of the mutant phenotype [29]. On the other hand, an ancient somatic mutation producing a stable stenospermocarpy phenotype likely emerged in a white-berried oriental cultivar known as 'Kishmish,' also known as 'Sultanina' or 'Thompson Seedless' [77]. Since the mutation responsible for stenospermocarpy has a lower impact than parthenocarpy in berry size and does not lead to sterility (pollen is fertile and embryos can also be rescued from seed traces), it has become the major source of seedlessness in table grape breeding [13, 78, 79] (**Figure 4**).

The stenospermocarpy phenotype has been associated with abnormal development of the inner ovule integument [80], which ends in impaired development and lack of lignification of maternal seed coat tissues [81]. Genetic analyses of seedlessness trait in several F1 progenies derived from at least one stenospermocarpic progenitor identified segregations that could be explained by the presence of a dominant locus named Seed Development Inhibitor (*SDI*) interacting with several recessive loci [82, 83]. Later, quantitative genetic analyses identified the *SDI* locus as a major QTL on linkage group 18, explaining up to 70% of the phenotypic variance for different seed variables [84–87]. Based on co-localization of this QTL with a grapevine homolog of the Arabidopsis MADS-box transcription factor gene *AGAMOUS-LIKE11* (*AGL11*), responsible for ovule morphogenesis and seed coat differentiation [88], *VviAGL11* was considered the best candidate gene for the *SDI* locus [86, 87]. More recently, using an independent positional study combined with targeted sequencing in a large collection of seeded and stenospermocarpic grapevine cultivars, a single nucleotide missense mutation in *VviAGL11* was identified as the causal origin of the dominant seedless phenotype [26]. This mutation causes the substitution of a conserved arginine 197 into leucine (**Figure 1**), which could disrupt the function of multimeric complexes containing *VviAGL11* proteins in a dominant manner. Interestingly, amino acid sequence variants of oil palm *AGL11* homologs have also been selected in this crop to reduce the level of seed coat lignification [89]. Apart from the relevant application of the identification of the causal point mutation in *VviAGL11* to develop efficient marker-assisted selection strategies for seedless grape breeding, this information paves the way to the development of targeted genome editing for the genetic improvement of seedless table grapes. Stenospermocarpic seedlessness could also be useful in black-berried wine grapes as a way to avoid the negative effects of unripe seeds in the sensory quality of red wines [90]. Ripening imbalance between pulp and seeds can become a problem under climate change conditions [3], what could be addressed with the use of

Figure 4.
Red globe and crimson seedless fruits.

seedless wine varieties. Finally, editing of *AGL11* homologs could also be useful to generate seedlessness in other fruit crops.

5. Final considerations on the use of somatic variation

The application of NGS to the study of somatic variation in grapevine is increasing our knowledge on the nucleotide sequence variation underlying phenotype variation. By direct comparison of somatic variants, this technology has the potential to identify causal candidates at the gene and gene variant levels. Regardless, genetic and molecular approaches are still required to confirm the role of those candidates. So far, NGS approaches have been used to unravel widely used classical phenotypes as those described along the chapter. When combined with genome edition technologies, they constitute new tools for the genetic improvement and adaptation of traditional elite grapevine wine cultivars.

The first conclusion that comes out from the review of currently available information in grapevine is that due to the essential heterozygous condition of emergent somatic mutations, only dominant mutations can generate somatic variant phenotypes. More frequently, these dominant mutations involve gains of function resulting from either SNV that generate nonneutral amino acid substitutions [26, 46, 47] or gene overexpression and misexpression caused by transposon insertions [42, 43] or recombinations [41, 69]. Loss of function mutations has also been described but so far only in the case of SV that unmasks the effect of recessive null alleles present at the color locus in cultivars that are heterozygous for functional and null alleles of the responsible *MYBA* genes [15]. Another interesting conclusion relates to the particular relevance that chimeric expression of the mutations can have in the generation of specific cultivars such Meunier or the gray-berried variants. These examples show once more how the same mutation can lead to different phenotypes depending on the meristem cell layers affected.

Dominant gain-of-function mutations identified in grapevine somatic variants exemplify how new gene functions can be created by mutations changing expression to different cell types, developmental stage, or transcription levels, o by the alteration of a key amino acid in functional protein domains. While the effects of loss of function mutations are generally easy to predict when the function of the affected genes is known, gain of function is much more unpredictable and represents a source of innovation that can create new possibilities for genetic improvement. Their dominant nature makes them especially useful not only for the improvement of traditional cultivars but also to breed new cultivars. Systematic screening of the large clonal germplasm hosted in old vineyards and collections of ancient accessions of traditional cultivars can unveil very relevant information and variant traits to be exploited in conventional, genomics-assisted, or genetic engineering-mediated breeding.

Acknowledgements

Current research activities in our group are funded by Project BIO2017-86375-R.

Conflict of interest

The authors declare no conflict of interest.

DOI: http://dx.doi.org/10.5772/intechopen.86443

Author details

Pablo Carbonell-Bejerano, Carolina Royo, Nuria Mauri, Javier Ibáñez and José Miguel Martínez Zapater*
Instituto de Ciencias de la Vid y del Vino (CSIC, University of La Rioja, Government of La Rioja), Logroño, Spain

*Address all correspondence to: zapater@icvv.es

References

[1] Wine Economics Research Centre. The University of Adelaide. 2010. Available from: http://www.adelaide.edu.au/wine-econ/databases/winegrapes/

[2] Mosedale JR, Abernethy KE, Smart RE, Wilson RJ, Maclean IM. Climate change impacts and adaptive strategies: Lessons from the grapevine. Global Change Biology. 2016;**22**:3814-3828. DOI: 10.1111/gcb.13406

[3] van Leeuwen C, Darriet P. The impact of climate change on viticulture and wine quality. Journal of Wine Economics. 2016;**11**:150-167. DOI: 10.1017/jwe.2015.21

[4] Caffarra A, Rinaldi M, Eccel E, Rossi V, Pertot I. Modelling the impact of climate change on the interaction between grapevine and its pests and pathogens: European grapevine moth and powdery mildew. Agriculture, Ecosystems and Environment. 2012;**148**:89-101. DOI: 10.1016/j.agee.2011.11.017

[5] Pertot I, Caffi T, Rossi V, Mugnai L, Hoffmann C, Grando MS, et al. A critical review of plant protection tools for reducing pesticide use on grapevine and new perspectives for the implementation of IPM in viticulture. Crop Protection. 2017;**97**:70-84. DOI: 10.1016/j.cropro.2016.11.025

[6] This P, Lacombe T, Thomas MR. Historical origins and genetic diversity of wine grapes. Trends in Genetics. 2006;**22**:511-519. DOI: 10.1016/j.tig.2006.07.008

[7] Bouquet A. Grapevines and viticulture. In: Adam-Blondon AF, Martínez Zapater JM, Kole C, editors. Genetics, Genomics, and Breeding of Grapes. Vol. 2011. Boca Raton, Florida, USA: CRC Press; 2011. pp. 1-29

[8] Toepfer R, Hausmann L, Eibach R. Molecular breeding. In: Adam-Blondon AF, Martínez Zapater JM, Kole C, editors. Genetics, Genomics, and Breeding of Grapes. Boca raton, Florida, USA: CRC Press; 2011. pp. 160-185

[9] Salmon JM, Ojeda H, Escudier JL. Disease resistant grapevine varieties and quality: The case of Bouquet varieties. OENO One. 2018;**52**:225-230. DOI: 10.20870/oeno-one.2018.52.3.2139

[10] Merdinoglu D, Schneider C, Prado E, Wiedemann-Merdinoglu S, Mestre P. Breeding for durable resistance to downy and powdery mildew in grapevine. OENO one. 2018;**52**:189-195. DOI: 10.20870/oeno-one.2018.52.3.2116

[11] Di Gaspero G, Cattonaro F. Application of genomics to grapevine improvement. Australian Journal of Grape and Wine Research. 2010;**16**:122-130. DOI: 10.1111/j.1755-0238.2009.00072

[12] Adam-Blondon AF, Zapater M, JM KC, editors. Genetics, Genomics, and Breeding of Grapes. CRC Press; 2011. 345 p

[13] Ibañez J, Carreño J, Yuste J, Martínez-Zapater JM. Grapevine breeding and clonal selection programs in Spain. In: Reynolds AG, editor. Grapevine Breeding Programs for the Wine Industry. Amsterdam, The Netherlands: Elsevier; 2015. pp. 183-209. ISBN: 978-1-78242-075-0

[14] Pelsy F. Molecular and cellular mechanisms of diversity within grapevine varieties. Heredity. 2010;**104**:331-340. DOI: 10.1038/hdy.2009.161

[15] Carbonell-Bejerano P, Royo C, Torres-Pérez R, Grimplet J, Fernandez L, Franco-Zorrilla JM, et al.

Catastrophic unbalanced genome rearrangements cause somatic loss of berry color in grapevine. Plant Physiology. 2017;**75**:1-16. DOI: 10.1104/pp.17.00715

[16] Torregrosa L, Fernandez L, Bouquet A, Boursiquot JM, Pelsy F, Martínez-Zapater JM. Origins and consequences of somatic variation in grapevine. In: Adam-Blondon AF, Martínez Zapater JM, Kole C, editors. Genetics, Genomics, and Breeding of Grapes. Boca Raton, Florida: CRC Press; 2011. pp. 68-92

[17] Acanda Y, Prado MJ, González MV, Rey M. Somatic embryogenesis from stamen filaments in grapevine (*Vitis vinifera* L. cv. Mencía): Changes in ploidy level and nuclear DNA content. In Vitro Cellular and Developmental Biology. Plant. 2013;**49**:276. DOI: 10.1007/s11627-013-9499-7

[18] Jaillon O, Aury JM, Noel B, Policriti A, Clepet C, et al. The grapevine genome sequence suggests ancestral hexaploidization in major angiosperm phyla. Nature. 2007;**449**:463-467. DOI: 10.1038/nature06148

[19] Thompson MM, Olmo HP. Cytohistological studies of cytochimeric and tetraploid grapes. American Journal of Botany. 1963;**50**:901-906. DOI: 10.1002/j.1537-2197.1963.tb06569.x

[20] Franks T, Botta R, Thomas MR. Chimerism in grapevines: Implications for cultivar identity, ancestry and genetic improvement. Theoretical and Applied Genetics. 2002;**104**:192-199. DOI: 10.1007/s001220100683

[21] Vezzulli S, Leonardelli L, Malossini U, Stefanini M, Velasco R, Moser C. Pinot blanc and pinot gris arose as independent somatic mutations of pinot noir. Journal of Experimental Botany. 2012;**63**: 6359-6369. DOI: 10.1093/jxb/ers290

[22] Velasco R, Zharkikh A, Troggio M, Cartwright DA, Cestaro A, et al. A high quality draft consensus sequence of the genome of a heterozygous grapevine variety. PLoS One. 2007;**2**:e1326. DOI: 10.1371/journal.pone.0001326

[23] Chin CS, Peluso P, Sedlazeck FJ, Nattestad M, Concepcion GT, et al. Phased diploid genome assembly with single-molecule real-time sequencing. Nature Methods. 2016;**13**:1050-1054. DOI: 10.1038/nmeth.4035

[24] Roach MJ, Johnson DL, Bohlmann J, van Vuuren HJJ, Jones SJM, Pretorius IS, et al. Population sequencing reveals clonal diversity and ancestral inbreeding in the grapevine cultivar Chardonnay. PLoS Genetics. 2018;**14**:e1007807. DOI: 10.1371/journal.pgen.1007807

[25] Xu Y, Gao Z, Tao J, Jiang W, Zhang S, Wang Q, et al. Genome-wide detection of SNP and SV variations to reveal early ripening-related genes in grape. PLoS One. 2016;**11**(2):e0147749. DOI: 10.1371/journal.pone.0147749

[26] Royo C, Torres-Pérez R, Mauri N, Diestro N, Cabezas JA, Marchal C, et al. The major origin of seedless grapes is associated with a missense mutation in the MADS-box gene *VviAGL11*. Plant Physiology. 2018;**177**:1234-1253. DOI: 10.1104/pp.18.00259

[27] Carrier G, Le Cunff L, Dereeper A, Legrand D, Sabot F, Bouchez O, et al. Transposable elements are a major cause of somatic polymorphism in *Vitis vinifera* L. PLoS One. 2012;7:e32973. DOI: 10.1371/journal.pone.0032973

[28] Gambino G, Dal Molin A, Boccacci P, Minio A, Chitarra W, Avanzato CG, et al. Whole-genome sequencing and SNV genotyping of 'Nebbiolo' (*Vitis vinifera* L.) clones. Scientific Reports. 2017;7:17294. DOI: 10.1038/s41598-017-17405-y

[29] Royo C, Carbonell-Bejerano P, Torres-Perez R, Nebish A, Martinez O, Rey M, et al. Developmental, transcriptome, and genetic alterations associated with parthenocarpy in the grapevine seedless somatic variant Corinto bianco. Journal of Experimental Botany. 2016;**67**:259-273. DOI: 10.1093/jxb/erv452

[30] Marroni F, Pinosio S, Morgante M. Structural variation and genome complexity: Is dispensable really dispensable? Current Opinion in Plant Biology. 2014;**18**:31-36. DOI: 10.1016/j.pbi.2014.01.003

[31] Li XQ. Natural attributes and agricultural implications of somatic genome variation. Current Issues in Molecular Biology. 2016;**20**:29-46

[32] Collins RL, Brand H, Redin CE, Hanscom C, Antolik C, et al. Defining the diverse spectrum of inversions, complex structural variation, and chromothripsis in the morbid human genome. Genome Biology. 2017;**18**:36. DOI: 10.1186/s13059-017-1158-6

[33] Liu P, Erez A, Nagamani SC, Dhar SU, Kolodziejska KE, et al. Chromosome catastrophes involve replication mechanisms generating complex genomic rearrangements. Cell. 2011;**146**:889-903. DOI: 10.1016/j.cell.2011.07.042

[34] McClintock B. The stability of broken ends of chromosomes in *Zea mays*. Genetics. 1941;**26**:234-282

[35] Sorzano CO, Pascual-Montano A, Sánchez de Diego A, Martínez AC, van Wely KH. Chromothripsis: Breakage-fusion-bridge over and over again. Cell Cycle. 2013;**12**:2016-2023. DOI: 10.4161/cc.25266

[36] Cardone MF, D'Addabbo P, Alkan C, Bergamini C, Catacchio CR, Anaclerio F, et al. Inter-varietal structural variation in grapevine genomes. Plant Journal. 2016;**88**:648-661. DOI: 10.1111/tpj.13274

[37] Mercenaro L, Nieddu G, Porceddu A, Pezzotti M, Camiolo S. Sequence polymorphisms and structural variations among four grapevine (*Vitis vinifera* L.) cultivars representing Sardinian agriculture. Frontiers in Plant Science. 2017;**8**:1279. DOI: 10.3389/fpls.2017.01279

[38] Zhou Y, Massonnet M, Sanjak JS, Cantu D, Gaut BS. Evolutionary genomics of grape (*Vitis vinifera* ssp. *vinifera*) domestication. Proceedings National Academy Sciences USA. 2017;**114**:11715-11720. DOI: 10.1073/pnas.1709257114

[39] Fedoroff NV. Transposable elements, epigenetics, and genome evolution. Science. 2012;**338**:758-767. DOI: 10.1126/science.338.6108.758

[40] Benjak A, Boue S, Forneck A, Casacuberta JM. Recent amplification and impact of MITEs on the genome of grapevine (*Vitis vinifera* L.). Genome Biology and Evolution. 2009;**1**:75-84. DOI: 10.1093/gbe/evp009

[41] Kobayashi S, Goto-Yamamoto N, Hirochika H. Retrotransposon-induced mutations in grape skin color. Science. 2004;**304**:982. DOI: 10.1126/science.1095011

[42] Fernandez L, Torregrosa L, Segura V, Bouquet A, Martinez-Zapater JM. Transposon-induced gene activation as a mechanism generating cluster shape somatic variation in grapevine. Plant Journal. 2010;**61**:545-557. DOI: 10.1111/j.1365-313X.2009.04090.x

[43] Fernandez L, Chaib J, Martinez-Zapater JM, Thomas MR, Torregrosa L. Mis-expression of a *PISTILLATA*-like MADS box gene prevents fruit development in grapevine. Plant Journal. 2013;**73**:918-928. DOI: 10.1111/tpj.12083

[44] Wegschneider E, Benjak A, Forneck A. Clonal variation in pinot noir revealed by S-SAP involving universal retrotransposon-based sequences. American Journal of Enology and Viticulture. 2009;**60**:104-109

[45] Castro I, D'Onofrio C, Martin JP, Ortiz JM, De Lorenzis G, Ferreira V, et al. Effectiveness of AFLPs and retrotransposon-based markers for the identification of Portuguese grapevine cultivars and clones. Molecular Biotechnology. 2012;**52**:26-39. DOI: 10.1007/s12033-011-9470-y

[46] Emanuelli F, Battilana J, Costantini L, Le Cunff L, This P, MS G. A candidate gene association study for Muscat flavor in grapevine *Vitis vinifera* L. BMC Plant Biology. 2010;**10**:241. DOI: 10.1186/1471-2229-10-241

[47] Boss PK, Thomas MR. Association of dwarfism and floral induction with a grape 'green revolution' mutation. Nature. 2002;**416**:847-850. DOI: 10.1038/416847a

[48] Pratt C. Reproductive anatomy in cultivated grapes: A review. American Journal of Enology and Viticulture. 1971;**22**:92-109

[49] Mullins MG. Regulation of inflorescence growth in cuttings of the grapevine (*Vitis vinifera* L.). Journal of Experimental Botany. 1968;**19**:532-543. DOI: 10.1093/jxb/19.3.532

[50] Peng J, Carol P, Richards DE, King KE, Cowling RJ, Murphy GO, et al. The Arabidopsis *GAI* gene defines a signaling pathway that negatively regulates gibberellin responses. Genes and Development. 1997;**11**:3194-3205. DOI: 10.1101/gad.11.23.3194

[51] Rézeau P. Dictionnnaire des noms de cépages en France. Lyons, France: CNRS Editions; 1998. 424 p. ISBN: 978-2-271-05526-2

[52] Ren C, Liu X, Zhang Z, Wang Y, Duan W, Li S, et al. CRISPR/Cas9-mediated efficient targeted mutagenesis in chardonnay (*Vitis vinifera* L.). Scientific Reports. 2016;**6**:32289. DOI: 10.1038/srep32289

[53] Chaïb J, Torregrosa L, Mackenzie D, Corena P, Bouquet A, Thomas MR. The grape microvine—A model system for rapid forward and reverse genetics of grapevines. Plant Journal. 2010;**62**:1083-1092. DOI: 10.1111/j.1365-313X.2010.04219.x

[54] Ribéreau-Gayon P, Boidron JN, Terrier A. Aroma of Muscat grape varieties. Journal of Agriculture and Food Chemistry. 1975;**23**:1042-1047. DOI: 10.1021/jf60202a050

[55] Crespan M, Milani N. The Muscats: A molecular analysis of synonyms, homonyms and genetic relationships within a large family of grapevine cultivars. Vitis. 2001;**40**:23-30

[56] Duchêne E, Legras JL, Karst F, Merdinoglu D, Claudel P, Jaegli N, et al. Variation of linalool and geraniol content within two pairs of aromatic and non-aromatic grapevine clones. Australian Journal of Grape and Wine Research. 2009;**15**:120-130. DOI: 10.1111/j.1755-0238.2008.00039.x

[57] Doligez A, Audiot E, Baumes R, This P. QTLs for Muscat flavour and monoterpenic odorant content in grapevine (*Vitis vinifera* L.). Molecular Breeding. 2006;**18**:109-125. DOI: 10.1007/s11032-006-9016-3

[58] Battilana J, Costantini L, Emanuelli F, Sevini F, Segala C, Moser S, et al. The 1-deoxy-D-xylulose 5-phosphate synthase gene co-localizes with a major QTL affecting monoterpene content in grapevine. Theoretical and Applied Genetics. 2009;**118**:653-669. DOI: 10.1007/s00122-008-0927-8

[59] Duchêne E, Butterlin G, Claudel P, Dumas V, Jaegli N, Merdinoglu D. A grapevine (*Vitis vinifera* L.)

deoxy-D-xylulose synthase gene collocates with a major quantitative trait locus for terpenol content. Theoretical and Applied Genetics. 2009;**118**:541-552. DOI: 10.1007/s00122-008-0919-8

[60] Battilana J, Emanuelli F, Gambino G, Gribaudo I, Gasperi F, Boss PK, et al. Functional effect of grapevine 1-deoxy-D-xylulose 5-phosphate synthase substitution K284N on Muscat flavour formation. Journal of Experimental Botany. 2011;**62**:5497-5508. DOI: 10.1093/jxb/err231

[61] Fischer BM, Salakhutdinov I, Akkurt M, Eibach R, Edwards KJ, Toepfer R, et al. Quantitative trait locus analysis of fungal disease resistance factors on a molecular map of grapevine. Theoretical and Applied Genetics. 2004;**108**:501-515. DOI: 10.1007/s00122-003-1445-3

[62] Doligez A, Adam-Blondon AF, Cipriani G, Di Gaspero G, Laucou V, Merdinoglu D, et al. An integrated SSR map of grapevine based on five mapping populations. Theoretical and Applied Genetics. 2006;**113**:369-382. DOI: 10.1007/s00122-006-0295-1

[63] Wong DC, Schlechter R, Vannozzi A, Höll J, Hmmam I, Bogs J, et al. A systems-oriented analysis of the grapevine R2R3-MYB transcription factor family uncovers new insights into the regulation of stilbene accumulation. DNA Research. 2016;**23**:451-466. DOI: 10.1093/dnares/dsw028

[64] Walker AR, Lee E, Bogs J, McDavid DA, Thomas MR, Robinson SP. White grapes arose through the mutation of two similar and adjacent regulatory genes. Plant Journal. 2007;**49**:772-785. DOI: 10.1111/j.1365-313X.2006.02997.x

[65] Kobayashi S, Ishimaru M, Hiraoka K, Honda C. Myb-related genes of the Kyoho grape (*Vitis labruscana*) regulate anthocyanin biosynthesis. Planta. 2002;**215**:924-933. DOI: 10.1007/s00425-002-0830-5

[66] Fournier-Level A, Lacombe T, Le Cunff L, Boursiquot JM, This P. Evolution of the *VvMybA* gene family, the major determinant of berry colour in cultivated grapevine (*Vitis vinifera* L.). Heredity. 2009;**104**:351-362. DOI: 10.1038/hdy.2009.148

[67] Fournier-Level A, Le Cunff L, Gomez C, Doligez A, Ageorges A, Roux C, et al. Quantitative genetic bases of anthocyanin variation in grape (*Vitis vinifera* L. ssp. *sativa*) berry: A quantitative trait locus to quantitative trait nucleotide integrated study. Genetics. 2009;**183**:1127-1139. DOI: 10.1534/genetics.109.103929

[68] Lijavetzky D, Ruiz-García L, Cabezas JA, De Andrés MT, Bravo G, Ibáñez A, et al. Molecular genetics of berry colour variation in table grape. Molecular Genetics and Genomics. 2006;**276**:427-435. DOI: 10.1007/s00438-006-0149-1

[69] Azuma A, Kobayashi S, Goto-Yamamoto N, Shiraishi M, Mitani N, Yakushiji H, et al. Color recovery in berries of grape (*Vitis vinifera* L.) 'Benitaka', a bud sport of 'Italia', is caused by a novel allele at the *VvmybA1* locus. Plant Science. 2009;**176**:470-478. DOI: 10.1016/j.plantsci.2008.12.015

[70] Walker AR, Lee E, Robinson SP. Two new grape cultivars, bud sports of cabernet sauvignon bearing pale-coloured berries, are the result of deletion of two regulatory genes of the berry colour locus. Plant Molecular Biology. 2006;**62**:623-635. DOI: 10.1007/s11103-006-9043-9

[71] Furiya T, Suzuki S, Sueta T, Takayanagi T. Molecular characterization of a bud sport of pinot gris bearing white berries. American

Journal of Enology and Viticulture. 2009;**60**:66-73

[72] Migliaro D, Crespan M, Muñoz-Organero G, Velasco R, Moser C, Vezzulli S. Structural dynamics at the berry colour locus in*Vitis vinifera* L. somatic variants. Australian Journal of Grape and Wine Research. 2014;**20**: 485-495. DOI: 10.1111/ajgw.12103

[73] Pelsy F, Dumas V, Bevilacqua L, Hocquigny S, Merdinoglu D. Chromosome replacement and deletion lead to clonal polymorphism of berry color in grapevine. PLoS Genetics. 2015;**11**:e1005081. DOI: 10.1371/journal.pgen.1005081

[74] Yakushiji H, Kobayashi S, Goto-Yamamoto N, Tae Jeong S, Sueta T, Mitani N, et al. A skin color mutation of grapevine, from black-skinned Pinot Noir to white-skinned Pinot Blanc, is caused by deletion of the functional VvmybA1 allele. Bioscience, Biotechnology, and Biochemistry. 2006;**70**:1506-1508. DOI: 10.1271/bbb.50647

[75] Vargas AM, Vélez MD, de Andrés MT, Laucou V, Lacombe T, Boursiquot J-M, Borrego J, Ibáñez J. Corinto bianco: A seedless mutant of Pedro Ximenes. American Journal of Enology and Viticulture; 2007;**58**:540-543. ISSN: 0002-9254

[76] Ledbetter CA, Ramming DW. Seedlessness in grapes. Horticultural Reviews. 1987;**11**:159-184. DOI: 10.1002/9781118060841.ch5

[77] Dangi GS, Mendum ML, Prins BH, Walker MA, Meredith CP, Simon CJ. Simple sequence repeat analysis of a clonally propagated species: A tool for managing a grape germplasm collection. Genome. 2001;**44**:432-438

[78] Adam-Blondon AF, Lahogue-Esnault F, Bouquet A, Boursiquot JM, This P. Usefulness of two SCAR markers for marker-assisted selection of seedless grapevine cultivars. Vitis. 2001;**40**:147-155

[79] Ibáñez J, Vargas AM, Palancar M, Borrego J, de Andrés MT. Genetic relationships among table-grape varieties. American Journal of Enology and Viticulture. 2009;**60**:35-42

[80] Pearson HM. Parthenocarpy and seedlessness in *Vitis vinifera*. Science. 1932;**76**:594. DOI: 10.1126/science.76.1982.594

[81] Malabarba J, Buffon V, Mariath JEA, Gaeta ML, Dornelas MC, Margis-Pinheiro M, et al. The MADS-box gene Agamous-like 11 is essential for seed morphogenesis in grapevine. Journal of Experimental Botany. 2017;**68**: 1493-1506. DOI: 10.1093/jxb/erx025

[82] Bouquet A, Danglot Y. Inheritance of seedlessness in grapevine, *Vitis vinifera* L. Vitis. 1996;**35**:35-42

[83] Lahogue F, This P, Bouquet A. Identification of a codominant scar marker linked to the seedlessness character in grapevine. Theoretical and Applied Genetics. 1998;**97**:950-959

[84] Doligez A, Bouquet A, Danglot Y, Lahogue F, Riaz S, Meredith P, et al. Genetic mapping of grapevine (*Vitis vinifera* L.) applied to the detection of QTLs for seedlessness and berry weight. Theoretical and Applied Genetics. 2002;**105**:780-795. DOI: 10.1007/s00122-002-0951-z

[85] Cabezas JA, Cervera MT, Ruiz-García L, Carreño J, Martínez-Zapater JM. A genetic analysis of seed and berry weight in grapevine. Genome. 2006;**49**:1572-1585. DOI: 10.1139/g06-122

[86] Costantini L, Battilana J, Lamaj F, Fanizza G, Grando MS. Berry and

phenology-related traits in grapevine (*Vitis vinifera* L.): From quantitative trait loci to underlying genes. BMC Plant Biology. 2008;**8**:38. DOI: 10.1186/1471-2229-8-38

[87] Mejía N, Soto B, Guerrero M, Casanueva X, Houel C, Miccono MA, et al. Molecular, genetic and transcriptional evidence for a role of VvAGL11 in stenospermocarpic seedlessness in grapevine. BMC Plant Biology. 2011;**11**:57. DOI: 10.1186/1471-2229-11-57

[88] Mizzotti C, Mendes MA, Caporali E, Schnittger A, Kater MM, Battaglia R, et al. The MADS box genes *SEEDSTICK* and Arabidopsis *Bsister* play a maternal role in fertilization and seed development. Plant Journal. 2012;**70**:409-420. DOI: 10.1111/j.1365-313X.2011.04878.x

[89] Singh R, Low ET, Ooi LC, Ong-Abdullah M, Ting NC, Nagappan J, et al. The oil palm *SHELL* gene controls oil yield and encodes a homologue of *SEEDSTICK*. Nature. 2013;**500**: 340-344. DOI: 10.1038/nature12356

[90] Casassa LF, Beaver CW, Mireles M, Larsen RC, Hopfer H, Heymann H, et al. Influence of fruit maturity, maceration length, and ethanol amount on chemical and sensory properties of merlot wines. American Journal of Enology and Viticulture. 2013;**64**:437-449. DOI: 10.5344/ajev.2013.13059

Chapter 3

Reconstruction of Parental SSR Haplotypes from a Single Grape Seed

Gizella Győrffyné Jahnke, Zóra Annamária Nagy, Csaba Németh, Gyöngyi Knolmajerné Szigeti and János Májer

Abstract

Microsatellite ('single sequence repeats', SSR) markers were widely used in the last decade for the identification of parents of a given grapevine variety or for pedigree reconstruction as well. By now the pedigree of the majority of the most important varieties is established. At the same time, knowing both of the parents gives information about which one was the mother plant and which one was the pollinator. Analyses of archaeological grapevine seeds can give new opportunities in the research of the evolution of varieties. In most of the angiosperms, the endosperm is triploid with two genome equivalents from the maternal line and one from the paternal line. Our presumption was that this numeral difference in the maternal and paternal alleles causes measureable difference in the amplification of SSR alleles from grapevine seeds. To validate our method, pre-experiments were carried out on 12 'Pinot gris' seeds, which verified our theory.

Keywords: grapevine, single sequence repeats (SSR), double fertilisation

1. Introduction

Microsatellite (SSR) markers were widely used in the last decade for the identification of parents of a given grapevine variety or for pedigree reconstruction as well. By now the pedigree of the majority of the most important varieties is established. For example, large-scale parentage analyses were carried out by Lacombe et al. [1]. At the same time, knowing both of the parents gives information about which one was the mother plant and which one was the pollinator only in special cases, for example, if one of the parents is female flowered. Based on only SSR analyses, no information is available about the time, when the given variety was born.

Analyses of archaeological grapevine seeds can give new opportunities in the research of the evolution of varieties. Over the analyses of seed morphology, the genetic analyses of the seeds can provide interesting data.

Angiosperms have a complex fertilisation process, the so-called double fertilisation (**Figure 1**). It is double, because two female cells, the egg cell and the central cells, are fertilised by two sperm cells forming the embryo and endosperm, respectively. This unique procedure was discovered parallel by Nawaschin [3] and Guignard [4]. Due to the double fertilisation, in such species, where the mass of the endosperm is significantly higher than the mass of the embryo, the haplotype of the

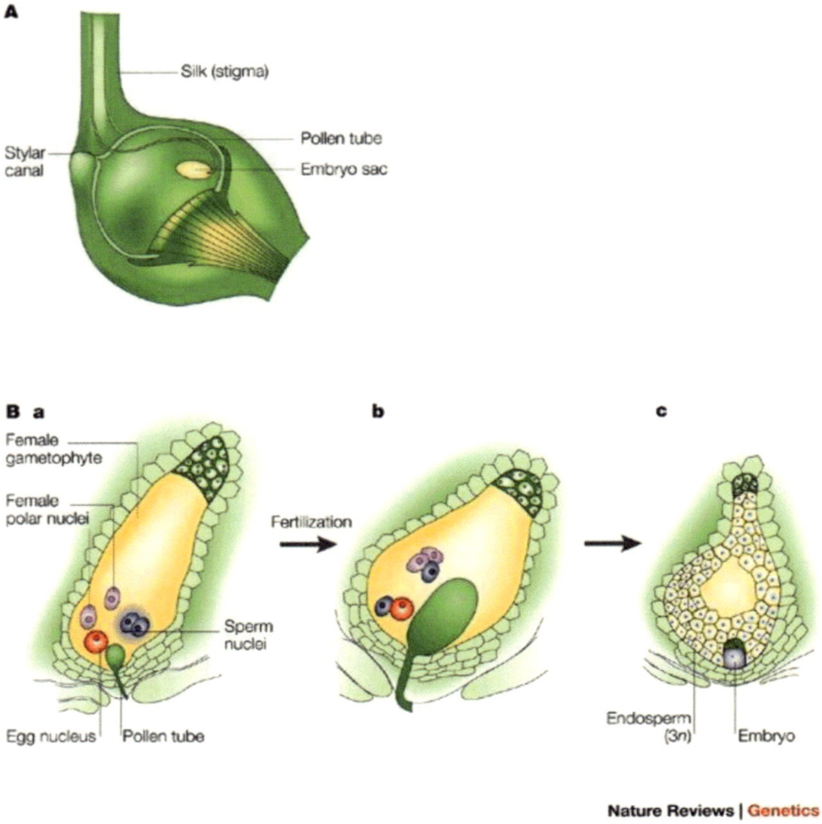

Figure 1.
Process of double fertilisation [2].

parents can be determinable. The main point of the idea is based on the fact that in most of the angiosperms, the endosperm is triploid with two genome equivalents from the maternal line and one from the paternal line [5]. Our presumption was that this numeral difference in the maternal and paternal alleles causes measureable difference in the amplification of SSR alleles.

To validate our method, pre-experiments were carried out on 12 Pinot gris seeds.

2. Materials and methods

Fully matured berries were collected from Pinot gris (clone B. 10) stocks in September 2016. Seeds were removed from the berries, washed in tap water and dried in room temperature. Only fully matured, brown-coloured seeds were used in the analyses. Pinot noir, Pinot gris (clone B. 10.), Chardonnay and Cabernet sauvignon DNA (extracted formerly from leaves) were also used as controls.

DNA was extracted from 12 single seeds. Prior to extraction the seeds were crushed in a mortar to powder; then, this powder was moved to a tube. Qiagen Plant Mini Kits were used for DNA extraction following the instructions of the manufacturer. The amount and quality of DNA were determined spectrophotometrically. The DNA was diluted to a concentration of 10 ng/ml.

The SSR analysis was performed at 19 loci (see **Table 1**). PCR reaction mix was the following: 0.2 mM of each primer, 12.5 ml of Hot Start Master Mix (Quiagen) and 50 ng of template DNA, completed to the total volume of 25 ml with DNA- and

Linkage group[a]	SSR locus	Annealing temp.
1	VMC8A7	64°C
2	VMC7G3	60°C
3	VVMD28	62°C
4	VrZag21	62°C
5	VrZag79	60°C
6	VMC4G6	50°C
7	VVMD7	50°C
8	VMC1F10	57°C
9	VMC1C10	60°C
10	VrZag25	67°C
11	VVS2	60°C
12	VMC2H4	57°C
13	VMC3D12	57°C
14	VMCNG1E1	58°C
15	VMC5G8	58°C
16	VVMD5	53°C
17	Scu06vv	60°C
18	VVIM10	57°C
19	VMC5E9	58°C

[a]*Linkage group numbers according to Adam-Blondon et al. [7].*

Table 1.
List of SSR loci [6].

RNA-free distilled water. The following thermal profile was used: (1) 94°C for 45 min; 35 cycles of: (2) 94°C for 1 min, (3) annealing temperature (see **Table 1**) for 1 min, 73°C for 1 min; and (4) 73°C for 7 min. Each forward primer of the primer pairs was fluorescently labelled with 6FAM on the 5′ end of the DNA chain. Separation of the amplified products was carried out through capillary electrophoresis in a PE-Applied Biosystems 3100 Automated Capillary DNA Sequencer; the molecular sizes of the products were determined using Peak Scanner Software (v. 1.0; Applied Biosystems) [8]. Allele sizes and peak area were also recorded for every single allele. Data were stored in Microsoft Excel [8].

3. Results

DNA extraction and amplification in 18 loci (out of the 19) were successful in all of the 12 seeds. In VVMD5 the amplification was weak, so the results were unevaluated. In all of the remaining 18 loci, the maternal and parental alleles were determined (**Figure 2**); according to our presumption, the allele with the higher area value was supposed to be maternal. Ratio of the quantity of maternal and paternal alleles was computed based on area values. In some loci (where 'Pinot gris' has a homozygote genotype and the majority of the seeds showed also homozygote genotype—VMC4G6 and VMCNG1E1), this ration was excluded from the further analyses.

Based on the remaining 16 loci, the average ratio of maternal and paternal alleles ranged from 1.89 (VMC5E9) to 2.58 (VrZag25), which confirms our presumption.

Maternal and paternal haplotypes were determined by the separation of maternal and paternal alleles based on the method described previously (**Table 2**).

Based on our results, it can be established that most of the seeds are originating from the selfing of 'Pinot gris', but cross-fertilisation is appearing in some cases, such as in the case of seed no. 4, where in six loci the paternal allele is absent from

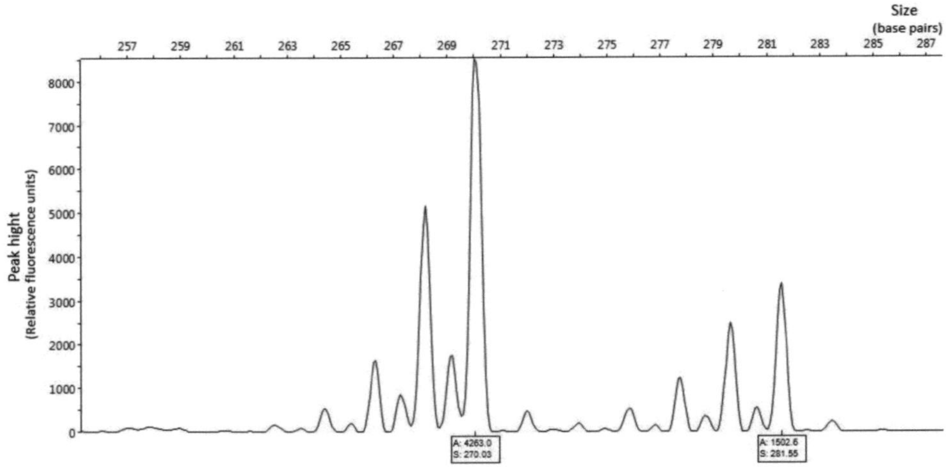

Figure 2.
Example for the determination of maternal and paternal allele in the case of seed no. 11 and VMC4A1 locus: maternal allele, 270; paternal allele, 282 (maternal genotype, 270:278).

Locus	Maternal haplotype	Paternal haplotype	'Pinot gris' genotype
VMC8A7	**158**	**158**	158:158
VVMD28	**236**	**236**	218:236
VrZag79	**242**	**248**	242:248
VVMD7	**240**	*238*	240:244
VMC1C10	**157**	**157**	157:157
VVS2	**150**	**134**	134:150
VMC3D12	**236**	*234*	200:236
VMC5G8	**315**	**311**	311:315
Scu06vv	**164**	**172**	164:172
VMC5E9	**220**	*218*	216:220
VMC7G3	**116**	*114*	116:116
VrZag21	**205**	*195*	199:205
VMC4G6	**122**	**122**	122:122
VMC1F10	**190**	**208**	190:208
VrZag25	**237**	**225**	225:237
VMC2H4	**206**	*204*	206:224
VMCNG1E1	**124**	**124**	124:124
VVIM10	**335**	**335**	335:339

Table 2.
Determination of parental haplotypes in the case of seed no. 4. (Alleles of 'Pinot gris' are red coloured, bold letters; alleles not in 'Pinot gris' genotype are blue coloured, italic letters) [9].

'Pinot gris' (blue coloured, italic letters in **Table 2**). These alleles were inherited from the pollinator, which is surely different from 'Pinot gris'.

4. Discussion

Such amount of DNA can be extracted from a single grape seed, which is suitable for SSR analyses. The amplification of DNA is successful in most of the loci; the PCR reaction optimised for other plant parts can be applied.

Our method is safely applicable for the determination of parental haplotypes. Based on the determined haplotypes, the parental identity could be determined by the use of databases.

The method could be suitable for the analyses of archaeological grapevine seeds, with the following limitations:

The quality and quantity of the extracted DNA could be poor because of the degradation; it depends on the age of the seeds and the environmental effects, e.g. carbonisation processes [10].

Mutations can occur in the SSR loci during the time, which can be manifested in the occurrence of the so-called null alleles. Null alleles result in amplification failures, which can be rid by the use of shortened primers in the PCR reactions ([11, 12]).

Licencing

The file from **Figure 1** is created under the Creative Commons Attribution licence [13].

Author details

Gizella Győrffyné Jahnke*, Zóra Annamária Nagy, Csaba Németh, Gyöngyi Knolmajerné Szigeti and János Májer
NARIC Research Institute for Viticulture and Enology, Hungary

*Address all correspondence to: jahnke.gizella@szbki.naik.hu

References

[1] Lacombe T, Boursiquot JM, Laucou V, Di Vecchi-Staraz M, Péros JP, This P. Large-scale parentage analysis in an extended set of grapevine cultivars (*Vitis vinifera* L.). Theoretical and Applied Genetics. 2013;**126**(2):401-414

[2] Double fertilization from Wikipedia, the free encyclopedia. Available from: https://commons.wikimedia.org/wiki/File:Double_Fertilization.jpg Description: Kiesselbach TA. The structure and reproduction of corn. Nebr. Agric. Exp. Stn Annu. Rep. 1949;**161**:1-96. (2019.01.15.)

[3] Nawaschin S. Resultate einer revision der befruchtungsvorgänge bei *Lilium martagon* und *Fritillaria tenella*. Bulletin de l'Académie impériale des sciences de St.-Pétersbourg. 1898;**9**:377-382

[4] Guignard ML. Sur les anthérozoides et la double copulation sexuelle chez les végétaux angiospermes. Revue Générale de Botanique. 1899;**11**:129-135

[5] Walbot V, Evans MM. Unique features of the plant life cycle and their consequences. Nature Reviews Genetics. 2003;**4**:369-379

[6] Kocsis L, Győrffyné Jahnke G. Teleki szőlőalanyok eredete. (Origin of the Teleki's grape rootstocks) LII. Georgikon Napok (52th Georgikon Scientific Conference-Proceedings). 2010. Available from: http://napok.georgikon.hu/upload/publications/2010-08-23_17-26-18__teleki-szoloalanyainak-eredete-lii-georgikon-napok.doc

[7] Adam-Blondon AF, Roux C, Claux D, Butterlin G, Merdinoglu D, This P. Mapping 245 SSR markers on the *Vitis vinifera* genome: A tool for grape genetics. Theoretical and Applied Genetics. 2004;**109**(5):1017-1027

[8] Jahnke G, Májer J, Lakatos A, Györffyné Molnár J, Deák E, Stefanovits-Bányai É, et al. Isoenzyme and microsatellite analysis of *Vitis vinifera* L. varieties from the Hungarian grape germplasm. Scientia Horticulturae. 2009;**120**(2):213-221

[9] Győrffyné Jahnke G, Nagy ZA, Májer J. Szülői haplotípusok rekonstruálása egyetlen szőlő magból. LIX Georgikon Napok (59th Georgikon Scientific Conference-Proceedings). 2017. Available from: www.napok.georgikon.hu/hu/cikkadatbazis/cikkek-2012/doc_download/411-gyorffyne-jahnke-gizella-nagy-zora-annamaria-majer-janos-szuloi-haplotipusok-rekonstrualasa-egyetlen-szolo-magbol

[10] Manen JF, Bouby L, Dalnoki O, Marinval P, Turgay M, Schlumbaum A. Microsatellites from archaeological *Vitis vinifera* seeds allow a tentative assignment of the geographical origin of ancient cultivars. Journal of Archaeological Science. 2003;**30**(6):721-729

[11] Dakin EE, Avise JC. Microsatellite null alleles in parentage analysis. Heredity. 2004;**93**:504-509

[12] Kwok S, Kellogg DE, Mckinney N, Spasic D, Goda L, Levenson C, et al. Effects of primer-template mismatches on the polymerase chain reaction: Human immunodeficiency virus type 1 model studies. Nucleic Acids Research. 1990;**18**(4):999-1005

[13] Creative Commons Attribution Licece: Available from: https://creativecommons.org/licenses/by/3.0/deed.en, https://creativecommons.org/licenses/by/3.0/legalcode

Chapter 4

Ampelographic and Genetic Characterization of Montenegrin Grapevine Varieties

Vesna Maraš

Abstract

Montenegro is a small country in Balkan Peninsula with very long tradition of grapevine growing and wine making that originate from the pre-Roman period. Dominant place in Montenegrin viticulture belongs to autochthonous grapevine varieties Vranac, Kratosija, and Krstac, while in minor part, the other varieties are presented. Among many literature sources, the oldest historical document that pointed out the importance of autochthonous varieties is The Medieval Statute of Budva from fifteenth century. In order to better present Montenegrin germplasm, this research provides an overview of literature, ampelographic, and genetic analysis on autochthonous and domesticated varieties. Achieved results showed an important breeding history of grapevine and a large number of unique DNA profiles. Montenegro has the richness of grapevine diversity that can significantly enrich the diversity of vines in Europe.

Keywords: autochthonous varieties, Montenegro, grapevine diversity, genetic characterization, ampelographic characterization

1. Introduction

Montenegro is a small country placed in the Balkan Peninsula with one part overlooking the Adriatic Sea, right across the Italian region Puglia. A long tradition of grapevine growing in Montenegro is very well known and it dates back before the Roman period [1]. On the Montenegrin territory, a large number of tombstones with grapevine and wine motives which originated in ancient era were found. Found decorations were in the vine form and were directly related to the Dionysus god cult [2]. Numerous archeological sites and found objects that originated from the Illyrian period indicate that the wine was much appreciated and was quite used as the beverage. In the middle century, vine growing and winemaking were well developed in Montenegro and one of the oldest written documents that point out the importance of grapevine cultivation and importance of autochthonous varieties on the territory of today's Montenegro is the Medieval Budva's Statute from fifteenth century. After all, more organized work started during the reign of the King Nikola Petrović (1860–1918), who introduced the grape-growing and winemaking regulations. Within the grape varieties that have a long tradition of growing in Montenegro, the dominant place belongs to autochthonous grape varieties Vranac

IntechOpen

and Kratošija that are used for making red wines, and for production of white wines, the Krstač was the dominant one [3, 4], while in a minor part, the other varieties were also presented. A major turning point in the development of Montenegrin viticulture was the realization of the project Ćemovsko polje since 1977–1982, during which 1500 ha of vineyards were planted and modern wine cellar with 2 million liters capacity was established. However, as autochthonous grape varieties were in that time the most important, they also now constitute the viticulture and winemaking sector of Montenegro. In fact, Vranac grape variety represents more than 70% of total production and promotes Montenegro as an important wine country. In order to better present the germplasm of grapevine varieties in Montenegro, beside literary research, ampelographic and genetic analysis of autochthonous and domesticated grapevine varieties was done in order of their secure identification.

2. Materials and methods

Multi-year research included work on autochthonous and domesticated grapevine varieties in Montenegro. A detailed review of available literature and writing of earlier and contemporary authors regarding autochthonous and domesticated grapevine varieties was done. For a better understanding, we did also ampelographic and genetic analysis of selected varieties. Analysis included 188 samples of old representative vines aged between 50 and 300 years that are grown in affirmed vineyards in Montenegro. These studies also included 17 biotypes of the Kratošija variety (**Figure 2**) that were collected in 1987 in an experimental field in Ljeskopolje-Podgorica.

2.1 Ampelographic description

Ampelographic analysis, that is, a method of describing characteristics of grapevine varieties, was done with codes—a descriptor prescribed by O.I.V. (*Office International de la Vigne et du Vin*)—International Wine and Wine Office [5]. Observations were made on young shoots (OIV-003 and -004), young leaves (OIV-051 and -053), mature leaves (OIV-067, 068, 070, 076, 079, 080, 084, and 087), flowers (OIV-151), shoots (OIV-155), bunches (OIV-202, 204, 206, and 208), berries (OIV-220, 223, 225, 235, and 236), and, when possible, on must quality (OIV-505, 506, and 508). Ampelographic description also was done for 17 Kratošija biotypes (with following OIV codes: 003, 004, 016, 065, 068, 076, 079, 084, 085,151,202, 203, 204, 206, 220, 223, 225, 231, 235, 236, and 241).

2.2 Genetic analysis

For genetic analysis, DNA was extracted from young leaves. In the first phase of research, genotyping was performed with 11 SSR loci for variety identification: VVS2 [6]; VVMD5, VVMD7, VVMD27 and VVMD28 [7, 8]; VrZAG62 and VrZAG79 [9]; ISV2, ISV3 and ISV4 [10]; and VMCNG4b9 [11], as described by Ref. [12].

During the second phase of research, genotyping was done with nine microsatellite loci: VVS2 [6]; VVMD5, VVMD7, VVMD25, VVMD27, VVMD28, and VVMD32 [7]; and ssrZAG62 and ssrZAG79 [9] as proposed by the GrapeGen06 consortium and by the European Vitis Database [13].

3. Results and discussion

3.1 Literature survey

The first mention of Montenegrin grapevine varieties was in fifteenth century [14], and later they were studied and described by many authors. Early mentioning of autochthonous Montenegrin grapevine variety was done by M. Plamenac [3]. He stated that in Montenegrin grape growing region Crmnica, Kratošija, Vranac, Krsmač, Sjerovina, Lisica, and Muskacelica varieties were grown. But the first more significant description of varieties Vranac and Kratošija was given by P. Plamenac [15]. All authors from the former Yugoslavia [16–29] reported Vranac and Kratošija as Montenegrin autochthonous grapevine varieties. Moreover, they stated that Vranac and Kratošija were grown only in Montenegro. From Montenegro, these varieties were spread to Macedonia [20] and Dalmatia [17] and to other countries in the former Yugoslavia. The Macedonian professor Nastev [20] states that Vranac is a Montenegrin autochthonous grapevine variety mostly cultivated in the Skadar lake region (Crmnica), but also in the Montenegrin seacoast. This author declares that Vranac has been transferred in the 1950s in Macedonia (experimental field Butel), from where it has been spread out through the former Yugoslavia. Montenegrin academic Ulicevic [18, 19] states that Vranac is a characteristic variety of vine growing area Crmnica in famous vine growing region Skadar lake, which occupies about 40% of the assortment. According to the same author, this is the only vine growing area where this variety is dominant and the growing area was not wider than 30 km.

The earliest reference of the Montenegrin variety Kratošija is reported in the Budva's Medieval Statute [14] in fifteenth century (1426–1431). In particular, it mentioned the "Kratošija's vineyards" indicating the importance of the Kratošija variety in that time in Budva (Montenegro). The Dalmatian ampelographer Bulić [17] described Kratošija (also considering the synonyms Gartošija, Grakošija, and Kratkošija) from nine municipalities of the Montenegrin cost (Budva, Grbalj, Luštica, Krtole, Kotor, Paštrovići, Prčanj, Tivat, and Herceg Novi). Moreover, the author stated that this variety was rarely found in the Dalmatia region where it was likely spread over from Montenegro. For the Kratošija variety, Ulicevic [18, 19] states that, it is strongly dominated in all plantations older than 60–70 years in that period and made 90% of the assortment in other regions. According to Ulicevic [18], the growing area of Kratošija was between 100 and 150 km through Montenegro and that is the main and probably the oldest Montenegrin variety.

In addition to very long Kratošija growing in Montenegro, there is also a huge heterogeneity of its population and it was described by many authors. M. Plamenac [3] for the first time mentioned biotypes of Kratošija and described some kind of Kratošija whose clusters are not compacted, but loose and it was called Reavica. Authors [16, 17, 19, 23, 24, 25, 30, 27, 28] also described different Kratošija's biotypes. Ulicevic [18] mentioned three types of Kratošija: Obična Kratošija, Slaborodna Kratošija, and Rehuljava Kratošija. Bozinovik et al. [30] stated that Kratošija has a high number of biotypes and described three of them (Kratošija standardna, Kratošija rehuljava, and Kratošija neoplodjena). The variability of the Kratošija population in Montenegro was also studied by Pejovic [24] and Maras [27]. Ampelographic analyses [27] were done on 17 biotypes of Kratošija, which are known under different names in viticultural areas in Montenegro: Velja Kratošija, Velji Vran, Crni Krstač, Vrančina, Bikača, Vran, Srednja Kratošija, Kratošija or Vran, Srednji Vranac, Velji Vranac, Vrančić, Ljutica, Kratošija, Čestozglavica, Kratošija mala, Kratošija sa dubokim urezima, and Rehuljača.

Krstač was dominant among the white grapevine varieties used for white wine production. Its name comes from the look and shape of the bunch that resembles a cross [15, 31]. Ulicevic [18] wrote that Krstač was believed to be autochthonous of Montenegro and probably originated from Beri (near Podgorica) with a growing area of 40–50 km. The same author stated synonyms for Krstač, in Doljani it was called Krstača bijela, in Vražegrmci Bijeli Krstač and in Beri Bijela vinogradarska.

Beside the most important grapevine varieties for viticulture in Montenegro Vranac, Kratošija, and Krstač there are also some literature data about minor grapevine varieties.

Žižak or Žižak bijeli [17] is considered another autochthonous variety of Montenegro and its origin is unknown. Individual vines can be found nearby Podgorica, but it is mostly grown on the Montenegrin seacoast (Boko-Kotorski sub region). Ulićević [18] also described Žižak as an important variety from which, in some places in Boka, are produced dessert wines called Prošek.

Ulicevic [18] states that Čubrica is used for red wine production and is represented in very small percentage in the vineyards of the Podgorica sub region (Doljani and Kuči). According to the author, vineyards in Doljani are quite old and

Figure 1.
Bunches of researched grapevine varieties.

none of the contemporary people, neither then nor now, did not know when these vineyards were planted.

Literature data about Muškaćelica were given by M. Plamenac [3], and he stated that it is the little grown white grapevine variety that has very strong smell while the variety Sjerovina is russet grapevine variety that has round berries.

The first mention of Lisica was in Grlica [3], where the author reported that this variety is grown in Crmnica. About Lisičina, Stojanović [16] reports the use of this cultivar for white wine production in Montenegro. Bulić [17] and Ulićević [18] wrote that Lisičina (synonyms are Lisica, Ružica, Sjemerava, and Sjeruša) was grown in Montenegro around Bar, Ulcinj, Podgorica, and Virpazar (Crmnica).

Zadrimka was a major variety grown in the Ulcinj viticultural region in the late 1800s until phylloxera, and later World War II devastated the vineyards and almost drove the cultivar to extinction [19].

According to Ulićević [18], large areas under vineyards were in Bokokotorski subregion. Kadarun was dominated, while Kratošija and Vranac were less represented.

Ulićević [18] states that the Razaklija cultivar from both the Skadar Lake and the seacoast region constituted 95% of total table grape production in Montenegro. Many authors from Yugoslavia consider that Razaklija originates from Asia Minor [19]. It is not known how and when it arrived in Montenegro, Macedonia, and other countries and how it was spread.

3.2 Ampelographic descriptions

Ampelographic descriptors for certain varieties (**Figure 1**) already existed and through these researches data with some descriptions of additional varieties were fulfilled. Ampelographic descriptions of researched varieties, except Trojka, are given in **Table 1**.

Ampelographic descriptions are available for all Kratošija accessions listed in **Table 2** and grown at the Experimental Estate in Podgorica-Lješkopolje [27]. The name of each Kratošija biotype is in connection with some of its particular characteristics. Cluster weight was highly variable among accessions and correlated with morphology and these traits showed to be stable within each biotype [27].

3.3 Genetic identification

The varietal identification was achieved by comparing the obtained SSR profiles with available molecular databases and literature data. The work on the genetic identification of Montenegrin autochthonous varieties began in collaboration with *Istituto sperimentale per la viticoltura-Susegana-Conegliano-TV*. The analysis confirmed that Vranac, Krstač, and Žižak have an original DNA profile. Kratošija, an old Montenegrin grape variety, has an identical genetic profile as the Zinfandel from California, Primitivo from Italy, and Crljenak kaštelanski from Croatia [32]. The same authors also suggest a probable first degree relationship between Vranac and Kratošija. Research on Montenegrin grapevine diversity continued and further analysis of 70 samples revealed 14 different genotypes. The results showed already identified genotypes: Vranac, Kratošija, Krstač, and Žižak and 10 new identified genotypes [33]. In **Table 3**, SSR markers of identified varieties are presented. Also, all 17 biotypes of Kratošija were confirmed to have the same genetic profile as Zinfandel/Primitivo [33]. According to Maras et al. [33] the variety Muškaćela is identical to a variety Muscat bianco–Muscat a petits grains. Trojka accession has the same profile as Muscat rouge de Madere (alias Moscato violetto), another important member of the Muscat family. Based on analysis, it can be concluded that Plavina is

	OIV	Japudžak	Čubrica	Kovačka bijela	Kratošija	Krstač	Lisičina	Sijer	Vranac	Zadrimka	Žižak	Muškačela
Young shoot: intensity of anthocyanin coloration on prostrate hairs of the shoot tip	3	5	3	1	3	3	5	1	3	7	3	3
Young shoot: density of prostrate hairs on the shoot tip	4	7	7	1	7	5	9	1	5	7	9	7
Young leaf: color of upper side of blade (fourth leaf)	51	3	1	2	3	3	3	3	3	3	3	3
Young leaf: density of prostrate hairs between main veins on lower side of blade (fourth leaf)	53	9	5	9	7	5	9	7	5	9	9	9
Mature leaf: shape of blade	67	2	2	4	3	2	4	2	2	2	2	2
Mature leaf: number of lobes	68	2	3	3	3	2	3	2	3	2	2	3
Mature leaf: area of anthocyanin coloration of main veins on the upper side of blade	70	4	1	1	2	1	1	1	1	1	4	2
Mature leaf: shape of teeth	76	2	2	2	2	2	2	2	2	2	2	2
Mature leaf: degree of opening/overlapping of petiole sinus	79	5	5	3	7	9	3	1	7	3	5	4
Mature leaf: shape of base of petiole sinus	80	1	1	3	1	1	1	3	1	1	1	1
Mature leaf: density of prostrate hairs between main veins on lower side of blade	84	5	5	5	5	5	7	3	5	5	5	1
Mature leaf: density of erect hairs on main veins on lower side of blade	87	3	7	7	7	1	3	1	3	7	3	1
Flower: sexual organs	151	3	3	3	3	3	3	3	3	3	3	3
Shoot: fertility of basal buds	155	5	5	5	5	5	1	1	5	5	5	5
Bunch: length (peduncle excluded)	202	5	5	5	7	9	5-7	9	7	7	5	5
Bunch: density	204	7	7	7	9	9	7	3	7	5	5	7
Bunch: length of peduncle of primary bunch	206	1	3	3	1	1	1	3	1	1	5	1
Bunch: shape	208	2	2	1	2	1	2	1	2	2	2	2
Berry: length	220	3-5	5	3-5	5	5	3-5	3-5	3-5	5	3-5	5
Berry: shape	223	3	3	2	2	3	3	2	2	3	3	3
Berry: color of skin	225	1	6	1	6	1	5	5	6	6	1	1
Berry: firmness of flesh	235	2	2	2	2	2	2	2	2	2	2	3
Berry: particular flavor	236	1	1	1	1	1	2	2	1	1	1	2
Sugar content of must (Oe)	505	7	7	7	9	7	7	7	9	7	7	7
Total acidity of must	506	3	3	3	3	3	5	3	3	3	3	3
Must -specific pH	508	3	5	5	7	3	3	3	7	7	7	5

	Young shoot: intensity of anthocyanin coloration on prostrate hairs of the shoot tip	Young shoot: density of prostrate hairs on the shoot tip	Young leaf: color of upper side of blade (fourth leaf)	Young leaf: density of prostrate hairs between main veins on lower side of blade (fourth leaf)	Mature leaf: shape of blade	Mature leaf: number of lobes	Mature leaf: area of anthocyanin coloration of main veins on the upper side of blade	Mature leaf: shape of teeth	Mature leaf: degree of opening/overlapping of petiole sinus	Mature leaf: shape of base of petiole sinus	Mature leaf: density of prostrate hairs between main veins on lower side of blade	Mature leaf: density of erect hairs on main veins on lower side of blade	Flower: sexual organs	Shoot: fertility of basal buds	Bunch: length (peduncle excluded)	Bunch: density	Bunch: length of peduncle of primary bunch	Bunch: shape	Berry: length	Berry: shape	Berry: color of skin	Berry: firmness of flesh	Berry: particular flavor	Sugar content of must (Oe)	Total acidity of must	Must -specific pH
	OIV	OIV	OIV	OIV	OIV	OIV	OIV	OIV	OIV	OIV	OIV	OIV	OIV	OIV	OIV	OIV	OIV	OIV	OIV	OIV	OIV	OIV	OIV	OIV	OIV	OIV
	3	4	51	53	67	68	70	76	79	80	84	87	151	155	202	204	206	208	220	223	225	235	236	505	506	508
Bioka	1	1	1	5	4	3	3	2	5	3	5	1	3	5	5-7	7	1	1	3	2	1	1	1	7	3	3
Crna loza	3	7	3	5	3	3	2	2	7	1	5	7	3	5	7	9	1	2	5	3	6	2	1	7	3	7
Razaklija	3	7	3	9	4	4	2	2	6	1	5	3	3	1	7	7	1	1	9	5	3	9	1	9	3	3
Kadarun	5	5	3	7	3	3	1	2	5	1	9	7	3	1	7	9	1	2	5	3	6	5	1	7	3	3
Plavina	3	5	3	7	3	2	3	2	5	1	1	7	3	3	7	7	1	1	5	3	4	5	1	7	3	7
Razaklija crna	5	7	3	7	4	4	1	2	7	1	9	3	3	5	9	7	3	1	9	9	9	7	4	9	3	3

Table 1.
Ampelographic description of research grapevine varieties using 26 OIV codes [5].

Figure 2.
Bunches of Kratošija variety biotypes.

a progeny of Kratošija. Bioka shares the same genotype as the Italian Francavidda and Croatian Zlatarica Vrgorska. Comparing data with the European Vitis database in the Vitis International Variety Catalog [34] Kadarun is Reported as a Turkish cultivar. The red berry Razaklija accession that was analyzed matches the SSR profile of Crven Drenok [35]. Crna Loza, Čubrica, Lisičina, and Razaklija crna show unique SSR profiles. Crna Loza was considered as a Kratošija synonym, but analysis shows a different SSR profile for this variety. Based on SSR allele sharing at all analyzed loci, Razaklija crna could really be a progeny of Drenak Crven [33]. Bearing in mind the importance of grapevine germplasm in Montenegro, the research was continued through two international projects SEEDNet and SEE.ERA NET. As result of SEEDNET project, from 16 considered samples, 6 different genotypes were identified [36]. The identified varieties are Vranac, Kratošija, Krstač, Čubrica, Lisičina, and Razaklija. The variety Razaklija has the identical SSR profile

	Young shoot: intensity of anthocyanin coloration on prostrate hairs of the shoot tip	Young shoot: density of prostrate hairs on the shoot tip	Shoot: attitude (before tying)	Mature leaf: size of blade	Mature leaf: number of lobes	Mature leaf: shape of teeth	Mature leaf: degree of opening/ overlapping of petiole sinus	Mature leaf: density of prostrate hairs on main veins on the lower side of blade	Mature leaf: density of erect hairs on main veins on the lower side of blade	Flower: sexual organs
	OIV	OIV	OIV	OIV	OIV	OIV	OIV	OIV	OIV	OIV
	3	4	6	65	68	76	79	86	87	151
Bikača	3	7	1	7	3-4	2	7	5	7	3
Crni krstač	3	7	1	7	3-4	2	6	5	7	3
Kratošija	3	7	1	7	3	2	7	5	7	3
Kratošija ili Vran	3	7	1	7	3	2	7	5	7	3
Kratošija mala	3	7	1	7	3-4	2	7	5	7	3
Kratošija sa dubokim urezom	3	7	1	7	3-4	2	7	5	9	3
Kratošija srednja	3	7	1	7	3	2	7	5	7	3
Ljutica	3	7	1	7	3-4	2	6	5	7	3
Rehuljača	3	7	1	5	3	2	7	5	7	3
Srednji Vranac	3	7	1	7	3	2	7	5	7	3

	Bunch: length (peduncle excluded)	Bunch: width	Bunch: density	Bunch: length of peduncle of primary bunch	Berry: length	Berry: shape	Berry: color of skin	Berry: intensity of flash anthocyanin coloration	Berry: firmness of flesh	Berry: particular flavor	Berry: formation of seeds
	OIV	OIV	OIV	OIV	OIV	OIV	OIV	OIV	OIV	OIV	OIV
	202	203	204	206	220	223	225	231	235	236	241
Bikača	5	5	7	1	5	3	6	3	3	1	3
Crni krstač	7	5	7	1	5	3	6	3	3	1	3
Kratošija	7	5	7	1	5	3	6	3	3	1	3
Kratošija ili Vran	7	5	7	1	5	3	6	3	3	1	3
Kratošija mala	5	5	9	1	5	3	6	3	3	1	3
Kratošija sa dubokim urezom	7	5	7	1	5	3	6	3	3	1	3
Kratošija srednja	5	5	7	1	5	3	6	3	3	1	3
Ljutica	7	5	7	1	5	3	6	3	3	1	3
Rehuljača	7	3	1	1	5	3	6	3	3	1	3
Srednji Vranac	7	5	7	1	5	3	6	3	3	1	3

	Young shoot: intensity of anthocyanin coloration on prostrate hairs of the shoot tip	Young shoot: density of prostrate hairs on the shoot tip	Shoot: attitude (before tying)	Mature leaf: size of blade	Mature leaf: number of lobes	Mature leaf: shape of teeth	Mature leaf: degree of opening/ overlapping of petiole sinus	Mature leaf: density of prostrate hairs on main veins on the lower side of blade	Mature leaf: density of erect hairs on main veins on the lower side of blade	Flower: sexual organs	Bunch: length (peduncle excluded)	Bunch: width	Bunch: density	Bunch: length of peduncle of primary bunch	Berry: length	Berry: shape	Berry: color of skin	Berry: intensity of flash anthocyanin coloration	Berry: firmness of flesh	Berry: particular flavor	Berry: formation of seeds
	OIV	OIV	OIV	OIV	OIV	OIV	OIV	OIV	OIV	OIV	OIV	OIV	OIV	OIV	OIV	OIV	OIV	OIV	OIV	OIV	OIV
	3	4	6	65	68	76	79	86	87	151	202	203	204	206	220	223	225	231	235	236	241
Velji Vranac	3	7	1	7	3	2	7	5	7	3	7	5	7	1	5	3	6	3	3	1	3
Vranac	3	7	1	7	3	2	7	5	7	3	7	5	7	1	5	3	6	3	3	1	3
Vran	3	7	1	7	3	2	7	5	7	3	7	5	9	1	5	3	6	3	3	1	3
Vrančić	3	7	1	7	3	2	7	5	9	3	7	5	7	1	5	3	6	3	3	1	3
Vrančina	3	7	1	7	3	2	6	5	7	3	7	5	7	1	5	3	6	3	3	1	3
Velja Kratošija	3	7	1	7	3-4	2	7	5	7	3	7	5	9	1	5	3	6	3	3	1	3
Čestozglavica	3	7	1	7	3	2	7	5	7	3	5	5	7	1	5	3	6	3	3	1	3

Table 2.
Ampelographic descriptions of Kratošija accessions [27].

DOI: http://dx.doi.org/10.5772/intechopen.85676

Variety	VVS2		VVMD5		VVMD7		VVMD27		VrZAG62		VrZAG79		VVMD28		ISV2 (VNC671)		ISV3 (VMC6F1)		ISV4 (VMC6G1)		VMC NG4B9	
Vranac	133	133	226	226	247	249	181	181	193	199	258	258	239	251	151	165	133	139	177	177	164	172
Kratošija	133	143	226	236	247	249	179	181	199	203	236	258	251	261	141	165	139	139	177	177	150	164
Krstač	133	139	232	240	239	239	185	185	187	195	250	258	247	261	145	151	133	139	169	177	158	166
Žižak	143	145	240	240	239	263	179	191	187	193	250	250	251	261	141	159	136	139	183	187	150	152
Trojka	133	133	226	228	247	249	179	183	185	203	244	254	249	261	141	161	133	133	169	177	158	158
Čubrica	133	143	236	246	239	249	179	181	187	199	236	258	239	261	141	141	133	139	177	193	164	172
Muškaćela	133	133	228	236	233	249	179	294	185	195	250	254	249	271	141	143	133	139	163	187	158	166
Razaklija	139	143	232	246	239	247	181	185	185	187	250	258	239	261	141	143	133	145	177	193	150	176
Kadarun	143	145	232	236	249	249	179	181	193	199	258	258	249	261	165	165	139	139	177	193	150	150
Lisičina	133	137	238	246	239	239	181	185	187	195	250	250	237	239	141	143	133	139	177	197	158	172
Bioka	135	143	226	232	239	249	179	181	187	193	250	258	249	251	137	165	133	139	169	187	166	176
Crna loza	139	143	226	232	247	249	179	181	185	199	236	258	261	261	141	141	133	139	177	177	150	164
Plavina	133	143	232	236	239	249	179	189	187	199	236	242	251	261	143	165	139	139	177	177	150	152
Razaklija crna	135	139	232	238	247	255	181	185	185	203	250	258	239	247	141	165	139	145	177	187	158	176

Table 3.
SSR profiles or researched grapevine varieties in Montenegro.

to Drenok crveni from Macedonia.Within SEE.ERaNet project on various viticulture areas, from different vine growing regions, 96 samples of vines were selected and marked for identification. From these samples, 15 different genotypes were revealed some already known (Vranac, Kratošija, Krstač, Žižak, Čubrica, and Lisičina) and some with original SSR profile (Kovačka bijela, Sijer, and Zadrimka). Out of the six remaining samples, three accessions were found to be misnomers, one coincides with a previously identified variety in another country, and two accessions showed the original SSR profile which did not match any of the known varieties [37]. Accession of Japudžak from Montenegro is identical to the Turkish variety Yapıncak [38]. As a result of this project, in 2012, Montenegro presented and included its autochthonous and domesticated grapevine varieties in the EU Vitis database (Vranac, Kratošija, Čubrica, Krstač, Žižak, Japudžak, Sijer, Lisičina, Zadrimka, and Kovačka bijela). To preserve grapevine germplasm, the National collection of identified varieties was planted in Ćemovsko polje.

4. Conclusions

According to available literature and obtained results of ampelographic and genetic identification, Montenegro has a very long tradition of grapevine growing and very rich grapevine germplasm. There are varieties whose identification was done, but there are a lot of varieties with unknown origin and identity. Ampelographic description of 18 identified varieties as well as of 17 Kratošija biotypes was done and presented. During multiple years of research, genetic identification of 188 samples was carried out and the results revealed the original DNA profile for Vranac, Krstač, Žižak, Crna Loza, Čubrica, Lisičina, Razaklija crna, Kovačka bijela, Zadrimka, and Sijer. Kratošija, and every of its 17 biotypes, have the same DNA profile as Italian Primitivo, Californian Zinfandel and Croatian Crljenak Kaštelanski. Muškaćela is Moscato bianco while Trojka is Moscato violeto. Montenegrin Bioka is the same as Italian Francavidda and Croatian Zlatarica Vrgorska. The variety Razaklija is the same as Drenak crveni. For Japudžak the same SSR profile as for Turkish Yampincak is discovered.

Research and work on autochthonous and domesticated grapevine varieties in Montenegro are of great importance for the viticulture and winemaking sector. Having in mind this and very interesting results achieved, it was necessary to continue with investigation of Montenegrin grapevine germplasm. Further research with partners from the Institute for Vine and Wine in La Rioja will be done with the aim of analyzing a large number of samples across Montenegro, and then determining its origin and genetic relationships (pedigree analysis).

Acknowledgements

This research was supported by the following projects: “Variability of Kratošija cultivar in Montenegrin viticultural regions” and “Clonal selection of Kratošija cultivar” funded by the Ministry of Science of Montenegro; “Identification and characterization of Montenegrin autochthonous grapevine cultivars” and “Clonal selection of autochthonous grapevine cultivars” funded by the company 13 jul Plantaze and the Ministry of Agriculture of Montenegro. Part of this work was financially supported by the Swedish International Development Agency (SIDA) within SEEDNET project: “Identification, characterization and conservation of old and autochthonous vine varieties in Eastern European countries” and by the respective national funding bodies and the European Commission represented by the Joint Call Secretariat (JCS) at the German Aerospace Center (DLR) within SEERA NET 91/01 project: "Preservation and establishment of true-to-type and virus free material of endangered grapevine cultivars in Croatia and Montenegro". The author gratefully acknowledges all partners for the participation and collaboration.

Conflict of interest

I declare I have no affiliations with or involvement in any organization or entity with any financial interest (such as honoraria; educational grants; participation in speakers' bureaus; membership, employment, consultancies, stock ownership, or other equity interest; and expert testimony or patent-licensing arrangements), or nonfinancial interest (such as personal or professional relationships, affiliations, knowledge, or beliefs) in the subject matter or materials discussed in this manuscript.

Author details

Vesna Maraš
Plantaže, Podgorica, Montenegro

*Address all correspondence to: vesnam@t-com.me

References

[1] Vuksanović P. Rejonizacija Vinogradarstva Crne Gore. Titograd: University "Veljko Vlahović" Agriculture Institute; 1977

[2] Gazivoda D. Antički kultovi na tlu Crne Gore. Beograd: Filozofski fakultet, Univerzitet u Beogradu; 2015

[3] Plamenac M. Grlica. Cetinje: Državna štamparija. p. 1891

[4] Mijušković Đ. Kalendar poljoprivrede i šumarstva za prestupnu 1948. Titograd. 1948:126-128

[5] Organisation Internationale de la Vigne et du Vin. 2nd Edition of the OIV Descriptor List for Grape Varieties and Vitis Species. Paris: Organization Intergouvernementale crée Par l'Accord International; April 2001

[6] Thomas MR, Scott NS. Microsatellite repeats in grapevine reveal DNA polymorphisms when analyzed as sequence-tagged sites (STSs). Theoretical and Applied Genetics. 1993; **86**(8):985-990. DOI: 10.1007/BF00211051

[7] Bowers JE, Dangl GS, Vignani R, Meredith CP. Isolation and characterization of new polymorphic simple sequence repeat loci in grape (*Vitis vinifera* L.). Genome. 1996;**39**(4): 628-633 Available from: https://link.springer.com/article/10.1007/s11032-004-1362-4

[8] Bowers JE, Dangl GS, Meredith CP. Development and characterization of additional microsatellite DNA markers for grape. American Journal of Enology and Viticulture. 1999;**50**:243-246. Available from: http://www.ajevonline.org/content/50/3/243

[9] Sefc KM, Regner F, Turetschek E, Glössl J, Steinkellner H. Identification of microsatellite sequences in *Vitis riparia* and their applicability for genotyping of different *Vitis* species. Genome. 1999; **42**(3):367-373. Available from: https://www.ncbi.nlm.nih.gov/pubmed/10382286

[10] Crespan M. The parentage of Muscat of Hamburg. Vitis. 2003;**42**(4): 193-197. Available from: https://www.vitis-vea.de/admin/volltext/e049265.pdf

[11] Welter LJ, Göktütk-Baydar N, Akkurt M, Maul E, Eibach R, Töpfer R, et al. Genetic mapping and localization of quantitative trait loci affecting fungal disease resistance and leaf morphology in grapevine (*Vitis vinifera* L.). Molecular Breeding. 2007;**20**(4): 359-374. DOI: 10.1007/s11032-007-9097-7

[12] Crespan M, Cabello F, Giannetto S, Ibáñez J, Kontić JK, Maletić E, et al. Malvasia delle lipari, malvasia di Sardegna, greco di gerace, malvasia de Sitges and malvasia dubrovačka—Synonyms of an old and famous grape cultivar. Vitis. 2006;**45**(2):69-73. Available from: https://www.vitis-vea.de/admin/volltext/e052325.pdf

[13] EVD. European Vitis Database [Internet]. Available from: http://www.eu-vitis.de

[14] Statuta et leges civitatis Buduae, civitatis scardonae et civitatis et insulae lesinae, opera prof. Simeonis Ljubić. XV Century

[15] Viala P, Vermorel V. Ampélographie I-IV. Paris: Massonet Cie; 1901-1910

[16] Stojanović M. Novo vinogradarstvo. Belgrade: "Sv. Sava" M. Sladekovića; 1929. 462 p

[17] Bulić S. Dalmatinska Ampelografija. Zagreb: Poljoprivredni Nakladni Zavod,

Tisak Nakladnog Hrvatskog Zavoda; 1949. 359 p

[18] Ulićević M. Prilog Rejonizaciji Vinogradarstva u Crnoj Gori. Naša poljoprivreda i šumarstvo, num.2/V. Titograd. 1959

[19] Ulićević M. Prilog Proučavanju Osobina Najvažnijih Sorata Vinove Loze Gajenih u SR Crnoj Gori. Archive of Agricultural Sciences. 1966;**X**(23): 1-100

[20] Nastev D. Specijalno Lozarstvo. Skopje: Izdanie na Universitetot vo Skopje; 1967

[21] Ćetković V. Uticaj đubrenja i navodnjavanja na biološke osobine i prinos grožđa sorte kratošija u ekološkim uslovima Titograda [doctoral dissertation]. Sarajevo; 1978

[22] Zirojević D. Poznavanje sorata vinove loze. Beograd: Nolit; 1979. 432 p

[23] Avramov L. Savremeno Gajenje Vinove Loze. Beograd: Nolit; 1988. 640 p

[24] Pejović L. Ampelografska Proučavanja Varijeteta Kratošije, N. 3-4. Beograd: Jugoslovensko Vinogradarstvo i Vinarstvo; 1988

[25] Burić D. Savremeno Vinogradarstvo. Beograd: Nolit; 1995. 522 p

[26] Božinovik Z. Ampelografija. Skopje: Agencija Akademik; 1996

[27] Maraš V. Ampelografske karakteristike varijeteti sorte vinove loze [doctoral dissertation]. Zemun-Beograd: Poljoprivredni fakultet; 2000. 155 p

[28] Maraš V, Milutinović M, Pejović L. Variability in the autochthonous vine variety Kratošija. Acta Horticulturae. 2004;(640):237-241. DOI: 10.17660/ActaHortic.2004.640.26

[29] Milosavljević M. Biotehnika vinove loze. Srbija, Beograd, Draganić, Zemun: Institut za istraživanja u poljoprivredi; 1998. 566 p

[30] Božinovik Z, Petkov M, Beleski K, Boškov K. Proizvodne i tehnološke osobine nekih varieteta sorte Kratošija u Republici Makedoniji; 1998

[31] Vujović M. Vrste Domaćih Loza. Naša poljoprivreda, Titograd. 1956;**1/11**

[32] Calò A, Costacurta A, Maraš V, Meneghetti S, Crespan M. Molecular correlation of zinfandel (primitivo) with Austrian, Croatian and Hungarian cultivars and Kratošija, an additional synonym. American Journal of Enology and Viticulture. 2008;**59**:205-209 Available from: http://www.ajevonline.org/content/59/2/205

[33] Maraš V, Božović V, Giannetto S, Crespan M. SSR molecular marker analysis of the grapevine germplasm of Montenegro. Journal International Des Sciences de La Vigne et Du Vin. 2014; **48**(2):87-97. DOI: 10.20870/oeno-one.2014.48.2.1562

[34] VIVC. European Vitis Database in the Vitis International Variety Catalogue. Available from: www.vivc.de

[35] Vitis-WBC. Western-Balkans Vitis Database. Available from: http://www.vitis.atcglabs.com

[36] Žulj Mihaljević M, Šimon S, Pejić I, Carka F, Sevo R, Kojic A, et al. Molecular characterization of old local grapevine varieties from South East European countries. Vitis. 2013;**52**(2): 69-76

[37] Žulj Mihaljević M, Anhalt UCM, Rühl E, Tomić Mugoša M, Forneck A, Zdunić G, et al. Cultivar identity, intravarietal variation, and health status of native grapevine varieties in Croatia and Montenegro. American Journal of

Enology and Viticulture. 2015;**66**: 531-541. DOI: 10.5344/ajev.2015.15023

[38] Lacombe T, Boursiquot JM, Laucou V, Di Vecchi-Staraz M, Péros JP, This P. Large-scale parentage analysis in an extended set of grapevine cultivars (*Vitis vinifera* L.). Theoretical and Applied Genetics. 2013;**126**(2):401-414. DOI: 10.1007/s00122-012-1988-2

Section 2

Vine Technology

Chapter 5

Effects of Vine Water Status on Yield Components, Vegetative Response and Must and Wine Composition

Pilar Baeza, Pedro Junquera, Emilio Peiro,
José Ramón Lissarrague, David Uriarte and Mar Vilanova

Abstract

Despite *Vitis vinifera* L. is a drought-tolerant species—rainfed traditionally grown in a very diversity of climates—irrigation has more and more become a usual practice aimed to obtain regular yields along seasons and to control must composition. Results on vineyard irrigation are dependent on the timing, length and intensity of the water deficit. From budbreak to flowering, shoot growth is very sensitive to water stress, while reproductive growth is almost unaffected. Severe water deficit during fruit set can reduce yield by affecting ovary cell multiplication and expansion. During maturation water stress induces yield reduction by limiting berry growth; along this phase must composition is also affected. There is a positive, linear relationship between must sugar content and available water; however, no relationship has been found to either total acidity or pH. Biosynthesis of anthocyanins and fruity aromas is enhanced by water deficit. Usually, wines from moderate irrigation treatments scored the highest. There is a general agreement that severe, long water deficits diminish must quality, leaf area, fertility and yield, and it has a negative carryover effect on the next seasons by limiting wood reserves to be used the following seasons.

Keywords: wine grapes, *Vitis vinifera* L., irrigation, water potential, yield, berry size, shoot growth, vigour, soluble solids, pH, total acidity, organic acids, polyphenols, anthocyanins, aroma compounds

1. Introduction

Wine grapes are grown over a very wide diversity of environmental conditions. Originally, wine grapes were confined around the Mediterranean basin, but as humans spread around the world, these plants were able to conquer new habitats. Vineyards now exist in areas with Mediterranean climatic conditions (i.e. with relatively long, dry summers) in Europe, South America and North America but also in Atlantic regions in Europe and North America and in places with a similar climate in New Zealand. In recent decades, *V. vinifera* vineyards have been established in North Beijing and Washington State (USA) where winters are cold and even tropical areas in Thailand. This highlights the plasticity of *V. vinifera* cultivars, which have become adapted to very different climatic conditions, producing reputable wines and table grapes under most of them.

Along history, growers have been forced to choose those cultivars best adapted to the local availability of water, seasonal temperatures, the dry periods they must face, etc., increasing experience allowing the most to be made of each situation. Different training systems and cultivation practices have also been developed, striking a balance between plant, vineyard management and the environment and giving rise to different viticultural landscapes, some now recognized as part of the world heritage. This balance, however, can be altered when priorities change, perhaps driven by the desire to produce more, or because of a change in market conditions. Thus, an area might need to increase yields or open up new areas of sustainable production. Areas naturally suited to raising white wine grape varieties might suddenly need to shift to red, or the variety habitually grown may need to be changed due to customer demand. Under certain circumstances, newly imposed conditions can only be met sustainably by modifying the vineyard agrosystem, perhaps by introducing a different kind of trellising or canopy management or by introducing irrigation.

For a long time, the drought tolerance of grapevines meant irrigation was not contemplated as a viticultural practice. Indeed, it took hard work to overcome the prejudice that grapevines are not well suited to it. By and large, vineyards in growing areas brought into production in the last 50 years have been irrigated. In some traditional areas, however, irrigation was banned until some decades ago. Irrigation results obtained from vineyards under regional regulations (geographical indications), with limitations either to yield or bud load, for example, may complicate the discussions of irrigation as it often happens that irrigated vines cannot express the most of them when we are limiting their optimal performance under those new conditions and when they are harvested at the same date. This turned out that part of the industry felt that the best wines were produced under situations of severe water stress. The aim of irrigating wine grapes is not always to produce higher yields but to ensure the quality required for different products. For example, some grapes are grown with the intention of producing young wines, others are raised to make wines for ageing and yet others for making spirits, etc.; as a result, they require different irrigation regimens and different optimal yields and different harvest time. In recent times, attitudes are changing as irrigation studies have increased and irrigation management becomes ever more technically friendly and controllable, and the consequences of global warming are felt.

In the following paragraphs, a review of the effects of water status on yield, vine growth and must and wine composition is exposed, and results are explained taken into account the phenological stage and the berry growth stage at which excess-optimal-severe available water took place.

2. Effects of plant water status on vine response

2.1 Effects of vine water status on yield components

Different components are taken into account when calculating the yield of a vineyard. The yield per hectare can be expressed as follows:

$$\text{Yield} = \frac{\text{N°vines}}{\text{ha}} \cdot \frac{\text{N°buds}}{\text{vine}} \cdot \frac{\text{N°shoots}}{\text{bud}} \cdot \frac{\text{N°clusters}}{\text{shoot}} \cdot \frac{\text{N°berries}}{\text{cluster}} \cdot \text{berry weight} \quad (1)$$

The vine spacing, training system and pruning level determine the number of potentially productive buds. In most viticultural regions, budburst follows its normal course since soil water is usually available. However, a strong water deficit at the beginning of the season negatively affects budburst since the mobilisation of

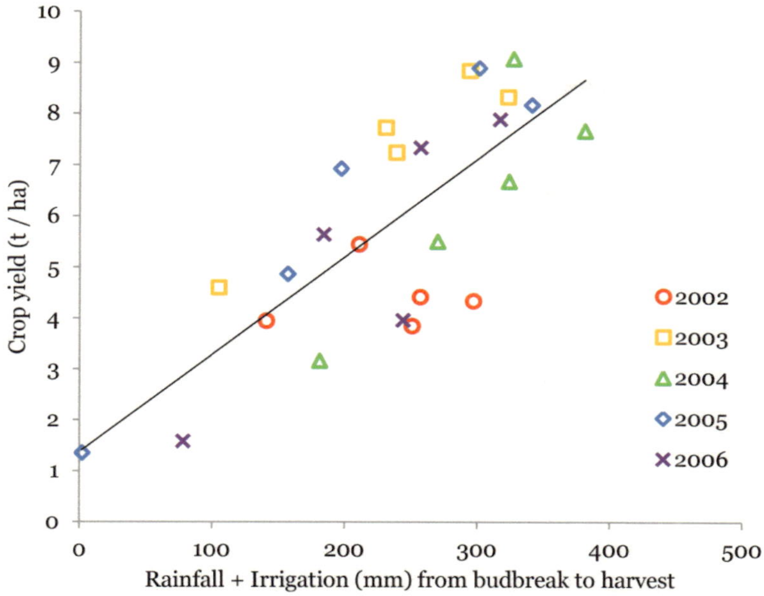

Figure 1.
Relationship between crop yield and water supply (rainfall + irrigation) from budbreak to harvest in a cv. Cabernet sauvignon vineyard in Madrid, Spain. Data correspond to five different irrigation treatments applied during 2002–2006 (adapted from Junquera et al. [6]).

nutrients from the reserve structures is reduced [1]. Once the number of potentially productive shoots is defined, the yield of a vineyard depends on a set of internal and external factors, and the interactions among them, all of which have an impact on the processes of floral induction and differentiation and the growth of the berries. These factors include the genotype of the vine (variety and rootstock), environmental conditions (climate and soil) and cultivation practices [2].

Water deficit is one of the main environmental factors limiting vegetative growth and berry yield [3, 4] (reproductive development is less sensitive to water shortages than vegetative growth [5]). The water status of a vineyard depends on the availability of water (soil water, rainfall and irrigation), atmospheric conditions (relative humidity, vapour pressure deficit, temperature, etc.) and leaf area as well as the ability of the vine to absorb and transport water to its organs.

Some studies have reported a direct relationship between the amount of water available during the growth cycle (rainfall + irrigation) and yield (**Figure 1**) [6, 7]. However, this relationship is not immediately obvious when data from different studies are brought together in the search for correlations. This is largely the consequence of differences in environmental conditions (soil and climate) and vineyard characteristics (genotype, training system, etc.), which generate differences in water use efficiency (kg fresh fruit/m^3 water applied) [8]. However, the meta-analysis conducted by Medrano et al. [8] clearly shows a positive linear relationship between yield and water use efficiency, even when an increase in the latter can only be achieved by reducing the total amount of water used—which generally involves a certain reduction in yield. Indeed, several studies have concluded that irrigation doses equivalent to 60–80% of crop evapotranspiration (ETc) are sufficient to maximise yield [9–11]. Irrigation doses exceeding 100% ETc might lower yield via reductions in fertility, and even in berry weight, perhaps due to competition between berry and vegetative growth [10].

Reproductive growth correlates with water availability, with this relationship dependent on the development stage of the vine. Generally, water deficit reduces yield, particularly when shortages occur early in the season [12]. However, the complexity and duration of the reproductive cycle of the vine make a more detailed analysis necessary. The reproductive cycle of the vine is completed after a 2-year period: the buds formed in the first season develop and give rise to fruiting shoots in the following season. This process includes numerous phenomena: induction and floral differentiation, flowering, pollination, fertilisation, fruit setting and berry growth [13]. Thus, there is a long period of time over which the yield is liable to alterations due to environmental conditions and/or vineyard management practices.

Intense and persistent water deficits usually reduce bud fertility via falls in the number and size of inflorescences [14]. Induction is particularly sensitive to water stress, with shortages during flowering normally leading to important reductions in bud fertility [15]. Vasconcelos et al. [1] reviewed the different means by which water status can affect floral induction and differentiation, and therefore bud fertility, reporting it to be influenced (1) directly, via the amount of water available to processes determining cell division and expansion, and (2) indirectly, via its effect on photosynthetic activity, nutrition, the microclimate of the renewal zone and hormonal balance. These authors also indicate that the many determining factors and possible interactions among them make it difficult to establish clear correlations between water status and bud fertility. Certainly, the potential for reduced fertility exists via excessive water availability leading to increased vigour and vegetative growth and therefore reduced light interception in the renewal zone [1, 16, 17]. This same excessive vigour and lack of illumination can, however, also favour primary bud necrosis and therefore a lack of primary bud growth at budbreak and reduced fertility [10, 18]. Fertility can thus be reduced by both limited and excessive water availability.

Shortly after budburst, reproductive growth is relatively unaffected by water deficit. In most viticultural regions, water deficit is not normally a problem during inflorescence development; the soil water content is generally sufficient throughout spring, supplied either by rain or irrigation. Moreover, at this point in the reproductive cycle, inflorescences are able to compete for photoassimilates against the vegetative structures of the shoots, with the production of carbohydrates by the former sufficient for self-supply. It is only later, during flowering, when vine requirements for photoassimilates exceed photosynthetic capacity and the sensitivity to water deficit increases [16]. Of course, there may be times when drought conditions occur even during early spring. Excessive water deficit at this time can cause the vine to loose whole inflorescences, reducing the eventual number of future clusters. This is particularly true when such drought conditions are combined with high temperatures and low vigour [13].

The reviews by McCarthy [19] and Keller [16] reveal the importance of vine water status during the flowering period. The male organs are more sensitive to this variable than the female organs; deficits near the time of flowering may limit ovary growth, leading to smaller berries, but the effects on pollen formation, germination and pollen tube growth are even more severe. Water deficit, like other stressors, can limit sugar uptake and starch accumulation in developing pollen grains, causing sterility and compromising the course of fertilisation and fruit set, even leading to the loss of whole inflorescences [2]. Severe water stress during fruit set can reduce the success of this stage via reductions in the photosynthetic rate and carbohydrate availability [17].

Once fruit set has taken place, and the final number of berries in the vineyard is determined, the last yield component to play a role in the yield is berry weight. Berry development follows a double sigmoid curve [20] that can be divided into three stages. In Stage I (the beginning of the green phase of berry development), berry growth is caused by cell division and enlargement. Stage II is the shortest stage; growth at this point is markedly reduced. At the end of Stage II, the berry colour starts

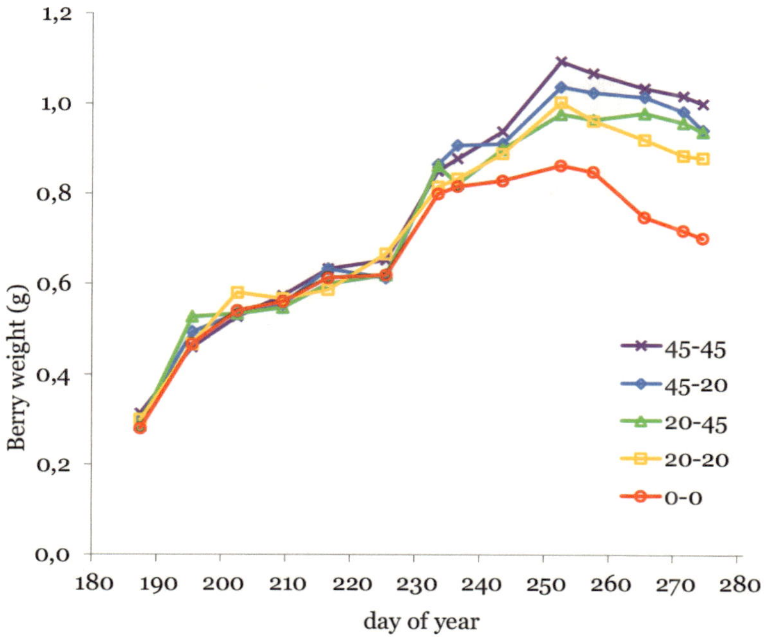

Figure 2.
Change in berry weight for five different irrigation treatments applied during 2004 in a cv. Cabernet sauvignon vineyard in Madrid, Spain. Numbers for each treatment correspond to the %ETc applied by irrigation before and after veraison. Unpublished data.

to change, and metabolic processes that trigger ripening take place. This moment in the cycle is called veraison. In Stage III, the so-called ripening, berry growth is restarted due to cell enlargement. During Stage I, both multiplication and cell growth can be affected by water stress, although multiplication is less sensitive than cell enlargement. Water stress at this time alters the properties of the cell wall, irreversibly restricting the capacity for cell enlargement [21]. Later on in the cycle, only cell expansion is affected by water stress, limiting berry and seed growth. However the effect here is never as significant as in the earlier stages. Berries become increasingly resistant to stress from veraison onward. In fact, the reduction in yield due to water deficit is much more important when this occurs before veraison, as made clear by numerous studies on regulated deficit irrigation (**Figure 2**) [6, 12, 22–25].

In their review, Chaves et al. [4] indicate the effect on photosynthesis to be the main cause of water availability-induced reductions in berry growth after veraison. During ripening, the berries take up water mainly via the phloem; uptake from the xylem is very limited. Occasionally, berry weight losses are observed in late ripening, reducing the final yield (**Figure 3**). Recent studies have shown that, in addition to possible water losses by transpiration (which are less severe at this point than during Stage I), water return via the xylem may occur. This return is dependent on grape variety and is determined by the late-ripening integrity of the cell membranes and the hydraulic conductivity of the xylem [26, 27]. Different grape varieties show either isohydric or anisohydric water regulation behaviours at the leaf and root level; the idea of variety-dependent water regulation strategies at the berry level cannot, therefore, be ruled out [28]. Illand et al. [17] hypothesize weight loss taking place during late ripening whenever berries continue to be vascularly connected to the vine and there is a loss in cell viability (shrinkage in Syrah). This suggests that weight loss would not occur if (a) cell viability is preserved (Thompson Seedless) or (b) the berries become vascularly disconnected from the vine (Chardonnay).

Figure 3.
Shrivelling and weight loss in cv. Graciano grapes during late ripening.

2.2 Effects of vine water status on vegetative growth

Plant growth is strongly affected by water availability. Freeman and Smart [29] reported increases in root growth under water deficit conditions, while Van Zyl [30] indicated that irrigating vines after 25 or 50% of the total water available had been used up by around the time of flowering led to 190 and 300 actively growing root tips/m^2, respectively. However, this was reduced to 40 root tips/m^2 when the soil was irrigated after 75% of the total water available had been consumed. Prolonged exposure to moderate water deficit thus increases the root-to-shoot ratio [4], but both severe water deficit and irrigation that keep the soil close to saturation for long periods negatively affect root growth.

There is, however, no evidence that vegetative growth in vines is increased under water deficit conditions. Indeed, water deficit negatively affects the vegetative growth of vine trunks, shoots and leaves. However, the limitation of vegetative growth depends on the timing, duration and severity of water deficit. The most active period of vegetative growth takes place between budbreak and veraison [12], with a maximum reached at the beginning of the growth cycle some 60 days after budbreak [6, 31–33]. Growth then progressively decreases until a vegetative standstill is reached close to the time of veraison.

After budbreak, shoot growth occurs at the expense of reserves stored during previous vegetative cycles [34]. In Mediterranean-type climates, it is uncommon for soil water deficits to be strong enough to inhibit the growth of shoots during this initial growth phase (rain usually falls during this period, and there is still winter rain stored in the soil [35]), although total needs up to veraison are rarely so easily met.

The effect of water deficit on the canopy has been widely studied. Many authors indicate it to reduce shoot growth and canopy development [6, 11, 36–40], a consequence of reduced activity in the shoot terminal meristem [41], reduced leaf area [42] and the senescence and fall of the basal leaves. Physiological changes in the vine, such as reduced leaf photosynthetic activity in response to stomatal closure due to water stress, have been studied in many experiments [4, 22, 43–52].

Post-veraison water stress has little or no effect on shoot growth [22, 53, 54]. Nevertheless, severe water stress during the ripening period can significantly diminish leaf area due to early senescence [55]. During the postharvest period, root growth and nutrient absorption contribute towards the accumulation of reserve carbohydrates. This period is important for the vigour and productivity of the vine in the following season. The soil water content aimed for at this time should ensure the accumulation of carbohydrates but avoid the regrowth of laterals [56].

Shellie [9] and Greenspan [57] reported a reduction in the main shoot growth from 20 days after budbreak when the midday leaf water potential approached—1.0 MPa in cvs. Merlot, Cabernet Sauvignon and Pinot Gris vines. Munitz et al. [58] observed a reduction in leaf area in cv. Merlot after continuous irrigation at 20, 35 and 50% of the ETc, reaching maximum ψ_{stem} values of −1.2, −1.3 and −1.4 MPa during cluster closure to veraison. However, in cv. Cabernet Sauvignon, Acevedo-Opazo [59] reported no differences neither in shoot length, number of stems, internode length or pruning weight between three treatments in which the midday ψ_{stem} was maintained at between −0.8 and −0.95 MPa, −1.0 and −1.2 MPa and −1.25 and −1.4 MPa, from post-setting to harvest—although in that work ψ_{stem} did reach values of −1.0 MPa close to veraison. Water stress induced at the beginning of the growing season (flowering to cluster closure or earlier) may thus result in a reduction in canopy size. Under field conditions, however, severe water stress might be hard to induce; soils will normally contain some stored water. In another experiment on cv. Cabernet France [36], less shoot growth was observed in an early water deficit treatment compared to control, although the leaf water potential values reached were similar (≈−0.8 MPa). The authors [36] suggested that this level of water deficit was not responsible for the reduced shoot length observed, but to the early limitation of photoassimilates, probably caused by a reduction in the hydraulic conductivity of the wood through prolonged exposure to early water deficit repeated over many seasons [60].

This high sensitivity of shoot growth to water deficit has sometimes been used as an early indicator of the latter (based on allometric measurements). Pellegrino et al. [61] analysed the effects of water deficit on certain components of shoot vegetative growth (the number of leaves to emerge on the first- and second-order laterals, leaf area, internode length of each phytomer on the first-order laterals and the frequency of second-order laterals) in cv. Shiraz. Sensitivity to water deficit was seen to increase as the second-order laterals emerged, i.e. the rate of emergence of second-order laterals decreases in response to water deficit. These authors also established a water deficit indicator (ratio of branching intensity between first- and second-order laterals) that was sensitive to slight water deficit—even more so than the stomatal response. A more recent study showed significant changes in the abundance of proteins involved in translation, energy production, antioxidant defence and steroid metabolism during early growth and indicates these changes to occur before any detectable reduction in shoot elongation, stomatal conductance or photosynthesis [62].

The availability of water in the soil leads to differences in hydraulic conductivity that leave permanent marks on vine plant organs. At the trunk level, high water availability early in the season results in wider xylem vessels (and therefore greater hydraulic conductivity) and greater trunk diameter, ring width and ring area. Also, when vines are subjected to late water deficit, they show more negative water potential values at the end of the season than do vines that receive low-level but relatively constant irrigation [60]. Thus, high water availability during the vegetative growth of *Vitis* increases vessel diameter and hydraulic conductivity, leaving plants more vulnerable to stress during the ripening period [60]. However, at the shoot level, Pagay et al. [63] reported xylem vessels with larger diameters to be more resistant to cavitation, concluding that they have less inter-vessel pitting. This would result in a hydraulic advantage allowing them to better maintain growth and productivity under water stress.

Finally, pruning weight is linearly related to the amount of water applied and is less influenced by the timing of the water supply than is leaf area (**Figure 4**) [33, 64]. Reductions in shoot weight are accentuated by long-term water deficit [6]. Thus,

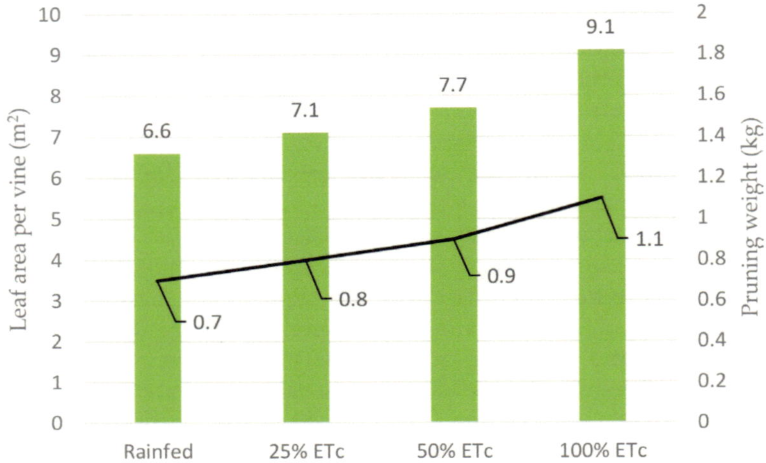

Figure 4.
Leaf area per vine at ripening (green bars) and winter pruning weight (blue line) from rainfed vines and three irrigation treatments applying 25, 50 and 100% of vine evapotranspiration (ETc) in averaged over four seasons in cv. Tempranillo. After [64].

water deficit has a cumulative effect, probably due to reduced starch and sucrose accumulation in the perennial organs [65–67]. This is important since the sugars accumulated in the trunk and roots are the first carbohydrates to be used during the following spring's growth.

2.3 Effects of vine water status on must and wine composition

2.3.1 Effects of irrigation on total soluble solids

Total soluble solids (TSS) accumulate in the berry during phase III of berry growth. The increase in sugar content (°Brix) is coupled with a resumption in berry growth, accompanied by a sharp increase in berry weight (**Figure 5**). At the beginning of ripening, sugar accumulation occurs through leaf photosynthesis and through the mobilisation of reserves [68] although this mobilisation soon ceases. Sugar is transported from leaves to the berry via phloem in the form of sucrose. Once in the berry, it is changed into glucose and fructose, the ratio between them remaining close to 1:1 throughout ripening [2]. Since the TSS content is directly related to leaf photosynthetic activity [2, 69], sugar content can be used to indirectly evaluate the plant photosynthesis activity. At the end of ripening, photoassimilates also divert to the fruit and to reserve tissues in the perennial parts of the plant [68].

A linear relationship exists between berry size and TSS (**Figure 6**). Thus, berry size provides a meaningful, inexpensive means of estimating plant activity and tracking ripening. At the end of the ripening period, the increase in berry weight levels off, and the discharge of sugar into the berry ends. When the berries reach maximum TSS due to photosynthesis, ripening is finished. The final °Brix may differ depending on variety, cultivation practices and climate.

The amount of available water influences both the sugar accumulation rate and berry size. Studies have shown that, under water deficit conditions, °Brix increases faster than under high soil water conditions (**Figures 5** and **6**). This means that for the same berry weight, the °Brix reached in rainfed vines (or grown under water deficit conditions) is higher than those reached in well-irrigated vines [33, 71]. However, sugar accumulation expressed on a per-berry basis is higher for irrigated vines. Since irrigated vines produce higher yields than either moderately water stressed or nonirrigated vines, the sugar concentration of the berries produced under the former conditions increases slower, but finally they can get the same concentration if they remain in the vine (**Figure 6**) [71].

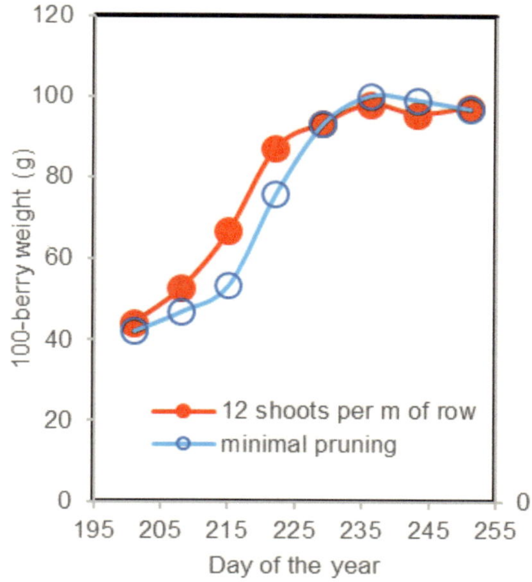

Figure 5.
Seasonal change in 100-berry weight in cv. Merlot in two shoot load treatments. Compact circles represent a shoot load of 12 shoots per metre of row; open circles represent minimal pruning conditions. Unpublished data.

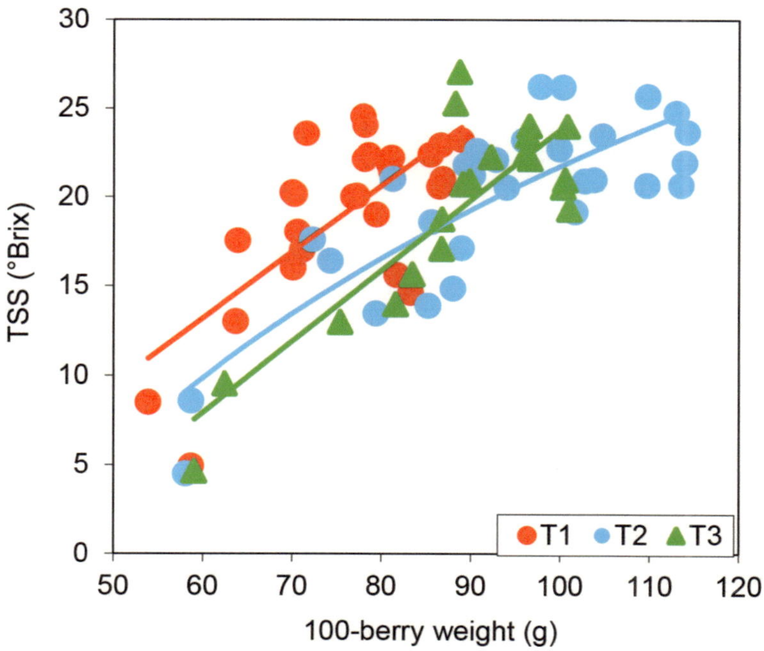

Figure 6.
Change in the relationship between berry size and TSS under three irrigation regimens in cv. Cabernet sauvignon in Madrid (Spain). T1: rainfed ($Y = 0.37x - 9.0R_2 = 0.50^*$), T2: irrigated 0.4·ETo ($Y = 23.2Ln(x) - 85.12R2 = 0.68^{**}$), T3: irrigated 0.2·ETo ($Y = 0.40x - 15.82R2 = 0.71^{**}$) [70].

Berries increase in weight according to the availability of soil water. In trials involving different irrigation treatments, seasonal berry weight trends run in accordance with the supply of water [33, 72]. However, other authors report that berry sugar concentration may not differ between irrigation treatments since the smaller photosynthetic rate reached during ripening may be compensated for by a smaller berry weight [6, 71, 73]. Under moderate water stress, berry weight is reduced, but ripening quality in terms of sugar content is unaffected [9, 21, 74–76]. When water deficit is very mild, neither berry size nor sugar content is affected [77].

Although berry sugar concentration may not be affected by an increase in water deficit, other must components—such as anthocyanins [6]—may be. When water deficit has been long and intense, photosynthetic rates become low, and leaf abscission can occur at mid ripening, collapsing the ripening process [6, 69, 72]. Thus, final sugar content depends on water deficit intensity and deficit timing; several authors [8, 9, 59, 69, 70, 72, 78] have examined the thresholds between moderate deficit and severe water stress; on the whole, a midday stem water potential (Ψ_S) of −1.2/−1.3 MPa is required to maintain the yield and must quality within the required range. A midday Ψ_s of ≤1.4 MPa has clearly detrimental effects on photosynthesis, quality and yield. Some authors report that over optimal irrigation can have a detrimental effect on sugar content and lead to delayed ripening, increased acidity, reduced berry colour intensity and a smaller yield [35, 79–81], but in some of these experiments, however, midday Ψ_s was kept at above −0.9 MPa [80].

2.3.2 Effects of irrigation on pH and titratable acidity

Must titratable acidity and pH are important quality variables in winemaking. They both affect wine perception in the mouth (including smoothness, freshness and stringency). pH also influences the colour of the anthocyanins (red-blue at pH 3.0, orange near pH 4.0 and transparent at around pH 7.0), conditions microbial stability and when low acts as a shield against oxidation in musts and wines.

Organic acids accumulate in the berries during pre-*veraison*, increasing in concentration from fruit set to the end of phase II of berry growth [2]. At the end of this phase, the berries contain many different acids, but tartaric and malic acids together account for 70–90% of the total acid content.

During ripening, the berry acid content decreases by (1) dilution as sap flows inwards from the phloem, (2) by malate being used as a carbon source in respiration and (3) by gluconeogenesis in the berry, although this is responsible for only a small amount. Citric acid transforms into malic acid which might then follow any of its degradation routes. During ripening, the total acid content of the berry decreases, and the pH increases (**Figure** 7). At the end of ripening, the berry acid content and pH depend on the balance between the acid content at pre-*veraison*, leaf photosynthetic activity during pre- and post-*veraison*, the vine microclimate during pre- and post-*veraison*, final berry size and the berry cation (Ca^{2+} and K^+) content, which transforms free acids into their corresponding salts. Variety and rootstock influence cation uptake, thus affecting the final pH too.

In trials, the effects of irrigation on total acidity and pH have been inconsistent. Irrigation has been reported to increase, reduce or not affect either variable. Esteban et al. [71] in a trial comparing nonirrigated and irrigated vines of cv. Tempranillo grafted onto 110 Richter rootstocks found that the grape must of the irrigated vines had lower pH and higher titratable acidity. In contrast, after a 5-year study of different irrigation regimens on cv. Cabernet Sauvignon/SO4, Junquera et al. [6] reported a positive relationship between water availability and total acidity but indicated that must pH was unaffected by irrigation. Differently again, after a 5-year of study involving cv. Tempranillo/161-49C, Intrigliolo and Castel [33] concluded that the

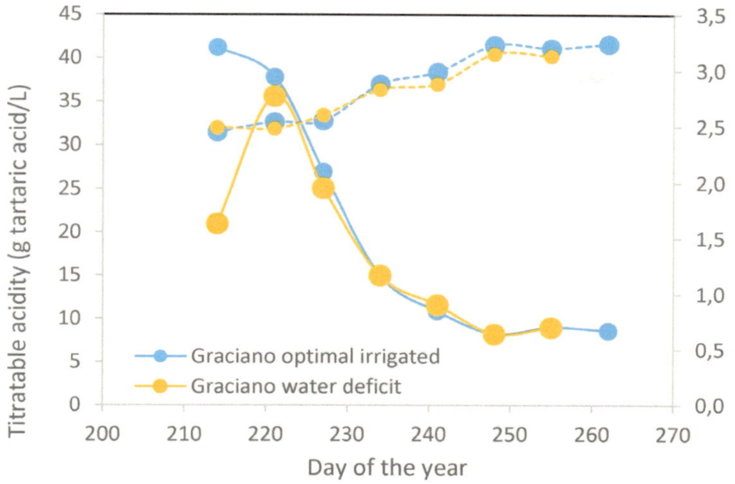

Figure 7.
Seasonal change in titratable acidity (compact lines) and pH (dashed lines) in cv. Graciano/41B under optimum water availability (compact circles) and water deficit conditions (empty circles).

only detrimental effect of irrigation was an increase in pH compared to no irrigation, with total acidity increasing. In another trial involving cv. Tempranillo in pots, [72] titratable acidity increased with increasing water stress regardless of the stage at which water stress was induced. Working in a very warm region with the white cultivar Doña Blanca, Uriarte et al. [73] reported higher tartaric acid and lower malic contents in the must of water deficit vines, regardless when water deficit was induced. However, after a 3-year trial on cv. Monastrell/1103 Pa [81], neither titratable acidity nor the malic or tartaric acid contents were altered by any irrigation treatment. They did indicate pH and K^+ to be significantly reduced in the highest irrigation treatment, but these differences were negligible from an oenological point of view; adding to the confusion, the tartaric and malic acid contents were inconsistent from 1 year to the next.

The results of other authors have further compounded the problem. For instance, neither Acevedo-Opazo [59], who ran a 3-year trial on cv. Cabernet Sauvignon in Chile, nor Munitz et al. [32], in their 4-year trial in Israel involving cv. Merlot, could found any differences in total acidity or pH between irrigation treatments. However, in a trial involving cv. Tempranillo/110R under a wide range of irrigation doses between budbreak and *veraison* and between *veraison* and harvest, Santesteban et al. [24] obtained higher titratable acidity values in the higher irrigation treatments before *veraison* (average predawn leaf water potential ~ −0.35 MPa). Regarding organic acid results, differences were significant 2 years out of 4 for the tartaric and malic acid contents which make us to state that irrigation effects on must acidity are still inconclusive.

These apparently very contradictory results do have some explanation, however, when examined taking into account vine physiology and factors that regulate the synthesis, accumulation and breakdown of these components [24, 69, 81, 82]: when water deficit is imposed from early in the season up to *veraison*, it negatively affects vigour, berry size and photosynthetic rate. If the photosynthetic rate is low, the acid and phenol contents accumulated in the berry during phase I are reduced [71, 83, 84]. On the contrary, optimal vine water status during this phase enhances photosynthetic activity, vigour and the acid and phenol content in the berries.

Esteban et al. [71] and other works obtained a tight positive, linear relationship between pH and K^+ in grape must. Potassium is meanly accumulated in berry during maturation [85]; grapevines suffering water deficit during maturation have lower berry potassium concentration due to reduced mobility in soil and impaired root uptake [85]. This could explain the reduced pH in lower irrigated vines with respect to irrigated ones resulted in some trials [33]. Therefore, final pH and titratable acidity will depend on the timing and intensity of vine water deficit.

The effect of water deficit on must composition during ripening depends on the plant's previous water status. Mild water stress (midday Ψ_{stem} = −1.2/−1.3 MPa) after no previous water stress favours an optimal photosynthetic rate and sugar accumulation by the berries while avoiding excess berry growth. Thus, acids and phenols are not diluted and reach an optimal concentration at the end of ripening. Severe water deficit after *veraison* slows sugar accumulation severely and can collapse ripening. At this point, water stress favours leaf abscission, and the berries remain unripe with a high acid content, with a low sugar content and with an unripe colour and immature seeds. Early leaf fall renders the clusters more exposed to direct sunlight, affecting the breakdown of malate and the synthesis of anthocyanins. Either overirrigation or excess of available soil water after *veraison* may cause the berry acids to be diluted due to excess berry growth [16]. If there is a high soil water content, laterals develop and compete with the berries for sugars from leaf photosynthesis. Ripening is then delayed, something that could be sought more often in warm growing areas.

2.3.3 Effects of vine irrigation on berry phenolic maturity and aroma compounds

Volatile and phenolic compounds are grapevine secondary metabolites critical to grape quality and wine sensory attributes. Viticultural practices can influence the concentration of these compounds and their precursors in grapes via plant stress responses. Deficit irrigation (moderate water restriction), for example, is an important vineyard management strategy used to alter grape composition and therefore improve the final organoleptic quality of wine [85]. Water deficit in the vineyard reduces vegetative growth, alters the canopy microclimate and increases the amount of intercepted light in the cluster zone [9, 33, 86]. This renders the fruit more susceptible to heat stress, especially when there are high levels of ambient solar radiation. However, increased exposure of the fruit to sunlight has been associated with improvements in must and wine quality [87]. Indeed, several authors have reported it to be associated with increases in the volatile compound contents of grapes and wines, especially monoterpenes and carotenoids [88, 89]. Light and temperature also influence norisoprenoid concentrations, which correlate directly with the concentrations of carotenoids in grapes under moderate water stress [90, 91].

Both light intensity and temperature also affect phenolic compound composition and berry colour. Sunlight favours the accumulation of polyphenols in the berries, mainly anthocyanins [92, 93], but increased temperatures from excessive exposure to sunlight may lead to reduced berry colour, especially in warm-climate regions [94–96].

Berry size is widely acknowledged to affect berry quality. Vine water deficits generally lead to smaller berries being produced and changes in fruit and wine composition [74]. Depending on the moment of induction of water stress and its severity, the proportion of skin surface area to mesocarp volume changes [74, 97] as does the rate of biosynthesis and degradation of volatile [98] and phenolic compounds [97]. Several authors have reported the effect of vine water status on grape and wine volatile compounds [98–106], while Chapman et al. [99] showed that water deficit influences berry composition and improves wine sensorial quality, increasing fruity aromas and reducing vegetation aromas.

Bindon et al. [101] report that deficit irrigation increases the concentration of some C13-norisoprenoids, such as β-damascenone and β-ionone, in cv. Cabernet Sauvignon berries at harvest. Other studies also report a positive effect of deficit irrigation on grape and wine volatile compounds in cv. Cabernet Sauvignon berries [100, 105]. Water deficit also affects cv. Merlot grape maturity and composition. Deficit irrigation reduces the concentration of negative compounds and increases the concentration of positive compounds [102]. Certainly, Qian et al. [98] observed that cv. Merlot wine produced from deficit-irrigated vines has increased vitispirane, β-damascenone, guaiacol, 4-methylguaiacol, 4-ethylguaiacol and 4-vinylguaiacol concentrations compared to wines produced from well-watered vines. Deficit irrigation had no effect on the concentrations of other measured volatile compounds such as esters and terpenes. Similarly, Talaverano et al. [103] suggest that low water supply has a negative effect on the aromatic potential (mainly related to ethyl esters) of wines at a similar ripening stage. However, this effect could be countered by harvesting at a later date. Recent work reported by Vilanova et al. [105] shows that volatile composition in cv. Verdejo wines is modified by the water regimen, with concentrations increasing under the most severe deficit irrigation regime.

In a study examining the effect of irrigation on the sensory profile of wines from Galicia, those made with grapes of the white cultivars Albariño and Godello were judged to be better when the vines were rainfed rather than irrigated. Wines made from cv. Treixadura grapes, however, were judged better when the vines were irrigated [107]. Balint and Reynolds [108] studied the effect of different irrigation strategies on cv. Cabernet Sauvignon aroma descriptors and reported regulated deficit irrigation (RDI) to improve wine quality over both full irrigation and no irrigation treatments. The 25-RDI (25% ETc) treatment especially returned higher scores for most of the positive sensory characteristics of cv. Cabernet wines. The former authors reported that soil and plant water status could be used to predict the flavour profile of these wines, reflecting the relationship of these variables with sensory descriptors. It was concluded that 100% water replacement was not recommendable at any phenological stage. However, 50 and 25% water replacement had overall positive effects on fruit composition and wine varietal typicity.

Water deficit can enhance the accumulation of anthocyanins by stimulating anthocyanin hydroxylation [4, 97]. Castellarin et al. [109] showed that water deficits accelerate anthocyanin accumulation and increase the expression of many genes responsible for the biosynthesis of anthocyanins. Moreover, the concentrations of different individual phenolic compounds are reported to change in response to plant water status, with those of flavonol and proanthocyanidin less affected than those of anthocyanins [97]. Ojeda et al. [110] report that severe water deficit before *veraison* reduces cv. Syrah anthocyanin synthesis. Similar results were reported by Romero et al. [69], with severe water stress associated with total grape phenolic compound concentration in cv. Monastrell grapes. However, Casassa et al. [111] recently reported that early and full deficit irrigation applied at pre-*veraison* produced cv. Cabernet Sauvignon grapes and wines with higher concentrations of phenolic compounds. In general, moderate water stress increases the concentrations of these compounds in red grapes, improving berry quality. However, when a certain threshold of water stress is surpassed, these positive effects are reported to disappear [69]. Similarly, Delgado et al. [112] report that the use of less water can increase the chemical and sensorial quality of wine; this is of some significance in a climate change context in which water supplies may decline. Niculea et al. [113] report phenolic compound accumulation and composition responses to sustained deficit irrigation during berry growth and ripening to be variety-dependent.

Finally, Herrera et al. [114] suggest that the interaction between water availability and weather conditions plays a crucial role in modulating berry composition. A meta-analysis performed by Mirás-Avalos and Intrigliolo [115], using published data for red and white varieties, concludes that cultivar, the timing of water restrictions and rootstock type have a great influence on must and wine composition. The effects of other factors, such as climate, the leaf surface/yield ratio and training systems, need to be examined in future research.

The effect of RDI on grape volatile and phenolic compounds remains incompletely understood, and further investigations are required to determine what compounds are influenced by irrigation. The timing, severity, duration of water stress, seasonal variations, the type of cultivar and the interaction of *genotype x environment* can all influence the response of vines to water stress, probably explaining the discrepancies seen in the results of different studies.

Author details

Pilar Baeza[1*], Pedro Junquera[2], Emilio Peiro[2], José Ramón Lissarrague[1,2], David Uriarte[3] and Mar Vilanova[4]

1 Department of Agronomy, Universidad Politécnica de Madrid, Madrid, Spain

2 Gestión Integral de Viticultura (GIVITI), Madrid, Spain

3 Centro de Investigaciones Científicas y Tecnológicas de Extremadura, Instituto de Investigaciones Agrarias Finca La Orden-Valdesequera, Badajoz, Spain

4 Spanish National Research Council (MBG-CSIC), Pontevedra, Spain

*Address all correspondence to: pilar.baeza@upm.es

References

[1] Vasconcelos MC, Greven M, Winefield CS, Trought MCT, Raw V. The flowering process of *Vitis vinifera*: A review. American Journal of Enology and Viticulture. 2009;**60**:411-434

[2] Keller M. The Science of Grapevines. 2nd ed. London: Academic Press; 2010. p. 509. DOI: 10.1016/C2009-0-01866-2

[3] Pearce I, Coombe BG. Grapevine phenology. In: Dry PR, Coombe BG, editors. Viticulture. 2nd ed. Vol. 1. Adelaide: Winetitles; 2005. pp. 150-166

[4] Chaves MM, Zarrouk O, Francisco R, Costa JM, Santos T, Regalado AP, et al. Grapevine under deficit irrigation: Hints from physiological and molecular data. Annals of Botany. 2010;**105**:661-676. DOI: 10.1093/aob/mcq030

[5] Williams LE, Dokoozlian NK, Wample R. Grape. In: Schafer B, Andersen PC, editors. Handbook of Environmental Physiology of Fruit Crops. Vol. I. Temperate Crops. Boca Raton: CRC Press; 1994. pp. 85-113. DOI: 10.1017/S0376892900034329

[6] Junquera P, Lissarrague JR, Jiménez L, Linares R, Baeza P. Long-term effects of different irrigation strategies on yield components, vine vigour and grape composition in cv. Cabernet-sauvignon (*Vitis vinifera* L.). Irrigation Science. 2012;**30**:351-361. DOI: 10.1007/s00271-012-0348-y

[7] Myburgh PA. Response of *Vitis vinifera* L. cv. merlot to low frequency irrigation and partial rootzone drying in the Western cape coastal region—Part II. Vegetative growth, yield, and quality. South African Journal of Enology and Viticulture. 2011;**32**:104-116. DOI: 10.21548/32-1-1370

[8] Medrano H, Tomás M, Martorell S, Escalona JM, Pou A, Fuentes S, et al. Improving water use efficiency of vineyards in semi-arid regions. A review. Agronomy for Sustainable Development. 2015;**35**:499-517. DOI: 10.1007/s13593-014-0280-z

[9] Shellie KC. Vine and berry response of merlot (*Vitis vinifera* L.) to differential water stress. American Journal of Enology and Viticulture. 2006;**57**:514-518

[10] Williams LE, Grimes DW, Phene CJ. The effects of applied water at various fractions of measured evapotranspiration on reproductive growth and water productivity of thompson seedless grapevines. Irrigation Science. 2010;**28**:233-243. DOI: 10.1007/s00271-009-0173-0

[11] Keller M, Romero P, Gohil H, Smithyman RP, Riley WR, Casassa LF, et al. Deficit irrigation alters grapevine growth, physiology, and fruit microclimate. American Journal of Enology and Viticulture. 2016;**67**: 426-435. DOI: 10.5344/ajev.2016.16032

[12] Williams LE, Matthews MA. Grapevine. In: Stewart BA, Nielson NR, editors. Irrigation of Agricultural Crops, Agronomy Monograph No 30. Madison: ASA-CSSA-SSSA; 1990. pp. 1019-1055

[13] May P. Flowering and Fruit Set in Grapevines. Adelaide: Lythrum Press; 2004. p. 119

[14] Buttrose MS. Fruitfulness in grape-vines: Effect of water stress. Vitis. 1974;**12**:299-305

[15] Jackson RS. Wine Science: Principles and Applications. 3rd ed. San Diego: Academic Press; 2008. p. 751. DOI: 10.1016/B978-0-12-373646-8.X5001-X

[16] Keller M. Deficit irrigation and vine mineral nutrition. American Journal of Enology and Viticulture. 2005;**56**:267-283

[17] Illand P, Dry P, Proffitt T, Tyerman S. The Grapevine from de Science to the Practice of Growing Vines for Wine. Adelaide: Patrick Illand Wine Promotions; 2011. p. 310

[18] Pérez J, Kliewer WM. Effect of shading on bud necrosis and bud fruitfulness of thompson seedless grapevines. American Journal of Enology and Viticulture. 1990;**41**:168-175

[19] McCarthy M. Water stress at flowering and effects on yield. In: Garis K, Dundon C, Johnstone R, Partridge S, editors. Transforming Flowers to Fruit. Adelaide: ASVO; 2005. pp. 35-37

[20] Coombe BG. Research on development and ripening of the grape berry. American Journal of Enology and Viticulture. 1992;**43**:101-110

[21] Ojeda H, Deloire A, Carbonneau A. Influence of water deficits on grape berry growth. Vitis. 2001;**40**:141-145

[22] Matthews MA, Anderson MM. Reproductive development in grape (*Vitis vinifera* L.): Responses to seasonal water deficits. American Journal of Enology and Viticulture. 1989;**40**:52-60

[23] McCarthy MG. The effect of transient water deficit on berry development of cv. Shiraz (*Vitis vinifera* L.). Australian Journal of Grape and Wine Research. 1997;**3**:102-108. DOI: 10.1111/j.1755-0238.1997.tb00128.x

[24] Santesteban LG, Miranda C, Royo JB. Regulated deficit irrigation effects on growth, yield, grape quality and individual anthocyanin composition in *Vitis vinifera* L. cv. 'Tempranillo'. Agricultural Water Management. 2011;**98**:1171-1179. DOI: 10.1016/j.agwat.2011.02.011

[25] Intrigliolo DS, Pérez D, Risco D, Yeves A, Castel JR. Yield components and grape composition responses to seasonal water deficits in Tempranillo grapevines. Irrigation Science. 2012;**30**:339-349. DOI: 10.1007/s00271-012-0354-0

[26] Keller M, Smith JP, Bondada BR. Ripening grape berries remain hydraulically connected to the shoot. Journal of Experimental Botany. 2006;**57**:2577-2587. DOI: 10.1093/jxb/erl020

[27] Rogiers SY, Greer DH, Hatfield JM, Orchard BA, Keller M. Solute transport into shiraz berries during development and late-ripening shrinkage. American Journal of Enology and Viticulture. 2006;**57**:73-80

[28] Tyerman SD, Chaves MM, Barrieu F. Water relations of the grape berry and aquaporins. In: Gerós H, Chaves MM, Delrot S, editors. The Biochemistry of the Grape Berry. Sharjah: Bentham Science Publishers; 2012. pp. 3-23. DOI: 10.2174/97816080536051120101

[29] Freeman BM, Smart RE. A root observation laboratory for studies with grapevines. American Journal of Enology and Viticulture. 1976;**27**(1):36-39

[30] Van Zyl JL. Response of grapevine roots to soil water regimes and irrigation systems. In: Van Zyl JL, editor. The Grapevine Root and Its Environment. Pretoria: Department of Agriculture and Water Supply. 1988. pp. 30-43

[31] Ben-Asher J, Tsuyuki I, Bravdo BA, Sagih M. Irrigation of grapevines with saline water: I. Leaf area index, stomatal conductance, transpiration and photosynthesis. Agricultural Water Management. 2006;**83**(1-2):13-21

[32] Munitz S, Schwartz A, Netzer Y. Evaluation of seasonal water use and crop coefficients for 'cabernet sauvignon' grapevines as the base for skilled regulated deficit irrigation. Acta Horticulturae. 2016;**1115**:33-40

[33] Intrigliolo DS, Castel JR. Response of grapevine cv. 'Tempranillo' to timing and amount of irrigation: Water relations, vine growth, yield and berry and wine composition. Irrigation Science. 2010;**28**(2):113-125

[34] Keller M, Mills LJ, Wample RL, Spayd SE. Cluster thinning effects on three deficit-irrigated *Vitis vinifera* cultivars. American Journal of Enology and Viticulture. 2005;**56**(2):91-103

[35] Dry PR, Loveys BR. Factors influencing grapevine vigour and the potential for control with partial rootzone drying. Australian Journal of Grape and Wine Research. 1998;**4**(3):140-148

[36] Matthews MA, Anderson MM, Schultz HR. Phenologic and growth-responses to early and late season water deficits in cabernet France. VITIS-Journal of Grapevine Research. 1987;**26**:147-160

[37] Williams LE, Grimes DW. Modelling vine growth-development of a data set for a water balance subroutine. In: Lee T, editor. Proceedings of the 6th Australian Wine Industry Technical Conference. 1987. pp. 169-174

[38] Hardie WJ, Martin SR. Shoot growth on de-fruited grapevines: A physiological indicator for irrigation scheduling. Australian Journal of Grape and Wine Research. 2000;**6**(1):52-58

[39] Olivo N, Girona J, Marsal J. Seasonal sensitivity of stem water potential to vapour pressure deficit in grapevine. Irrigation Science. 2009;**27**(2):175-182

[40] Lovisolo C, Schubert A. Effects of water stress on vessel size and xylem hydraulic conductivity in *Vitis vinifera* L. Journal of Experimental Botany. 1998;**49**(321):693-700

[41] Freeman BM, Kliewer MW. Effect of irrigation, crop level and potassium fertilization on carignane vines. II. Grape and wine quality. American Journal of Enology and Viticulture. 1983;**34**(3):197-207

[42] Fanizza G, Ricciardi L, Bagnulo C. Leaf greenness measurements to evaluate water stressed genotypes in *Vitis vinifera*. Euphytica. 1991;**55**(1):27-31

[43] Flexas J, Bota J, Escalona J. Effects of drought on photosynthesis in grapevines under field conditions: An evaluation of stomatal and mesophyll limitations. Functional Plant Biology. 2002;**29**(4):461-471

[44] Lovisolo C, Perrone C, Carra I, Ferrandino A, Flexas J, Medrano H, et al. Drought-induced changes in development and function of grapevine (*Vitis* spp.) organs and in their hydraulic and non-hydraulic interactions at the whole-plant level: A physiological and molecular update. Functional Plant Biology. 2010;**37**(2):98-116

[45] Bota J, Flexas J, Medrano H. Genetic variability of photosynthesis and water use in Balearic grapevine cultivars. The Annals of Applied Biology. 2001;**138**(3):353-361

[46] Poni S, Gatti M, Palliotti A, Dai Z, Duchêne E, Truang TT, et al. Grapevine quality: A multiple choice issue. Scientia Horticulturae. 2018;**234**:445-462

[47] Cole J, Pagay V. Usefulness of early morning stem water potential as a sensitive indicator of water status of deficit-irrigated grapevines (*Vitis vinifera* L.). Scientia Horticulturae. 2015;**191**:10-14

[48] Cuevas E, Baeza P, Lissarrague JR. Variation in stomatal behaviour and gas exchange between mid-morning and mid-afternoon of north–south oriented grapevines (*Vitis vinifera* L. cv. Tempranillo) at different levels

of soil water availability. Scientia Horticulturae. 2006;**108**(2):173-180

[49] Gómez-del-Campo M, Ruiz C, Lissarrague JR. Effect of water stress on leaf area development, photosynthesis, and productivity in chardonnay and Airén grapevines. American Journal of Enology and Viticulture. 2002;**53**(2):138-143

[50] Flexas J, Escalona JM, Medrano H. Down-regulation of photosynthesis by drought under field conditions in grapevine leaves. Australian Journal of Plant Physiology. 1998;**25**(8):893-900

[51] Schultz HR. Water relations and photosynthetic responses of two grapevine cultivars of different geographical origin during water stress. Acta Horticulturae. 1996;**427**:251-266

[52] Downton JS, Loveys R, Grant JR. Non-uniform stomatal closure induced by water stress causes putative non-stomatal inhibition of photosynthesis. The New Phytologist. 1988;**110**(4):503-509

[53] Poni S, Lakso N, Turner JR, Melious RE. Interactions of crop level and late season water stress on growth and physiology of field-grown Concord grapevines. American Journal of Enology and Viticulture. 1994;**45**(2):252-258

[54] Naor A, Bravdo B, Hepner Y. Effect of post-veraison irrigation level on sauvignon blanc yield, juice quality and water relations. South African Journal of Enology and Viticulture. 1993;**14**(2):19-25

[55] O'Neill SD. Role of osmotic potential gradients during water stress and leaf senescence in Fragaria virginiana. Plant Physiology. 1983;**72**(4):931-937

[56] Singh Z. Water Relations and Irrigation Scheduling in Grapevine. Hoboken, New Yersey, United States: John Wiley and Sons, Inc; 2001

[57] Greenspan M. Integrated irrigation of California wine grapes. Practical Winery and Vineyard. 2005;**27**(3):21-79

[58] Munitz S, Netzer Y, Schwartz A. Sustained and regulated deficit irrigation of field-grown merlot grapevines. Australian Journal of Grape and Wine Research. 2017;**23**(1):87-94

[59] Acevedo-Opazo C, Ortega-Farias S, Fuentes S. Effects of grapevine (*Vitis vinifera* L.) water status on water consumption, vegetative growth and grape quality: An irrigation scheduling application to achieve regulated deficit irrigation. Agricultural Water Management. 2010;**97**(7):956-964

[60] Munitz S, Netzer Y, Shtein I, Schwartz A. Water availability dynamics have long-term effects on mature stem structure in *Vitis vinifera*. American Journal of Botany. 2018;**105**(9):1443-1452

[61] Pellegrino A, Lebon E, Simmoneau T, Wery J. Towards a simple indicator of water stress in grapevine (*Vitis vinifera* L.) based on the differential sensitivities of vegetative growth components. Australian Journal of Grape and Wine Research. 2005;**11**(3):306-315

[62] Cramer GR, Van Sluyter SC, Hopper DW, Pascovici D, Keighley T, Haynes PA. Proteomic analysis indicates massive changes in metabolism prior to the inhibition of growth and photosynthesis of grapevine (*Vitis vinifera* L.) in response to water deficit. BMC Plant Biology. 2013:13-49

[63] Pagay V, Zufferey V, Lakso AN. The influence of water stress on grapevine (*Vitis vinifera* L.) shoots in a cool, humid climate: Growth, gas exchange and hydraulics. Functional Plant Biology. 2016;**43**(9):827

[64] Uriarte D, Intrigliolo DS, Mancha LA, Picón-Toro J, Valdés E, Prieto MH. I nteractive effects of irrigation and crop level on tempranillo vines in a semiarid climate. American Journal of Enology and Viticulture. 2015;**66**(2):101-111

[65] Rogiers SY, Holzapfel BP, Smith JP. Sugar accumulation in roots of two grape varieties with contrasting response to water stress. The Annals of Applied Biology. 2011;**159**(3):399-413

[66] Zufferey V, Murisier F, Vivin P, Belcher S, Lorenzini F, Spring JL, et al. Carbohydrate reserves in grapevine (*Vitis vinifera* L. 'Chasselas'): The influence of the leaf to fruit ratio. Vitis. 2012;**51**(3):103-110

[67] Rossouw GC, Smith JP, Barril C, Deloire A, Holzapfel BP. Implications of the presence of maturing fruit on carbohydrate and nitrogen distribution in grapevines under postveraison water constraints. Journal of the American Society for Horticultural Science. 2017;**142**(2):71-84

[68] Mullins MA, Bouquet A. Biology of the Grapevine. Oxford, England, United Kingdom: Oxford University Press; 1993. p. 239

[69] Romero P, Fernández-Fernández JI, Martínez-Cutillas A. Physiological thresholds for efficient regulated deficit-irrigation management in wine grapes grown under semiarid conditions. American Journal of Enology and Viticulture. 2010;**61**:300-312

[70] Baeza P, Sanchez de Miguel P, Centeno A, Junquera P, Linares R, Lissarrague JR. Water relations between leaf water potential, photosynthesis and agronomic vine response as a tool for establishing thresholds in irrigation scheduling. Scientia Horticulturae. 2007;**114**:151-158

[71] Esteban MA, Villanueva MJ, Lissarrague JR. Relationships between different berry components in tempranillo (*Vitis vinífera* L) grapes from irrigated and non-irrigated vines during ripening. Journal of the Science of Food and Agriculture. 2002;**82**:1136-1146

[72] Girona J, Marsal J, Mata M, Del Campo J, Basile B. Phenological sensitivity of berry growth and composition of tempranillo grapevines (*Vitis vinifera* L) to water stress. Australian Journal of Grape and Wine Research. 2009;**15**:268-277

[73] Uriarte D, Mancha LA, Moreno D, Bejarano D, Valdés E, Talaverano I, et al. Effects of timinf of water deficit induction on Doña Blanca white grapevine under semiarid growing conditions of South-Western Spain. In: Marsal J, Girona J, editors. ISHS Acta Horticulturae 1150: Proceedings VIII International Symposium on Irrigation of Horticultural Crops. 2017. DOI: 10.17660/actahorticulturae.2017.1150.68

[74] Kennedy JA, Matthews MA, Waterhouse AL. Effect of maturity and vine water status on grape skin and wine flavonoids. American Journal of Enology and Viticulture. 2002;**53**:268-274

[75] Van Leeuwen C, Friant P, Choné X, Tregoat O, Koundouras S, Dubourdieu D. Influence of climate, soil and cultivar on terroir. American Journal of Enology and Viticulture. 2004;**55**:207-217

[76] Keller M, Smithyman RP, Mills LJ. Interactive effects of deficit irrigation and crop load on cabernet sauvignon in an arid climate. American Journal of Enology and Viticulture. 2008;**59**:221-234

[77] Tarricone L, Alba V, diGenaro D, Amendolagine AM, Gentilesco G, Masi G. Grape and wine quality of *Vitis vinifera* "Nero di Troia" in response to moderate deficit irrigation. Acta Horticulturae. 2017;**1150**:485-491

[78] Deloire A, Carbonneau A, Wang Z, Ojeda H. Wine and water: A short review. Journal International des Sciences de la Vigne et du Vin. 2004;**38**:1-13

[79] Jackson DI, Lombard PB. Environmental and management practices affecting grape composition and wine quality. A review. American Journal of Enology and Viticulture. 1993;**44**:409-430

[80] Mirás-Ávalos JM, Trigo-Córdoba E, Bouzas-Cid Y, Orriols-Fernández I. Response of grapevine cv. "Brancellao" and "Sousón" to supplementary irrigation: Water relations, vine growth, yield and berry and wine composition. Ciência e Técnica Vitivinícola. 2016;**31**:98-113

[81] Romero P, Gil-Muñoz R, Del Amor FM, Valdés E, Fernández JI, Martinez-Cutillas A. Regulated deficit irrigation based upon optimun water status improves phenolic composition in monastrell grapes and wines. Agricultural Water Management. 2013;**121**:85-101

[82] Conde C, Silva P, Fontes N, Diaz ACP, Tavares RM, Sousa MJ, et al. Biochemical changes throughout grape berry development and fruit and wine quality. Food. 2007;**1**:1-22

[83] Salon JL, Chirivella C, Castel JR. Response of cv Bobal to timing of deficit irrigation in Requena, Spain: Water relations, yield and wine quality. American Journal of Enology and Viticulture. 2005;**56**:1-8

[84] Souza CR, Marocco JP, Santos T, Rodriguez ML, Lopes C, Pereira S, et al. Grape berry metabolism in field-grown grapevines exposed to different irrigation strategies. Vitis. 2005;**44**:103-109

[85] Rogiers SY, Coetzee ZA, Walker RR, Deloire A, Tyreman SD. Potassium in the grape (*Vitis vinifera* L.) berry: Transport and function. Frontiers in Plant Science. 2017;**8**:1-19

[86] Santos TP, Lopes CM, Rodrigues ML, Souza CR, Silva JR, Maroco JP, et al. Effects of deficit irrigation strategies on cluster microclimate for improving fruit composition of Moscatel field-grown grapevines. Scientia Horticulturae. 2007;**112**:321-330

[87] Morrison JC, Noble AC. The effects of leaf and cluster shading on the composition of cabernet sauvignon grapes and on fruit and wine sensory properties. American Journal of Enology and Viticulture. 1990;**41**:193-200

[88] Reynolds AG, Wardle DA. Influence of fruit microclimate on monoterpene levels on Gewerztraminer. American Journal of Enology and Viticulture. 1989;**40**:149-154

[89] Belancic A, Agosin E, Ibacache A, Bordeu E, Baumes R, Razungles A, et al. Influence of sun exposure on the aromatic composition of Chilean Muscat grape cultivars Moscatel de Alejandría and Moscatel rosada. American Journal of Enology and Viticulture. 1997;**48**:181-186

[90] Lee SH, Seo MJ, Riu M, Cotta JP, Block DE, Dokoozlian NK, et al. Vine microclimate and norisoprenoid concentration in cabernet sauvignon grapes and wines. American Journal of Enology and Viticulture. 2007;**58**(3):291-301

[91] Oliveira C, Silva Ferreira AC, Pinto MM, Hogg T, Alves F, Guedes de Pinho P. Carotenoid compounds in grapes and their relationship to plant water status. Journal of Agricultural and Food Chemistry. 2003;**51**:5967-5971

[92] Spayd SE, Tarara JM, Mee DL, Ferguson JC. Separation of sunlight and temperature effects on the composition of *Vitis vinifera* cv. Merlot berries.

American Journal of Enology and Viticulture. 2002;**53**(3):171-182

[93] Downey MO, Dokoozlian NK, Krstic MP. Cultural practice and environ- mental impacts on the flavonoid composition of grapes and wine: A review of recent research. American Journal of Enology and Viticulture. 2006;**57**(3):257-268

[94] Winkler AJ, Cook JA, Kliewer WM, Lider LA. General Viticulture. Berkeley: University of California Press; 1974

[95] Haselgrove L, Botting D, van Heeswijck R, Hoj P, Dry P, Ford C, et al. Canopy microclimate and berry composition, the effect of bunch exposure on the phenolic composition of *Vitis vinifera* L. cv shiraz grape berries. Australian Journal of Grape and Wine Research. 2000;**6**:141-149. DOI: 10.1111/j.1755-0238.2000.tb00173.x

[96] Bergqvist J, Dokoozlian N, Ebisuda N. Sunlight exposure and temperature effects on berry growth and composition of cabernet sauvignon and Grenache in the Central San Joaquin Valley of California. American Journal of Enology and Viticulture. 2001;**52**(1):1-7

[97] Grimplet J, Deluc LG, Tillet RL, Whetley MD, Schlauch KA, Cramer GR, et al. Tissue-specific mRNA expression profiling in grape berry tissues. BMC Genomics. 2007;**8**:187

[98] Qian M, Fang Y, Shellie K. Volatile composition of merlot wine from different vine water status. Journal of Agricultural and Food Chemistry. 2009;**57**(16):7459-7463

[99] Chapman DM, Roby G, Ebeler SE, Guinard JX, Matthews MA. Sensory attributes of cabernet sauvignon wines made from vines with different water status. Australian Journal of Grape and Wine Research. 2005;**11**:339-347

[100] Koundouras S, Marinos V, Gkoulioti A, Kotseridis Y, Van Leeuwen C. Influence of vineyard location and vine water status on fruit maturation of nonirrigated cv. Agiorgitiko (*Vitis vinifera* L.) effects on wine phenolic and aroma components. Journal of Agricultural and Food Chemistry. 2006;**54**:5077-5086

[101] Bindon KA, Dry PR, Loveys BR. Influence of plant water status on the production of C13-norisoprenoid precursors in *Vitis vinifera* L. cv. Cabernet-sauvignon grape berries. Journal of Agricultural and Food Chemistry. 2007;**55**:4493-4500

[102] Song J, Shellie KC, Wang H, Qian MC. Influence of deficit irrigation and kaolin particle film on grape composition and volatile compounds in merlot grape (*Vitis vinifera* L.). Food Chemistry. 2012;**134**:841-850

[103] Talaverano I, Valdés E, Moreno D, Gamero E, Mancha L, Vilanova M. The combined effect of water status and crop level on tempranillo wine volatiles. Journal of the Science of Food and Agriculture. 2017;**97**:1533-1542

[104] Bouzas-Cid Y, Díaz-Losada E, Trigo-Córdoba E, Falqué E, Orriols I, Garde-Cerdán T, et al. Effects of irrigation over three years on the amino acid composition of Albariño (*Vitis vinifera* L.) musts and wines in two different terroirs. Scientia Horticulturae. 2018;**227**:313-325

[105] Ju Y, Liu M, Tu T, Zhao X, Yue X, Zhang J, et al. Effect of regulated deficit irrigation on fatty acids and their derived volatiles in cabernet sauvignon grapes and wines of Ningxia, China. Food Chemistry. 2018;**245**:667-675

[106] Vilanova M, Rodríguez-Nogales JM, Vila-Crespo J, Yuste J. Influence of water regime on yield components, must composition and wine volatile

compounds of *Vitis vinifera* cv. Verdejo. Australian Journal of Grape and Wine Research. 2019;**25**(1):83-91

[107] Trigo-Córdoba E, Bouzas-Cid Y, Orriols-Fernández I, Mirás-Avalos JM. Irrigation effects on the sensory perception of wines from three white grapevine cultivars traditional from Galicia (Albariño, Godello and Treixadura). Ciência e Técnica Vitivinícola. 2014;**129**(2):71-80

[108] Balint G, Reynolds AG. Effect of different irrigation strategies on vine physiology, yield, grape composition and sensory profiles of *Vitis vinifera* L. cabernet sauvignon in a cool climate area. Journal International des Sciences de la Vigne et du Vin. 2014;**48**:269-292

[109] Castellarin SD, Matthews MA, Di Gaspero G, Gambetta GA. Water deficits accelerate ripening and induce changes in gene expression regulating flavonoid biosynthesis in grape berries. Planta. 2007;**227**:101-112

[110] Ojeda H, Andary C, Kraeva E, Carbonneau A, Deloire A. Influence of pre- and postveraison water deficit on synthesis and concentration of skin phenolic compounds during berry growth of *Vitis vinifera* cv. Shiraz. American Journal of Enology and Viticulture. 2002;**53**:261-267

[111] Casassa LF, Keller M, Harbertson JF. Regulated deficit irrigation alters anthocyanins, tannins and sensory properties of cabernet sauvignon grapes and wines. Molecules. 2015;**20**:7820-7844

[112] Delgado P, Salgado E, Ribalta-Pizarro C, Olaeta JA, López E, Pastenes C, et al. Phenolic composition and sensory characteristics of cabernet sauvignon wines: Effect of water stress and harvest date. International Journal of Food Science and Technology. 2018;**53**:1726-1735

[113] Niculea M, Martínez-Lapuente L, Guadalupe Z, Sánchez-Diaz M, Ayestarán B, Antolín MC. Characterization of phenolic composition of *Vitis vinifera* L. 'Tempranillo' and 'Graciano' subjected to deficit irrigation during berry development. Vitis. 2015;**54**:9-16

[114] Herrera JC, Hochberg U, Degu A, Sabbatini P, Lazarovitch N, Castellarin SD, et al. Grape metabolic response to postveraison water deficit is affected by interseason weather variability. Journal of Agricultural and Food Chemistry. 2017;**65**(29):5868-5878

[115] Mirás-Avalos JM, Intrigliolo DS. Grape composition under abiotic constrains: Water stress and salinity. Frontiers in Plant Science. 2017;**8**:851

Section 3

Wine Biotechnology

Chapter 6

Contribution of the Microbiome as a Tool for Estimating Wine's Fermentation Output and Authentication

Dimitrios A. Anagnostopoulos, Eleni Kamilari and Dimitrios Tsaltas

Abstract

Wine is the alcoholic beverage which is the product of alcoholic fermentation, usually, of fresh grape must. Grape microbiome is the source of a vastly diverse pool of filamentous fungi, yeast, and bacteria, the combination of which plays a crucial role for the quality of the final product of any grape must fermentation. In recent times, the significance of this pool of microorganisms has been acknowledged by several studies analyzing the microbial ecology of grape berries of different geographical origins, cultural practices, grape varieties, and climatic conditions. Furthermore, the microbial evolution of must during fermentation process has been overstudied. The combination of the microbial evolution along with metabolic and sensorial characterizations of the produced wines could lead to the suggestion of the microbial terroir. These aspects are today leading to open a new horizon for products such as wines, especially in the case of PDO-PGI products. The aims of this review is to describe (a) how the microbiome communities are dynamically differentiated during the process of fermentation from grape to ready-to-drink wine, in order to finalize each wine's unique sensorial characteristics, and (b) whether the microbiome could be used as a fingerprinting tool for geographical indication, based on high-throughput sequencing (HTS) technologies. Nowadays, it has been strongly indicated that microbiome analysis of grapes and fermenting musts using next-generation sequencing (NGS) could open a new horizon for wine, in the case of protected designation of origin (PDO) and protected geographical indication (PGI) determination.

Keywords: grape, wine, microbiome, terroir, fermentation, next-generation sequencing

1. Introduction

Fermented products are generated as a result of metabolic activities conducted by functional microbes, leading to the biochemical and organoleptic modification of the substrates in order to meet the requirements of the consumers [1]. The dynamic interaction between the members of the microbial communities guiding the process of fermentation has great influence in the nutritional, hygienic, safety, and organoleptic characteristics of the final product [2]. In a large number of fermented products, the formation of microbial biodiversity existing in the initial substrate is affected by a large number of factors, including the geographic origin, the cultural practices, differences among varieties, or the climatic conditions [3]. The contribution of the microbial community configuration, which is governed by spatial factors, land topography, environmental factors, etc. that sustain the spatial structure of the inhabitants, and their potential relation with the metabolic and sensorial characterizations of the final product, has been under deep research, leading to the suggestion of the microbial terroir [4]. The perspective of analyzing the microbial communities' dynamics as progressively differentiated during the process of fermentation for the determination of microbial terroir has been applied in grapes and consequently its final fermented product, the wine [5, 6].

Traditionally winemaking process relies on spontaneous fermentation, which is conducted without the addition of chemical compounds or supplementary microbes at the beginning of the fermentation process. Under spontaneous fermentation conditions, the microbial community participating in fermentation and which is responsible for the quality of the final product is considered to be quite unpredictable. At the initial stages of fermentation, the microbial communities are comprised by a rich biodiversity of several yeast and mold species, including *Metschnikowia*, *Candida*, *Hanseniaspora*, *Pichia*, *Lachancea*, *Kluyveromyces*, and *Saccharomyces* [7–9]. During must fermentation, the alcoholic fermentation conducted elevates the ethanol content and establishes the basic fermenters, such as *Saccharomyces cerevisiae,* among the predominant species [9]. Their dominant presence during the fermentation process has led to the isolation of several *S. cerevisiae* strains, which have been extensively studied for the potential application of their technological characteristics [9–11].

The development of high-throughput sequencing technologies has allowed the evaluation of the microbial consortium comprising grapes' microbiome in terms of revealing the concept of the microbial terroir [12–16]. The contribution of origin-associated factors of grape varieties, including climate and microclimate, region site, as well as grape cultivar, in the microbial community formation and the final metabolic profiles, has been recently investigated [12, 17–20]. These studies have led to an improved spatial and temporal determination of the wine grapes' microbiome and brought new insights into its dynamics and biodiversity, revealing a new horizon for the better characterization of this product, especially in the case of PDO and PGI wines' designation. These labels were established by the European Union (EU) to guarantee the authentication of the local products produced in distinct geographic origin, applying traditional specialties. Metagenomic studies have been recently applied to identify the microbial communities that influence the original sensorial characteristics of PDO wines [14, 16].

The aim of this chapter is to extensively review all latest literature in the scope to investigate (a) how the microbiome communities are dynamically differentiated from grape to ready-to-drink wine, in order to finalize each wine's unique sensorial characteristics ,and (b) whether the microbiome could be used as a fingerprint tool for regional characterization, based on high-throughput sequencing (HTS) technologies.

2. Methods to identify grape microbial species

Grapes are comprised by a complex microbiome, the members of which share different physiological characteristics and effects upon wine production. Some of them are present only in grapes and soil, such as parasitic fungi and environmental bacteria, while others have the ability to survive and grow during wine fermentation, constituting the wine microbial consortium. Several studies over the last years have reported that the biodiversity and the quantity of the microorganisms present on the surface of the grape berry are highly dependent on many factors, including the health state of the grapes, the temperature, the microclimate conditions, and the pesticide treatments [21–23]. Recently, the "terroir" idea was proposed to be extended to the microbiological aspect, indicating that the geographical distribution of the grape and soil microbiota is not randomly dispersed but is dependent on the cultivar, the location of the vineyard, and the vintage [17].

The application of culture-dependent methods is considered weak to support the terroir perspective, since less than 1% of the total population can be detected [24], and these methods also fail to detect viable but non-culturable organisms [25–27]. Additionally, the stressful environment shaped during winemaking due to the addition of SO_2, high ethanol concentration, etc. forces a number of bacteria and yeast to enter a viable but non-culturable state (VBNC) [28, 29]. Even though still viable and maintaining a detectable metabolic activity, the microbial cells are unable to grow on culture media during VBNC status [30]. Examples of such microorganisms include *Candida stellata*, *Brettanomyces bruxellensis*, *S. cerevisiae*, and *Zygosaccharomyces bailii* [27]. In order to study the existence of bacteria during VBNC, microbiologists have applied alternative culture-independent techniques. Three of the main culture-independent techniques applied include quantitative real-time PCR (qPCR), restriction fragment length polymorphism (RFLP), and denaturing gradient gel electrophoresis (DGGE) [24, 27, 31–33]. Still, the detection sensitivities of these techniques remain limited due to the predominance of certain yeast such as *C. zemplinina* and *S. cerevisiae* during fermentation, which restrict the detection of low-abundant species.

The introduction of next-generation sequencing (NGS) technologies has significantly enhanced the information elicited from microbiological studies, allowing the distinction of the high-abundant species from the low-abundant, with detection sensitivities greatly higher than the previously used molecular techniques [24]. For instance, analysis of the microbial communities' formation existing on grape and during Carignan and Grenache must fermentation from three vineyards in Priorat (Spain) highlighted the ability of NGS to detect an increased amount of species compared to DGGE [34]. Undoubtedly, NGS provides a new powerful tool, with elevated capabilities to enhance the understanding of the complexities of microbial communities as dynamically differentiated from grapes and its close environment to ready-to-drink fermented wine, in terms of diagnostic, monitoring, and traceability [16, 21, 35–38]. Understanding the progressive alterations of the microbial diversity during fermentation using HTS technologies is considered a promising approach to reveal correlations between microbiomes and geographical origin.

3. Identification of the microbial communities

Terroir is characterized by a multi-complex ecosystem where the vine (genetic material and cultural practices) interacts with the environmental factors (i.e., soil, climate, microclimate, humans, etc.) affecting the quality and typicity of the wine produced in a particular location. The understanding of the microbial terroir involves the identification of the microbes shaping grapes' environmental communities and the evaluation of their diversity dynamical evolution throughout the different stages of fermentation, until wine production. During natural fermentation the complex microbial communities that comprise the grape microbiome, including, yeasts, yeast-like fungi, and bacteria, are under the selective pressure of the alterations in the must microenvironment, caused by microbial interactions, as well as chemical and physical factors [39]. The must microbes have to handle stressful factors that affect their survival, including reduced oxygen, high ethanol and sulfur dioxide (SO_2) levels, and low pH [40]. Moreover, the amounts of sugar existing in must favor for particular species, and high sugar content sweet wines select for osmotolerant species [41, 42]. As a consequence of this stressful microenvironment, numerous environmental species become unable to survive, while others, which are able to perform alcoholic fermentation and were detected in reduced relative abundance before fermentation, such as *Saccharomyces cerevisiae*, become dominant by the end of fermentation [16]. Apart from alcoholic fermentation, malolactic fermentation (MLF) (conversion of malic acid into lactic acid) is also involved in the metabolic transformation of grape juice into wine, conducted mostly by lactic acid bacteria (LAB), including the genera *Oenococcus*, *Lactobacillus*, *Pediococcus*, and *Leuconostoc*, leading to must deacidification, a process that affects organoleptic characteristics' formation [43]. By the end of fermentation, the microbial diversity is limited to selected microbial species [12, 35]. As revealed by several studies, some species were found to decline rapidly at the initial or the middle stages of fermentation, such as *Cryptococcus carnescens*, *Paraburkholderia terricola*, *Aureobasidium pullulans*, and *Metschnikowia pulcherrima*, while others exist until the end of fermentation, including *Saccharomyces cerevisiae, Torulaspora delbrueckii*, *Lachancea thermotolerans*, and *Streptomyces bacillaris* [44–47].

Overall, the fungal population at a phylum level is very similar and mainly comprised by *Ascomycota*, the most abundant phylum, followed by *Basidiomycota* [3, 18, 19, 24, 35, 48]. Additional phyla frequently detected but in limited concentrations include *Zygomycota* and *Chytridiomycota*. The most commonly found filamentous fungi genera include *Aspergillus*, *Erysiphe*, *Alternaria*, *Cladosporium*, *Penicillium*, *Davidiella*, *Lewia*, *Botrytis*, as well as the yeast-like fungus *Aureobasidium pullulans*. Further yeast genera commonly found include *Issatchenkia*, *Candida*, *Hanseniaspora*, *Pichia*, *Rhodotorula*, *Metschnikowia*, *Lachancea*, *Filobasidiella*, *Cryptococcus*, *Torulaspora*, and *Sporobolomyces* [3, 18, 19, 24, 34, 35, 48, 49].

High-throughput sequencing studies have been applied to evaluate the bacterial communities associated with the vineyard. The most frequently detected phyla in vineyard soils and grapevine roots include *Proteobacteria*, *Bacteroidetes*, *Acidobacteria*, *Verrucomicrobia*, *Planctomycetes*, *Actinobacteria*, *Chloroflexi*, *Gemmatimonatedes*, and *Firmicutes* [21, 50–52]. High-throughput analysis of the grapevine phyllosphere, flowers, and grape berry surface indicated that the bacterial communities were predominated by *Proteobacteria* followed by *Firmicutes*, *Actinobacteria*, *Acidobacteria*, and *Bacteroidetes* [13, 38, 53, 54]. The relative abundances of the groups may vary depending on the plant tissue or organ. The dominant taxa include members of the genera *Pseudomonas*, *Sphingomonas*, *Frigoribacterium*, *Curtobacterium*, *Bacillus*, *Enterobacter*, *Acinetobacter*, *Erwinia*, *Citrobacter*, *Pantoea*, and *Methylobacterium* [3, 13, 21, 48, 53, 54]. In contrast, the endophytic community in grape berries is mainly comprised by *Ralstonia*, *Burkholderia*, *Pseudomonas*, *Staphylococcus*, *Mesorhizobium*, *Propionibacterium*, *Dyella*, and *Bacillus species* [35].

4. Factors affecting the microbial communities' formation

Grapevines' associated microbial communities originated from distinct geographic regions exhibit different profiles [13, 18, 34, 36, 55]. Each region is differentiated by the dominance of a few species per region. Indicatively, *Aspergillus* and *Penicillium* spp. were largely associated with the Chardonnay in Napa, while *Actinobacteria*,

Bacteroides, *Saccharomycetes*, and *Erysiphe necator* dominated in Central Coast, as well as *Proteobacteria* and *Botryotinia fuckeliana* in Sonoma [3]. Additionally, the prevalence of *Lachancea* in the Alentejo appellation was reported by Pinto et al. [13] while of *Rhodotorula* and *Botryotinia* was shown in the Estremadura appellation. Finally *Ramularia* and *Hanseniaspora* were the dominant genera in Bairrada, *Rhodotorula* and *Lachancea* in Dão, *Rhodotorula* and *Erysiphe* in Douro, and *Rhodotorula* and *Alternaria* in Minho appellation. Furthermore, the fungal grapes' associated diversity is also affected by agronomic practices. Vineyards that employed conventional, integrated pest management systems, organic, biodynamic, and ecophyto practices were shown to harbor different fungal communities [19, 23, 24, 44, 46, 48, 56–59]. However, the fact that these studies were carried out in vineyards from different countries (Austria, France, Italy, Spain, and Slovenia), subjected to different climates, pesticides, and regulatory constraints, may explain the contradictory results.

Many studies suggested that yeast diversity is dependent on climatic and microclimatic conditions. Higher yeast diversity has been described for vintages with high rainfall [40, 57] probably due to substantial fungal proliferation. Dry wines are produced by grapes submitted to prolonged withering in order to become moderately dried. The climate, as well as the extent of the withering period, was found to affect the formation of the fungal microbiome on grape skins in *V. vinifera* L. cv. Corvina, influencing the relative abundances of the fungal genera and consequently the secreted metabolites shaped in the must of Amarone red dry wine [57]. Grapes collected during a rainy season had increased bacterial biodiversity and enriched volatile compound (VOC) profile compared to a "dry" season collection, although some common microbial populations and VOC profiles maintained over the different vintages in grapes and musts samples, probably indicative of the typicity of Amarone.

Vineyard factors such as grape variety and berry chemical components are often described to influence microbial diversity [11, 43, 61, 62]. For instance, in similar soil and climatic conditions, *Cryptococcus* was the genera most frequently isolated (90% of all isolates) from Grenache grapes, whereas *Hanseniaspora* was the genus most frequently isolated from Carignan (75%) [58].

The health status of berries can also affect the diversity of yeasts. The ascomycete *Botrytis cinerea* is considered one of the most damaging fungi in low temperature viticulture [60]. It causes *Botrytis* bunch rot, alternatively gray mold in grapes, affecting the physiochemical condition of grapes dramatically. Botrytized wine fermentations were found to contain increased abundance of acetic acid bacteria (AAB) in comparison with unaffected wines [61]. The elevated presence of AAB was additionally shown in botrytized wine fermentations obtained from the Dolce Winery, Oakville, California, analyzed via HTS [36]. Interestingly, the lactic acid bacteria (LAB) community was comprised mostly by *Leuconostoc* and *Lactococcus*, whereas *Oenococcus* was completely absent. Berries affected by *Botrytis cinerea* indicated increased development of the genus *Metschnikowia* [62]. Additionally, the bacterial community structure may vary depending on the grape cultivars or the agronomic practices [13, 35, 48, 52, 53].

One of the factors found to contribute to microbial communities' formation is the amount of SO_2. Comparison of the bacterial community dynamics following the fermentation process of hand-harvested organically grown Riesling grapes following organic and conventional *pied-de-cuve* (PDC) indicated that the species *Gluconobacter oxydans* was significantly affected by the addition of SO_2 prior to PDC and bulk fermentation [37]. The ability of SO_2 to prevent the growth of *Gluconobacter* at concentrations $\geq$25 mg/L was also shown by Bokulich and colleagues [63]. The elevated presence of this spoilage bacterium in organic fermentation highlights the susceptibility of the organic fermentation procedures to wine spoilage.

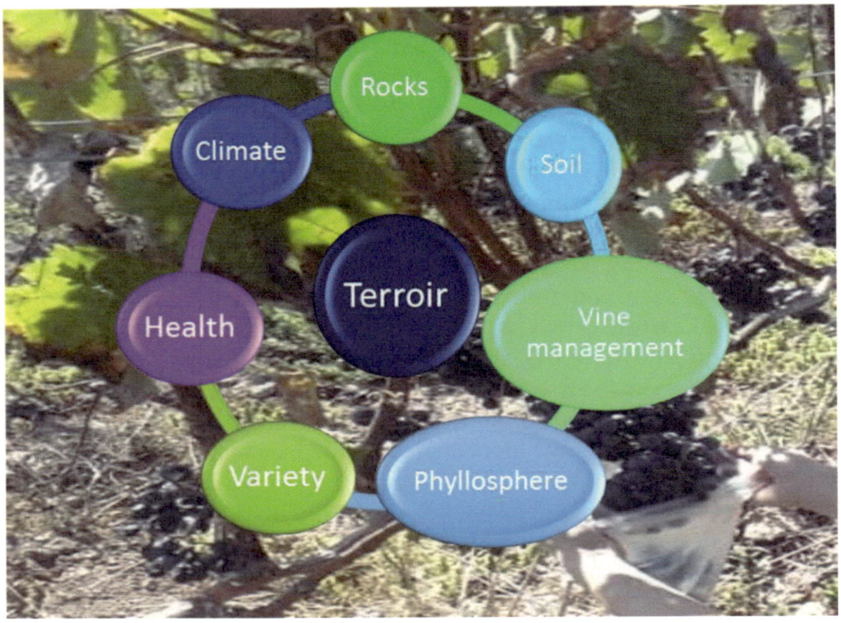

Figure 1.
Variables related to the microbiome formation.

Generally, many of these variables (e.g., climatic conditions or cultivar) are interdependent and may be clustered into broad groups of effects **(Figure 1)**. The study of Bokulich and Mills [17] has shown that grape-associated microbial region is totally related with varietal, biogeographical, and climatic factors across multiscale viticultural zones. According to other study [20], the distribution of yeast species promotes significantly intra-vineyard spatial fluctuations. Continuously, the heterogeneity of grape samples harvested from single vineyards at the same stage of ripeness might be related, at least in part, to differing microbial communities in different sections of the vineyard. The biodiversity of yeast species in grapes is affected by numerous biotic and abiotic factors, as well as the interactions among the resident populations. However, more studies need to be performed in order to confidently elucidate the vineyard and grapevine phyllosphere microbiome.

5. Microbial evolution of must during spontaneous fermentation process

High-throughput sequencing techniques have allowed the discrimination of the microbial diversity as dynamically formed from the initiation of fermentation until wine production, identifying also the non-culturable microorganisms, as well as the limited represented species [12–16] (**Figure 2**). During the process of fermentation, the microbial community is reshaped and become dominated by the fermentative organisms. These alterations, however, are to a large extent dependent from the origin of the must/wine, including the winery and the grape variety [12]. Metagenomic analysis of the microbial communities' structure fluctuations formed throughout the fermentation of grapes obtained from American Viticultural Areas (AVA), for Cabernet and Chardonnay wines production, combined with metabolomic analysis, indicated that the characteristic microbial signatures of grapes and soil disappeared during fermentation to become replaced by characteristic fermentative microbes, but still, the microbial and wine metabolite profiles were able to distinguish the

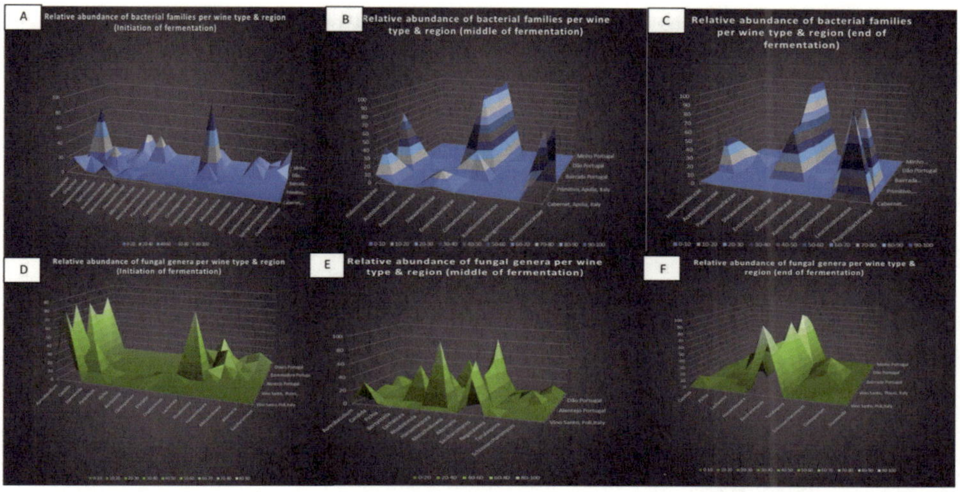

Figure 2.
Spatial distribution of the microbial communities shaping from the initiation of fermentation until wine production regarding the studies of Stefanini et al. [16], Marzano et al. [14], Wei et al. [15], and Pinto et al. [13]. (A) Representation of the relative abundance of bacterial families at the beginning of fermentation. (B) Representation of the relative abundance of bacterial families at the middle of fermentation. (C) Representation of the relative abundance of bacterial families at the end of fermentation. (D) Representation of the relative abundance of fungal genera at the beginning of fermentation. (E) Representation of the relative abundance of fungal genera at the middle of fermentation. (F) Representation of the relative abundance of fungal genera at the end of fermentation.

individual vineyards and the viticultural area, as revealed by random forest machine learning models [12]. Markedly, a negative association among the fermentation rate as well as bacterial richness with various taxa, such as *Lactobacillus* spp., *H. uvarum*, and *Gluconobacter*, was observed, indicative of the ability of some bacteria to prevent alcoholic fermentation, probably due to antagonism for available nutritional sources with the alcoholic fermentation fermenters, such as *S. cerevisiae*, while others, such as *Pseudomonas*, were positively correlated in both wines. The malolactic fermentation (MLF) conducted in Cabernet limits the bacterial biodiversity of wines to the presence of members of the family *Leuconostocaceae* (*Oenococcus oeni*), whereas the fungal biodiversity, as well as the microbial diversity of Chardonnay wines, remained enriched throughout fermentation and wine production, possibly responsible for the more distinct both regional and vineyard discriminations of Chardonnay wines compare to Cabernet Sauvignon wines.

In order to understand the association among the biogeographic distribution of wineries and wine microbiome of six different Portuguese wine appellations, HTS analysis was applied to reveal the dynamics of microbial communities' formation following the different stages of spontaneous wine fermentations [13]. The presence of an increased average microbial biodiversity dissimilarity among the grape microbiome from the different wine appellations (60.16 and 57.36% for eukaryotes and prokaryotes, respectively) indicated the elevated contribution of the vineyard environment in microbial communities' shaping and consequently the influence of the initial microbiome to the uniqueness of the different appellation-derived wines. During the process of fermentation, the average microbial dissimilarity was reduced, due to alterations in the microbial biodiversity and dominance of specific, able to perform fermentation species, leading to the loss of the biogeographic profile, but still each wine was distinguished by its unique pattern of microbial biodiversity.

The high detection sensitivities of HTS technologies have allowed the identification of the rich bacterial biodiversity implicated in Cabernet, Negroamaro, and

Primitivo Apulian red wines' production process, highlighting the alterations in the bacterial population during vinification [14]. Although a common microbiome core was identified among the three wine varieties, comprised by the genera *Candidatus liberibacter*, *Gilliamella*, *Gluconobacter*, *Halomonas*, *Halospirulina*, *Komagataeibacter*, *Pseudomonas*, and *Shewanella*, each wine was discriminated by a unique taxonomic signature. During malolactic fermentation *Shewanella*, *Halomonas*, and *Oenococcus* became the dominant genera, whereas at the end of fermentation, *Oenococcus*, with the species *Oenococcus oeni*, became the abundant bacterium of the three wines' microbiome. Similarly, HTS analysis of Cabernet Sauvignon samples from three different winery regions in Xinjiang province, China, from Fukang area, identified a common core microbiome composed mostly by the fungal genera *Aureobasidium*, *Pleosporaceae*, *Cryptococcus*, and *Dothideales* and the bacterial genera *Pseudomonas*, *Acinetobacter*, *Kaistobacter*, *Arthrobacter*, and *Sphingomonas* in all grape and grape juice samples analyzed, even though the relative abundances of those genera were different [15]. However, following malolactic fermentation, the microbial biodiversity was gradually reduced and limited mostly to the fungal genera *Aspergillus*, *Penicillium*, and *Alternaria*, while the slow-growing, necessary for malolactic fermentation, lactic acid bacterium *Oenococcus* appeared to be the dominant genus in all wine samples.

Metagenomic analysis, applied to reveal the spatial distribution of the microbial communities shaped in Vino Santo Trentino sweet wine, produced by Nosiola grapes from three wineries (Poli, Pedrotti, and Pisoni in the Italian Alps), indicated that a winery-specific "microbial-terroir" contributed mostly to the wines' microbial community shaping, rather than a regional "terroir" [16]. As a result of the spontaneous fermentation, the complex microbial diversity which composed the grapes' microbiome, including *Aureobasidium pullulans*, *Starmerella meliponinorum* MS 2010, *Penicillium polonicum*, *Pichia membranifaciens*, *Candida zemplinina*, *Penicillium bialowiezense*, and *Candida ethanolic*, was limited to some specific wine yeast species, which existed in limited relative abundance before fermentation, such as *Saccharomyces cerevisiae*, *Pichia membranifaciens*, and *Hanseniaspora osmophila*. Even though the must from the different wineries had significantly different mycobiome, the dominant presence of *Saccharomyces* at the end of fermentation was observed in all must tested, except from the Poli must, in which *Hanseniaspora osmophila* was also dominant.

6. Combination of microbial evolution studies with metabolism analysis could provide indications of the microbial terroir

The different varieties of grapevine (*Vitis vinifera* L.) are differentiated by a unique pool of compounds or chemical precursors that influence the aromatic composition of the produced wines. For instance, linalool is a typical characteristic aroma of Muscat varieties, while methoxypyrazine derivatives characterize the varieties Sauvignon blanc and Cabernet Sauvignon [64]. Apart from the grapevine variety, the degree of ripening, as well as the agronomic and oenological techniques applied, influence also wine's aromatic profile [65–71]. The metabolic reactions performed in wines, due to the specific enzymatic activity of selective wine yeasts that assist to the catabolism of sugar molecules and other ingredients, in order for the aroma compounds to be released have been reviewed extensively [72–74]. Indicatively, the basic yeast enzymes implicated in flavor compounds' secretion from the catabolism of grape components include: (a) glycosidases, such as α-L-arabinofuranosidase, α-L-rhamnosidase or β-D-apiosidase and β-D-glucosidase, which lead to the release of aromatic compounds found in the bound aroma sections of diglycosides, glucosides and chemical compounds including terpene diols, terpenols, C_{13}-norisoprenoids [72, 75, 76]. These enzymes are produced mainly by the genera *Saccharomyces, Debaryomyces, Candida, Hanseniaspora/Kloeckera, Metschnikowia, Zygosaccharomyces Kluyveromyces, Pichia, Schizosaccharomyces and Saccharomycodes, Brettanomyces, Torulaspora* and *Trichosporon* [70, 77–91]. *(b) Carbon-sulfur lyases, that catalyze the release of volatile or varietal thiols from glutathionated thiol precursors produced by yeasts, including S. cerevisiae, Pichia kluyveri, Candida zemplinina, Metschnikowia pulcherrima, Hanseniaspora uvarum, Kluyveromyces thermotolerans and Torulaspora delbrueckii* [92–94].

A great influence on the pool of the VOCs released in wine is due to the metabolic activities performed mostly by predominant yeasts, leading to secondary metabolites' production during fermentation [92]. These secondary aroma compounds include ethanol, CO_2, and glycerol, as well as volatile fatty acids, such as acetic acid and propanoic and butanoic acid esters, higher alcohols and aldehydes, and volatile derivatives of fatty acids and nitrogen- and sulfur-comprising compounds, which have greater contribution to the secondary aroma profile [96–99]. The spontaneous fermentation is conducted by autochthonous yeasts, which exist naturally on the surface of grapes. Increased biodiversity of yeast strains leads to elevated content of VOCs in wine [57]. The majority of the fermentative aroma metabolites are characterized by elevated sensory thresholds [70]. As a result, their combination shapes the characteristic aroma of wines. Importantly, some metabolic reactions performed by must microbiota are considered undesirable, since they spoil the quality of wine, such as by the acetic acid production [95]. Botrytized wine fermentations were found to contain increased abundance of acetic acid bacteria (AAB) in comparison with unaffected wines [36, 64]. Based on that, the selective microbial communities which are related to specific grape varieties, originated from particular locations, may extract distinctive metabolites, the combination of which could provide a characteristic *terroir* to the region [57].

The understanding of the contribution of the microbial communities in the sensorial characteristics of the wine requires the combination of metagenomic studies that will allow the identification of the wine's microbiome, with transcriptomics or metabolomics, which will reveal the volatile profile of the produced metabolites. Bokulich and colleagues [12] proposed that by identifying the microbial pool which composes grapes, and based on the existed knowledge, a great amount of the produced in the wine metabolites could be predicted. Indeed, by applying metabolomics and associating them with microbial communities—metagenomics—they discovered marker metabolites able to differentiate AVAs. Additionally, through a statistical model, they suggested that the grape must microbial conformation is able to predict the metabolites comprising the produced wine, proposing that regional microbial composition patterns may be able to characterize the wine physiognomies. Similarly, Belda and co-workers [96] suggested that the enzymatic activities of the wine-related microbial species population may predict the influence of the produced metabolites on wine aroma and establish region-derived clusters, via combination of metagenomics with information extracted by species-related enzymatic profiles analysis. Through gathering numerous non-*Saccharomyces* yeasts derived from three wine appellations in Spain and relating phylogenetic data with specific wine-associated enzymatic capabilities from glycosidases (β-glucosidase, α-L-arabinofuranosidase and β-D-xylosidase), β-lyases, pectinases, proteases, cellulases and sulfite reductases, indicated distinct origin-associated clusters for species such as *A. pullulans, T. delbrueckii, W. anomalus, H. uvarum and L. thermotolerans*.

Importantly, genetic variations among microbial strains may alter the overall profile of the wine's volatiles, proposing the influence of another contributing factor to regional characteristic terroir. Genetic variances between *S. cerevisiae* strains lead to alterations in the wines' metabolic profile affecting their sensory qualities [100–105]. Fluctuations in the expression levels of key enzymes affecting wine's aroma among different *S. cerevisiae* strains isolated from diverse geographic areas of New Zealand indicated correlations among geographic region and genetic background as well as the phenotypic profile of *S. cerevisiae* [103]. However, the phenotypic plasticity of *S. cerevisiae* to produce altered phenotypes based on the fermentation microenvironment was found to affect the metabolic profile of wines [104].

Moreover, genotypic characterization of different strains of *O. oeni*, *isolated from diverse geographic regions* during the process of malolactic fermentation, *revealed a highly diverse genetic background among the strains derived from different locations, but also strains categorized in the same phylogenetic group* were detected in diverse regions, adapted in the same type of wine [105]. Noteworthy, the genomic, transcriptomic, and proteomic profile of various *O. oeni* strains was found to be strongly influenced by microenvironmental conditions during winemaking [106–108].

Brettanomyces bruxellensis (or *Dekkera bruxellensis*), a yeast implicated in wine spoilage producing volatile phenols that create unpleasant flavors, was found to be composed by strains with differences in their genetic background that affected their adaptation in the wine-producing environment [109–112]. Microsatellite analysis of 1488 *B. bruxellensis* strains isolated from diverse geographic locations identified that the *B. bruxellensis* population was differentiated not only based on ploidy level, culture method, and fermentation environment but also on the origin of isolation [112], highlighting again the influence of geographic region in combination with additional influencing factors to microbial terroir formation.

7. Conclusion

Regional characteristics such as climate, agronomic practices, grape variety, and soil chemistry may influence the composition of the local microbial communities creating a characteristic regional microbial profile described with the term "microbial terroir." The composition of a particular variety grape microbiome, beyond its dynamic fluctuations during fermentation, was found to be able to provide indications regarding the chemical composition and the sensorial characteristics of the produced wines. The existence of specific regional microbial biomarkers, able to predict the metabolic composition of the wine, is a powerful indication of the existence of a clear association between region and local microbiome. Future studies based on the combination of HTS technologies with metabolomic studies may provide more enhanced evidence regarding the contribution of the regional microbial communities to wines' sensorial characteristics.

Acknowledgements

The authors would like to acknowledge the financial contribution from project AGROID funded by the INTERREG Greece-Cyprus 2014–2020 Program which is co-funded by the European Union (ERDF) and National Resources of Greece and Cyprus.

Conflict of interest

None of the authors has a financial or personal relationship with other people or organizations that could inappropriately influence or bias this publication.

Author details

Dimitrios A. Anagnostopoulos†*, Eleni Kamilari† and Dimitrios Tsaltas
Department of Agricultural Sciences, Biotechnology and Food Science,
Cyprus University of Technology, Lemesos, Cyprus

*Address all correspondence to: dimitris.tsaltas@cut.ac.cy
† These authors are equally contributed to this work.

References

[1] Tamang JP, Watanabe K, Holzapfel WH. Review: Diversity of microorganisms in global fermented foods and beverages. Frontiers in Microbiology. 2016;**7**:377

[2] Tamang JP, Shin DH, Jung SJ, Chae SW. Chae Functional properties of microorganisms in fermented foods. Frontiers in Microbiology. 2016;**7**:578

[3] Bokulich NA, Thorngate JH, Richardson PM, Mills DA. PNAS plus: From the cover: Microbial biogeography of wine grapes is conditioned by cultivar, vintage, and climate. Proceedings of the National Academy of Sciences of the United States of America [Internet]. 2014;**111**(1):E139-E148. DOI: 10.1073/pnas.1317377110

[4] Felder D, Burns D, Chang D. Defining microbial terroir: The use of native fungi for the study of traditional fermentative processes. International Journal of Gastronomy and Food Science. 2012;**1**(1):64-69

[5] Hachiya J. Studies of histological and molecular biological changes after graded periods of ischemia-reperfusion in mouse skeletal muscle. Journal of The Aichi Medical University Association. 1996;**24**(1):109-120

[6] Torriani S, Zapparoli G, Suzzi G, Agrarie I, Policattedra I. Genetic and phenotypic diversity of *Saccharomyces sensu stricto* strains isolated from Amarone wine. Antonie Van Leeuwenhoek [Internet]. 1999;**75**(3):207-215. DOI: 10.1023/A:1001773916407

[7] Cocolin L, Bisson LF, Mills DA. Direct profiling of the yeast dynamics in wine fermentations. FEMS Microbiology Letters. 2000;**189**(1):81-87

[8] Mills DA, Johannsen EA, Cocolin L. Yeast diversity and persistence in *Botrytis*-affected wine fermentations. Applied and Environmental Microbiology. 2002;**68**(10):4884-4893

[9] Fleet GH. Wine yeasts for the future. In: FEMS Yeast Research. 2008. pp. 979-995

[10] Isak SP. Tailoring wine yeast for the new millennium: Novel approaches to the ancient art of winemaking. Yeast [Internet]. 2000;**16**(8):675-729. DOI: 10.1002/1097-0061(20000615)16:8%3C675::AID-YEA585%3E3.0.CO;2-B

[11] Nisiotou AA, Spiropoulos AE, Nychas GJE. Yeast community structures and dynamics in healthy and *Botrytis*-affected grape must fermentations. Applied and Environmental Microbiology. 2007;**73**(21):6705-6713

[12] Bokulich NA, Collins TS, Masarweh C, Allen G, Heymann H, Ebeler SE, et al. Associations among wine grape microbiome, metabolome, and fermentation behavior suggest microbial contribution to regional wine characteristics. MBio. 2016;**7**:e00631-e00616. doi: 10.1128/mBio.00631-16

[13] Pinto C, Pinho D, Cardoso R, Custódio V, Fernandes J, Sousa S, et al. Wine fermentation microbiome: a landscape from different Portuguese wine appellations. Frontiers in Microbiology. 2015;**6**:905

[14] Marzano M, Fosso B, Manzari C, Grieco F, Intranuovo M, Cozzi G, et al. Complexity and dynamics of the winemaking bacterial communities in berries, musts, and wines from apulian grape cultivars through time and space. Plos One. 2016;**11**(6):e0157383

[15] Wei YJ, Wu Y, Yan YZ, Zou W, Xue J, Ma WR, et al. High-throughput sequencing of microbial community diversity in soil, grapes, leaves, grape

juice and wine of grapevine from China. PLoS One. 2018;**13**:e0193097

[16] Stefanini I, Albanese D, Cavazza A, Franciosi E, De Filippo C, Donati C, et al. Dynamic changes in microbiota and mycobiota during spontaneous "Vino Santo Trentino" fermentation. Microbial Biotechnology. 2016;**9**(2):195-208

[17] Bokulich NA, Mills DA. Facility-specific "house" microbiome drives microbial landscapes of artisan cheesemaking plants. Applied and Environmental Microbiology. 2013;**79**(17):5214-5223

[18] Taylor MW, Tsai P, Anfang N, Ross HA, Goddard MR. Pyrosequencing reveals regional differences in fruit-associated fungal communities. Environmental Microbiology. 2014;**16**:2848-2858

[19] Setati ME, Jacobson D, Bauer FF. Sequence-based analysis of the *Vitis vinifera L.* cv Cabernet Sauvignon grape must mycobiome in three South African vineyards employing distinct agronomic systems. Frontiers in Microbiology. 2015;**6**:1358

[20] Setati ME, Jacobson D, Andong UC, Bauer F. The Vineyard Yeast Microbiome, a Mixed Model Microbial Map. PLoS One. 2012;7(12):e52609

[21] Zarraonaindia I, Owens SM, Weisenhorn P, West K, Hampton-Marcell J, Lax S, et al. The soil microbiome influences grapevine-associated microbiota. MBio. 2015;**6**:e02527-e02514. DOI: 10.1128/mBio.02527-14

[22] Barata A, Loureiro V. International Journal of Food Microbiology The microbial ecology of wine grape berries. International Journal of Food Microbiology. 2012;**153**(3):243-259

[23] Martins G, Lauga B, Miot-Sertier C, Mercier A, Lonvaud A, Soulas ML, et al. Characterization of epiphytic bacterial communities from grapes, leaves, bark and soil of grapevine plants grown, and their relations. PLoS One. 2013;**8**(8):e73013

[24] David V, Terrat S, Herzine K, Claisse O, Rousseaux S, Tourdot-Maréchal R, et al. High-throughput sequencing of amplicons for monitoring yeast biodiversity in must and during alcoholic fermentation. Journal of Industrial Microbiology & Biotechnology. 2014;**41**(5):811-821

[25] Davey HM, Kell DB. Flow cytometry and cell sorting of heterogeneous microbial populations: The importance of single-cell analyses. Microbiological Reviews [Internet]. 2008;**60**(4):1-56. Available from: papers2://publication/uuid/EDF3C8F5-D5D9-4659-91A1-AD884E7A105F

[26] Quirós C, Herrero M, Garcáa LA, Dáaz M. Quantitative approach to determining the contribution of viable-but-nonculturable subpopulations to malolactic fermentation processes. Applied and Environmental Microbiology. 2009;**75**(9):2977-2981

[27] Salma M, Rousseaux S, Sequeira-Le Grand A, Divol B, Alexandre H. Characterization of the viable but nonculturable (VBNC) state in *Saccharomyces cerevisiae*. PLoS One. 2013;**8**(10):2-12

[28] Millet V, Lonvaud-Funel A. The viable but non-culturable state of wine micro-organisms during storage. Letters in Applied Microbiology. 2000;**30**(2):136-141

[29] Divol B, Lonvaud-Funel A. Evidence for viable but nonculturable yeasts in botrytis-affected wine. Journal of Applied Microbiology. 2005;**99**(1):85-93

[30] Yamamoto H. Viable but nonculturable state as a general

phenomenon of non-sporeforming bacteria, and its modeling. Journal of Infection and Chemotherapy [Internet]. 2000;**6**(2):112-114. Available from: http://linkinghub.elsevier.com/retrieve/pii/S1341321X00713250

[31] Andorrà I, Esteve-zarzoso B, Guillamón JM, Mas A. Determination of viable wine yeast using DNA binding dyes and quantitative PCR. International Journal of Food Microbiology. 2010;**144**(2):257-262

[32] Tofalo R, Schirone M, Corsetti A, Suzzi G. Detection of Brettanomyces spp. in red wines using real-time PCR. Journal of Food Science. 2012;**77**(9):545-549

[33] Neilson JW, Jordan FL, Maier RM. Analysis of artifacts suggests DGGE should not be used for quantitative diversity analysis. Journal of Microbiological Methods. 2013;**92**(3):256-263

[34] Wang C, García-Fernández D, Mas A, Esteve-Zarzoso B. Fungal diversity in grape must and wine fermentation assessed by massive sequencing, quantitative PCR and DGGE. Frontiers in Microbiology. 2015;**6**(Oct):1-8

[35] Campisano A, Antonielli L, Pancher M, Yousaf S, Pindo M, Pertot I. Bacterial endophytic communities in the grapevine depend on pest management. PLoS One. 2014;**9**(11):e112763

[36] Bokulich NA, Joseph CML, Allen G, Benson AK, Mills DA. Next-generation sequencing reveals significant bacterial diversity of botrytized wine. PLoS One. 2012;7(5):3-12

[37] Piao H, Hawley E, Kopf S, DeScenzo R, Sealock S, Henick-Kling T, et al. Insights into the bacterial community and its temporal succession during the fermentation of wine grapes. Frontiers in Microbiology. 2015;**6**:809

[38] Portillo C, Mas A. Analysis of microbial diversity and dynamics during wine fermentation of Grenache grape variety by high-throughput barcoding sequencing. LWT—Food Science and Technology. 2016;**72**:317-321

[39] Zott K, Miot-Sertier C, Claisse O, Lonvaud-Funel A, Masneuf-Pomarede I. Dynamics and diversity of non-Saccharomyces yeasts during the early stages in winemaking. International Journal of Food Microbiology. 2008;**125**:197-203

[40] Combina M, Elía A, Mercado L, Catania C, Ganga A, Martinez C. Dynamics of indigenous yeast populations during spontaneous fermentation of wines from Mendoza, Argentina. International Journal of Food Microbiology. 2005;**99**(3):237-243

[41] Tofalo R, Chaves-lópez C, Di F, Schirone M, Felis GE, Torriani S, et al. Molecular identification and osmotolerant pro fi le of wine yeasts that ferment a high sugar grape must. International Journal of Food Microbiology. 2009;**130**(3):179-187

[42] Williams KM, Liu P, Fay JC. Evolution of ecological dominance of yeast species in high-sugar environments. Evolution (NY). 2015;**69**(8):2079-2093

[43] Renouf V, Claisse O, Lonvaud-Funel A. Understanding the microbial ecosystem on the grape berry surface through numeration and identification of yeast and bacteria. Australian Journal of Grape and Wine Research. 2005;**11**(3):316-327

[44] Cordero-bueso G, Arroyo T, Serrano A, Tello J, Aporta I, Dolores M, et al. Influence of the farming system and vine variety on yeast communities associated with grape berries. International Journal of Food Microbiology. 2011;**145**(1):132-139

[45] Gobbi M, Comitini F, Domizio P, Romani C, Lencioni L, Mannazzu I, et al. *Lachancea thermotolerans* and *Saccharomyces cerevisiae* in simultaneous and sequential co-fermentation: A strategy to enhance acidity and improve the overall quality of wine. Food Microbiology. 2013;**33**(2):271-281

[46] Milanović V, Comitini F, Ciani M. Grape berry yeast communities: Influence of fungicide treatments. International Journal of Food Microbiology. 2013;**161**(3):240-246

[47] Bagheri B, Bauer FF, Setati ME. The diversity and dynamics of indigenous yeast communities in grape must from vineyards employing different agronomic practices and their influence on wine fermentation. South African Journal of Enology and Viticulture. 2015;**36**(2):243-251

[48] Kecskeméti E, Berkelmann-Löhnertz B, Reineke A. Are epiphytic microbial communities in the carposphere of ripening grape clusters (*Vitis vinifera L.*) different between conventional, organic, and biodynamic grapes? PLoS One. 2016;**11**(8):1-23

[49] Rodríguez-Gómez F, Ruiz-Bellido MA, Romero-Gil V, Benítez-Cabello A, Garrido-Fernández A, Arroyo-López FN. Microbiological and physicochemical changes in natural green heat-shocked Aloreña de Málaga table olives. Frontiers in Microbiology. 2017;**8**

[50] Burns KN, Kluepfel DA, Strauss SL, Bokulich NA, Cantu D, Steenwerth KL. Vineyard soil bacterial diversity and composition revealed by 16S rRNA genes: Differentiation by geographic features. Soil Biology and Biochemistry. 2015;**91**:232-247

[51] Faist H, Keller A, Hentschel U. Grapevine (*Vitis vinifera*) crown galls host distinct microbiota. Applied and Environmental Microbiology. 2016;**82**(18):5542-5552

[52] Calleja-Cervantes ME. From ground pollution to soil fertilization: An environmental assessment of soil amendments derived from organic wastes [Internet]. Universidad Pública de Navarra; 2015. Available from: https://academica-e.unavarra.es/bitstream/handle/2454/20850/Tesis%2BM.E.%2BCallejaMarcAg.pdf?sequence=1&isAllowed=y

[53] Perazzolli M, Antonielli L, Storari M, Puopolo G, Pancher M, Giovannini O, et al. Resilience of the natural phyllosphere microbiota of the grapevine to chemical and biological pesticides. Applied and Environmental Microbiology. 2014;**80**(12):3585-3596

[54] Portillo C, Franquès J, Araque I, Reguant C, Bordons A. International Journal of Food Microbiology Bacterial diversity of Grenache and Carignan grape surface from different vineyards at Priorat wine region (Catalonia, Spain). International Journal of Food Microbiology. 2016;**219**:56-63

[55] Bokulich NA, Collins T, Masarweh C, Allen G, Heymann H, Ebeler SE, et al. Fermentation behavior suggest microbial contribution to regional. MBio Journal. 2016;7(3):1-12

[56] Čadež N, Zupan J, Raspor P. The effect of fungicides on yeast communities associated with grape berries. FEMS Yeast Research. 2010;**10**(5):619-630

[57] Stefanini I, Carlin S, Tocci N, Albanese D, Donati C, Franceschi P, et al. Core microbiota and metabolome of *Vitis vinifera* L. cv. Corvina grapes and musts. Frontiers in Microbiology. 2017;**8**:457. DOI: 10.3389/fmicb.2017.0045

[58] Sabate J, Cano J, Esteve-zarzoso B, Guillamon JM. Isolation and

identification of yeasts associated with vineyard and winery by RFLP analysis of ribosomal genes. Microbiological Research. 2002;**157**:267-274

[59] Guerzoni E, Marchetti R. Analysis of yeast flora associated with grape sour rot and of the chemical disease markers. Applied and Environmental Microbiology. 1987;**53**(3):571-576

[60] Dean R, Van Kan JAL, Pretorius ZA, Hammond-Kosack KE, Di Pietro A, Spanu PD, et al. The top 10 fungal pathogens in molecular plant pathology. Molecular Plant Pathology. 2012;**13**:414-430

[61] Joyeux A, Lafon-Lafourcade S, Tibereau-Gayon P. Evolution of acetic acid bacteria during fermentation and storage of wine. Applied and Environmental Microbiology. 1984;**48**:153-156

[62] Sipiczki M. *Metschnikowia* strains isolated from botrytized grapes antagonize fungal and bacterial growth by iron depletion. Applied and Environmental Microbiology. 2006;**72**(10):6716-6724

[63] Bokulich NA, Swadener M, Sakamoto K, Mills DA, Bisson LF. Sulfur dioxide treatment alters wine microbial diversity and fermentation progression in a dose-dependent fashion. American Journal of Enology and Viticulture. 2015;**66**:73-79

[64] Berger RG. Flavours and Fragrances: Chemistry, Bioprocessing and Sustainability. Berlin, Germany: Springer; 2007

[65] Ewart A, Brien C, Soderlund R, Smart R. The effects of light pruning, irrigation and improved soil-management on wine quality of the *Vitis vinifera* CV riesling. VITIS—Journal of Grapevine Research. 1985;**24**:209-217

[66] Spayd SE, Tarara JM, Mee DL, Ferguson JC. Separation of sunlight and temperature effects on the composition of *Vitis vinifera* cv. Merlot berries. American Journal of Enology and Viticulture. 2002;**53**:171-181

[67] Hernández-Orte P, Cersosimo M, Loscos N, Cacho J, Garcia-Moruno E, Ferreira V. The development of varietal aroma from non-floral grapes by yeasts of different genera. Food Chemistry. 2008;**107**:1064-1077

[68] Hernandez-Orte P, Concejero B, Astrain J, Lacau B, Cacho J, Ferreira V. Influence of viticulture practices on grape aroma precursors and their relation with wine aroma. Journal of the Science of Food and Agriculture. 2014;**95**:688-701. DOI: 10.1002/jsfa.6748

[69] Padilla B, Gil JV, Manzanares P. Past and future of non-Saccharomyces yeasts: From spoilage microorganisms to biotechnological tools for improving wine aroma complexity. Frontiers in Microbiology. 2016;**7**:411. DOI: 10.3389/fmicb.2016.00411

[70] Belda I, Ruiz J, Esteban-Fernández A, Navascués E, Marquina D, Santos A, et al. Microbial contribution to Wine aroma and its intended use for wine quality improvement. Molecules. 2017;**22**:189. DOI: 10.3390/molecules22020189

[71] Capozzi V, Garofalo C, Chiriatti MA, Grieco F, Spano G. Microbial terroir and food innovation: The case of yeast biodiversity in wine. Microbiological Research. 2015;**181**:75-83. DOI: 10.1016/j.micres.2015.10.005

[72] Ebeler SE, Thorngate JH. Wine chemistry and flavor: Looking into the crystal glass. Journal of Agricultural and Food Chemistry. 2009;**57**:8098-8108

[73] Winterhalter P, Skouroumounis GK. Glycoconjugated aroma compounds: Occurrence, role and

biotechnological transformation. Advances in Biochemical Engineering/ Biotechnology. 1997;**55**:73-105

[74] Charoenchai C, Fleet GH, Henschke PA, Todd BEN. Screening of non-Saccharomyces wine yeasts for the presence of extracellular hydrolytic enzymes. Australian Journal of Grape and Wine Research. 1997;**3**:2-8

[75] Sabel A, Martens S, Petri A, König H, Claus H. Wickerhamomyces anomalus AS1: A new strain with potential to improve wine aroma. Annales de Microbiologie. 2014;**4**:483-491

[76] López MC, Mateo JJ, Maicas S. Screening of β-glucosidase and β-xylosidase activities in four non-Saccharomyces yeast isolates. Journal of Food Science. 2015;**80**:1696-1704

[77] Cordero-Bueso G, Esteve-Zarzoso B, Cabellos JM, Gil-Díaz M, Arroyo T. Biotechnological potential of non-Saccharomyces yeasts isolated during spontaneous fermentations of Malvar (*Vitis vinifera* cv. L.). European Food Research and Technology. 2013;**236**:193-207

[78] Fia G, Giovani G, Rosi I. Study of β-glucosidase production by wine-related yeasts during alcoholic fermentation. A new rapid fluorimetric method to determine enzymatic activity. Journal of Applied Microbiology. 2005;(99):509-517

[79] Villena MA, Iranzo JFÚ, Pérez AIB. β-Glucosidase activity in wine yeasts: Application in enology. Enzyme and Microbial Technology. 2007;**40**:420-425

[80] Wang Y, Kang W, Xu Y, Li J. Effect of different indigenous yeast β-glucosidases on the liberation of bound aroma compounds. Journal of the Institute of Brewing. 2011;**117**:230-237

[81] McMahon H, Zoecklein BW, Fugelsang K, Jasinski Y. Quantification of glycosidase activities in selected yeasts and lactic acid bacteria. Journal of Industrial Microbiology & Biotechnology. 1999;**23**:198-203

[82] Manzanares P, Rojas V, Genovés S, Vallés S. A preliminary search for anthocyanin-β-D-glucosidase activity in non-Saccharomyces wine yeasts. International Journal of Food Science and Technology. 2000;**35**:95-103

[83] Strauss MLA, Jolly NP, Lambrechts MG, Van Rensburg P. Screening for the production of extracellular hydrolytic enzymes by non-Saccharomyces wine yeasts. Journal of Applied Microbiology. 2001;**91**:182-190

[84] Spagna G, Barbagallo RN, Palmeri R, Restuccia C, Giudici P. Properties of endogenous β-glucosidase of a *Saccharomyces cerevisiae* strain isolated from Sicilian musts and wines. Enzyme and Microbial Technology. 2002;**31**:1036-1041

[85] Cordero Otero RR, Ubeda Iranzo JF, Briones-Perez AI, Potgieter N, Villena MA, Pretorius IS, et al. Characterization of the β-glucosidase activity produced by enological strains of non-Saccharomyces yeasts. Journal of Food Science. 2003;**68**:2564-2569

[86] Fernández-González M, Di Stefano R, Briones A. Hydrolysis and transformation of terpene glycosides from muscat must by different yeast species. Food Microbiology. 2003;**20**:35-41

[87] Rodríguez ME, Lopes CA, Van Broock M, Valles S, Ramón D, Caballero AC. Screening and typing of Patagonian wine yeasts for glycosidase activities. Journal of Applied Microbiology. 2004;**96**:84-95

[88] González-Pombo P, Pérez G, Carrau F, Guisán JM, Batista-Viera F,

Brena BM. One-step purification and characterization of an intracellular β-glucosidase from *Metschnikowia pulcherrima*. Biotechnology Letters. 2008;**30**:1469-1475

[89] Swiegers JH, Kievit RL, Siebert T, Lattey KA, Bramley BR, Francis IL, et al. The influence of yeast on the aroma of Sauvignon Blanc wine. Food Microbiology. 2009;**26**:204-211

[90] Anfang N, Brajkovich M, Goddard MR. Co-fermentation with *Pichia kluyveri* increases varietal thiol concentrations in sauvignon blanc. Australian Journal of Grape and Wine Research. 2009;**15**:1-8

[91] Zott K, Thibon C, Bely M, Lonvaud-Funel A, Dubourdieu D, Masneuf-Pomarede I. The grape must non-Saccharomyces microbial community: Impact on volatile thiol release. International Journal of Food Microbiology. 2011;**151**:210-215

[92] Romano P, Fiore C, Paraggio M, Caruso M, Capece A. Function of yeast species and strains in wine flavour. International Journal of Food Microbiology. 2003;**86**:169-180

[93] Rapp A, Versini G. Influence of nitrogen compounds in grapes on aroma compounds of wines. In: Rantz. editor. Proceedings of the International symposium on nitrogen in grapes and wines. Davis, CA: American Society for Enology and Viticulture; 1995. pp. 156-164

[94] Boulton RB, Singleton VL, Bisson LI, Kunkee RE. The maturation and ageing of wines. III. In: Boulton RB, Singleton VL, Bisson LF. Kunkee RE. editors. Principles and practices of winemaking. New York: Chapman and Hall; 1996. p. 382-426

[95] Zoecklein BW, Williams JM, Duncan SE. Effect of sour rot on the composition of white riesling (*Vitis vinifera* L.) grapes. Small Fruits Review. 2000;**1**:63-77

[96] Belda I, Ruiz J, Alastruey-Izquierdo A, Navascués E, Marquina D, Santos A. Unraveling the enzymatic basis of wine "Flavorome": A phylo-functional study of wine related yeast species. Frontiers in Microbiology. 2016;**7**:12

[97] Lopandic K, Gangl H, Wallner E, Tscheik G, Leitner G, Querol A, et al. Genetically different wine yeasts isolated from Austrian vine-growing regions influence wine aroma differently and contain putative hybrids between *Saccharomyces cerevisiae* and *Saccharomyces kudriavzevii*. FEMS Yeast Research. 2007;**7**:953-965

[98] Barrajón N, Capece A, Arévalo-Villena M, Briones A, Romano P. Co-inoculation of different *Saccharomyces cerevisiae* strains and influence on volatile composition of wines. Food Microbiology. 2011;**28**:1080-1086

[99] Callejon RM, Clavijo A, Ortigueira P, Troncoso AM, Paneque P, Morales ML. Volatile and sensory profile of organic red wines produced by different selected autochthonous and commercial *Saccharomyces cerevisiae* strains. Analytica Chimica Acta. 2010;**660**:68-75

[100] Patrignani F, Chinnici F, Serrazanetti DI, Vernocchi P, Ndagijimana M, Riponi C, et al. Production of volatile and sulfur compounds by 10 *Saccharomyces cerevisiae* strains inoculated in trebbiano must. Frontiers in Microbiology. 2016;**7**:243

[101] Swain-Lenz D, Nikolskiy I, Cheng J, Sudarsanam P, Nayler D, Staller MV, et al. Causal genetic variation underlying metabolome differences. Genetics. 2017;**206**(4):2199-2206

[102] Capece A, Granchi L, Guerrini S, Mangani S, Romaniello R, Vincenzini M, et al. Diversity of *Saccharomyces*

cerevisiae strains isolated from two Italian wine-producing regions. Frontiers in Microbiology. 2016;**1**:1018

[103] Knight S, Klaere S, Fedrizzi B, Goddard MR. Regional microbial signatures positively correlate with differential wine phenotypes: Evidence for a microbial aspect to terroir. Scientific Reports. 2015;**5**:14233

[104] Peltier E, Sharma V, Martí Raga M, Roncoroni M, Bernard M, Jiranek V, et al. Dissection of the molecular bases of genotype x environment interactions: A study of phenotypic plasticity of *Saccharomyces cerevisiae* in grape juices. BMC Genomics. 2018;**19**:772

[105] El Khoury M, Campbell-Sills H, Salin F, Guichoux E, Claisse O, Lucas PM. Biogeography of *Oenococcus oeni* reveals distinctive but nonspecific populations in wine-producing regions. Applied and Environmental Microbiology. 2017;**83**:1-11

[106] Costantini A, Rantsiou K, Majumder A, Jacobsen S, Pessione E, Svensson B, et al. Complementing DIGE proteomics and DNA subarray analyses to shed light on *Oenococcus oeni* adaptation to ethanol in wine-simulated conditions. Journal of Proteomics. 2015;**123**:114-127

[107] Olguín N, Champomier-Vergès M, Anglade P, Baraige F, Cordero-Otero R, Bordons A, et al. Transcriptomic and proteomic analysis of *Oenococcus oeni* PSU-1 response to ethanol shock. Food Microbiology. 2015;**51**:87-95. DOI: 10.1016/j.fm.2015.05.005

[108] Romero J, Ilabaca C, Ruiz M, Jara C. *Oenococcus oeni* in chilean red wines: Technological and genomic characterization. Frontiers in Microbiology. 2018;**9**:90. DOI: 10.3389/FMICB.2018.00090

[109] Borneman AR, Zeppel R, Chambers PJ, Curtin CD. Insights into the *Dekkera bruxellensis* genomic landscape: Comparative genomics reveals variations in ploidy and nutrient utilisation potential amongst wine isolates. PLoS Genetics. 2014:e1004161

[110] Curtin CD, Borneman AR, Chambers PJ, Pretorius IS. De-novo assembly and analysis of the heterozygous triploid genome of the wine spoilage yeast *Dekkera bruxellensis* AWRI1499. PLoS One. 2012:e33840

[111] Albertin W, Panfili A, Miot-Sertier C, Goulielmakis A, Delcamp A, Salin F, et al. Development of microsatellite markers for the rapid and reliable genotyping of *Brettanomyces bruxellensis* at strain level. Food Microbiology. 2014;**42**:188-195

[112] Avramova M, Cibrario A, Peltier E, Coton M, Coton E, Schacherer J, et al. *Brettanomyces bruxellensis* population survey reveals a diploid-triploid complex structured according to substrate of isolation and geographical distribution. Scientific Reports. 2018;**8**:4136. DOI: 10.1038/s41598-018-22580-7

Chapter 7

Biodiversity of *Saccharomyces cerevisiae* Yeasts in Spontaneous Alcoholic Fermentations: Typical Cellar or Zone Strains?

Pilar Santamaría, Rosa López, Maria del Patrocinio Garijo, Rocío Escribano, Lucía González-Arenzana, Isabel López-Alfaro and Ana Rosa Gutiérrez

Abstract

Spontaneous fermentation is the most traditional way and a low-intervention method for conducting alcoholic fermentation in wineries, giving rise to the most complex wine profiles. However, inoculation with single culture inocula of *Saccharomyces cerevisiae* strains has become widespread in the modern wine industry. Nevertheless, some authors have pointed out that the use of the same yeasts in all the winegrowing regions of the world can cause a loss of typicity and have a standardizing effect on the wines. For this reason, many wineries and regions are carrying out programs of isolation and selection of yeasts that are typical of their vineyards/wineries. The aim of this work was to study the ecology of spontaneous fermentations in 11 wineries from all over the Rioja qualified designation of origin (Spain) during 3–4 consecutive years in order to establish the existence of typical strains belonging to wineries, sub-zones, or regional ecosystems. The results obtained showed a great diversity of strains of *Saccharomyces cerevisiae* in each fermentation studied. These strains were different each year in each winery, and hardly any common strains were detected between neighboring wineries, which would indicate that there are no representative strains from the winery or the area.

Keywords: Rioja qualified designation of origin, alcoholic fermentation, *saccharomyces cerevisiae*, diversity, ecology, typical strains

1. Introduction

Spontaneous fermentation is the most traditional way and a low-intervention method for conducting alcoholic fermentation in wineries, giving rise to the most complex wine profiles. This complexity develops because of the large number of different yeast species involved (*Saccharomyces* spp. and non-*Saccharomyces*) [1]. However, the presence of unknown microbiota makes it a risky and unpredictable practice. For this reason, the inoculation with single culture inocula of *Saccharomyces cerevisiae* has become widespread in the modern wine industry to reduce the risk of wine spoilage. Nevertheless, some authors have pointed out that the use of the same yeasts in all the winegrowing regions of the world can cause a loss of typicity and have a negative effect on the biodiversity of natural yeasts present in the wineries [2].

S. cerevisiae is the predominant yeast species in alcoholic fermentation and the main element responsible for the characteristics of the wines. Many surveys carried out with spontaneous fermentations in different wine-producing regions have demonstrated that there is high genetic diversity within this species in each vinification [3]. However, in most cases, only a small number of strains of *S. cerevisiae* are dominant, mainly in the tumultuous and final fermentation stages, representing a high percentage out of the total number of strains identified.

Earlier studies have shown that some strains of *S. cerevisiae* have been isolated in several consecutive years in the same winery, which is why some authors have suggested the term "winery effect" [4], and also some strains of these species were detected in different wineries of the same wine-producing area, suggesting that they were representative of a specific enological ecosystem [5, 6]. Knight et al. [7] even found specific genotypes from a particular region. These findings suggest that specific native *S. cerevisiae* strains could be associated with a *terroir* and have an influence on terroir-associated wine characteristics [8].These authors found a correlation between genotypic and phenotypic groups and the geographical origin of the strains, supporting the concept that there can be a microbial aspect to *terroir*.

Nowadays, many wineries and regions are carrying out strain selection programs with yeasts isolated from their vineyard/winery ecosystems, based on the idea that these yeasts are better adapted to their musts, which have characteristics determined by the grape varieties and the *terroir* [9, 10]. Thus, the use of these typical strains as starter yeasts could provide wines with distinctive characteristics of a particular winery or region. For this reason, studying the existence of strains which are specific to one winery or enological area is very interesting for the wine industry [11].

Rioja is a wine region in Spain with qualified designation of origin status. It is subdivided into three sub-zones: Rioja Alta, Rioja Oriental, and Rioja Alavesa. Rioja Alta is located on the western edge of the region and at higher elevations with an Atlantic climate. Rioja Oriental, the eastern section, is strongly influenced by a Mediterranean climate which makes this area the warmest and driest part of the region. Rioja Alavesa, with a similar climate to Rioja Alta, produces different wines due to the relatively poor condition of the soil. Each sub-zone has its own character, which results in different wines derived from the different compositions and origins of the soils and the climate conditions.

The study of the ecology and biodiversity of the yeast population during alcoholic fermentation is an interesting and important step in the research into, and understanding of, a winegrowing area and should be a step prior to the selection and subsequent employment of the yeasts isolated from that area as starters. The aim of this work was to study the ecology of spontaneous fermentations in 11 wineries from all over the Rioja Designation of Origin during 3–4 consecutive years in order to establish the existence of typical strains belonging to wineries, sub-zones, and regional ecosystems. The wineries under study were distributed throughout the three sub-zones of the Rioja designation of origin.

2. Study on the ecology of S. *cerevisiae* in the Rioja qualified designation of origin

2.1 Material and methods

2.1.1 Sample collection

Samples were taken from 11 wineries (A–K) located in three different sub-zones of the Rioja designation of origin (Rioja Alta, Rioja Oriental, and Rioja Alavesa) over a period of 3 or 4 consecutive years (**Figure 1**). None of the wineries studied had ever used a commercial starter yeast. Wineries A and B were new; wineries C, E, and K were about 20 years old, while the others were over 50 years old.

Alcoholic fermentation (AF) was carried out by the destemming and crushing method in stainless steel (wineries A, B, C, D, and F) or wooden vats (winery H). In the other wineries, AF was carried out following the traditional carbonic maceration method (whole grape) in open concrete vats (wineries E, G, I, J, and K). The wines underwent spontaneous AF with the indigenous microbiota in all cases.

In each winery, one fermentation tank was monitored in each year studied. The sampling was carried out 24 hours after vatting, in tumultuous AF (density 1025 g/L) and final AF.

2.1.2 Microbial analyses: strain typing of S. cerevisiae

Samples collected in sterile bottles were taken to the laboratory and processed as follows: serial decimal dilutions were performed, and the samples were seeded onto plates containing a chloramphenicol glucose agar medium (CGA). The plates were incubated at 28°C for 48 h. Plates containing between 30 and 300 colonies were examined, and 10 colonies were randomly isolated from each CGA plate. The colonies were analyzed in order to identify *Saccharomyces* and non-*Saccharomyces* yeast and the clonal distribution of the *S. cerevisiae* by mitochondrial DNA (mtDNA) restriction analysis. Yeast cells were grown overnight in a culture of 5 ml YPD. DNA extraction and mtDNA restriction were determined by the method described by Querol and Barrio [12]. The DNA was digested with the restriction endonucleases Alu I, RSa I, and Hinf I, in accordance with the supplier's instructions (Boehringer Mannheim). The restriction fragments were separated by electrophoresis in agarose 1% gels and visualized on a UV transilluminator after ethidium bromide staining.

The different clones isolated in each fermentation were named with the letter of the cellar (A–K), followed by a Roman numeral and year of harvest (1–4).

The clonal variability of each fermentation was determined as a percentage of different *S. cerevisiae* genotypes compared to the total colonies *S. cerevisiae* identified.

The index of diversity (I.D.) [13] was calculated with the different *S. cerevisiae* strains identified in the tumultuous and final stages of fermentation, according to the following equation:

$$\text{I.D.} = 1 - \frac{\Sigma 3\, n_j\,(n_j - 1)}{\text{N}\,(\text{N} - 1)} \tag{1}$$

where N is the total number of *S. cerevisiae* strains and n_j is the number of *S. cerevisiae* strains with the same electrophoretic profile.

2.2 Results and discussion

2.2.1 Ecology of the yeast population during alcoholic fermentation in Rioja Oriental sub-zone

In this sub-zone, four wineries were sampled (A–D). In all of them, the vinification took place after destemming and crushing the grapes.

As expected, and concurring with previous studies on the ecology of alcoholic fermentation (FA) [14, 15], non-*Saccharomyces* yeasts were detected mainly in the first stages of the fermentation (data not shown). The rest of the yeasts isolated in this sub-zone (316 colonies) were identified as belonging to the *S. cerevisiae* species.

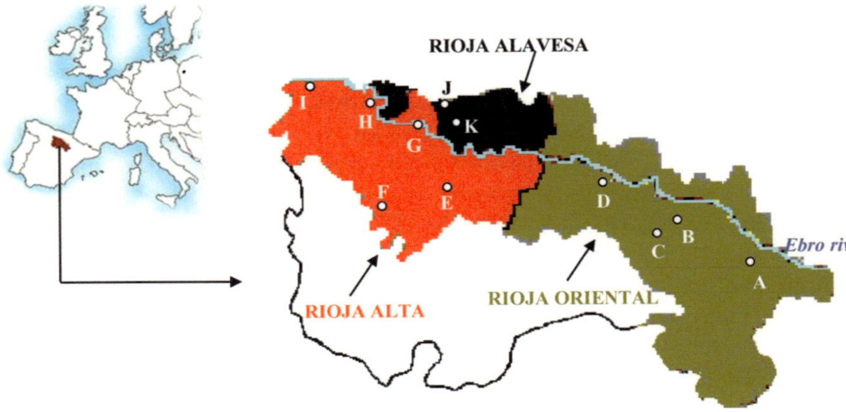

Figure 1.
Location of the wineries in the three different sub-zones of the Rioja designation of origin.

The clonal variability in this species was different depending on the winery and the year studied within each winery (**Table 1**). Thus, while in cellars A and B the total clonal variability was 13.8 and 21.8%, respectively, in the other two (C and D), it was around 35%. Because of the lower variability in the first two wineries, the index of diversity was also smaller.

A percentage of the isolated microbiota in each harvest had already been identified in previous years (**Table 1**). This percentage was higher in wineries A and B, while in the others, this proportion was lower. However, the evolution in the population of the strains during fermentation was similar in all the wineries, since in all of them the fermentation in general was carried out by one or two majority yeasts (**Table 2**) associated with a variable number of minority strains (**Table 3**). This majority strain(s) represented between 50 and 82% of the total isolated in each vinification (**Table 2**). The only exception was found in vintage 2 of cellar C, in which there were three major strains with a percentage of 14.3% each. Previous research on the *S. cerevisiae* population in spontaneous fermentations also showed the existence of a great diversity of strains. Among them, one or two represented more than 50% of the total [16, 17].

Nonetheless, only in wineries A and B, there was one common majority strain in the vinifications studied in the 3–4 years (it is highlighted in the same color in **Table 2**). The fact that the same strains were predominant in successive campaigns in the same winery has already been described in other works [4, 18], for which reason they considered that these yeasts were representative of the winery or the processing area. In contrast, in other studies, common yeasts were not found from one campaign to another or were only found in small proportions [19, 20].

The results obtained could suggest that in wineries A and B, there are yeasts characteristic of the winery that lead the fermentation, in proportions that vary depending on the characteristics of the vintage. However, in the other two wineries, the fermentations were carried out by a succession of strains, among which one or two were majority, but different from one vintage to another.

The main difference between the four wineries studied in the Rioja Oriental sub-zone was their age. While wineries A and B were new (this study began in its third and first harvest, respectively), cellar C had been producing wine for 15 years and D for 50. This means that the latter two had more complex ecosystems, which have been formed over successive harvests, as evidenced by the greater number of clones that participated in the fermentations and the higher diversity index (**Table 1**).

Winery	Years	Isolates of *S. cerevisiae* analyzed	Number of different clones	Clonal variability (%)	Index of diversity	Genotypes detected in previous years	
						Number	%
A	1	29	6	20.7	0.65	–	–
	2	28	7	25.0	0.77	3	71.4
	3	22	4	18.2	0.33	2	86.3
	4	30	6	20.0	0.46	3	83.2
	Total	**109**	**15**	**13.8**	**0.64**		
B	1	–	–	–	–	–	–
	2	20	5	25.0	0.76	–	–
	3	30	9	30.0	0.84	2	36.6
	4	28	8	28.6	0.77	3	63.8
	Total	**78**	**17**	**21.8**	**0.83**		
C	1	–	–	–	–	–	–
	2	21	11	53.4	0.94	–	–
	3	22	8	36.4	0.77	2	13.6
	4	24	11	45.8	0.87	3	45.8
	Total	**67**	**25**	**37.3**	**0.92**		
D	1	–	–	–	–	–	–
	2	20	10	50.0	0.86	–	–
	3	22	5	22.7	0,58	1	9.1
	4	20	8	40.0	0.82	2	10.0
	Total	**62**	**20**	**32.3**	**0.91**		

Table 1.
Clonal diversity of S. cerevisiae yeasts in Rioja Oriental wineries in 4 consecutive years (1–4).

In a later study, carried out by our research group in cellar B [21], 72 colonies of the species *S. cerevisiae* were isolated in three fermentation tanks analyzed. The results showed 41 different clones, which provided a clonal variability of 56.9% and a biodiversity index of 0.98. These data would indicate that the fermentations were carried out by the succession of different clones and that there were no dominant ones. On the other hand, the comparison of the restriction profiles of this year's clones with those identified in the years specified in **Tables 1** and **2** showed that there was no correspondence with any of them, and therefore the majority strain (**Table 2**) that was isolated in the first three vintages as representative of the winery did not appear years later. Therefore, the ecology of the fermentations of this last year in cellar B was similar to those found in cellars C and D in the period shown previously. The situation in winery A could be similar, but it could not be proven since in the following years, commercial yeasts were inoculated, and the study could not be carried out. Taking into account the results shown, the presence of a dominant strain in the elaborations of wineries A and B could be due to the fact that they are new wineries and not to the existence of representative yeasts.

On the other hand, the comparison of the different restriction profiles obtained in the four wineries during the 3–4 years of study showed that only two yeasts were common among the different wineries (**Table 3**). Therefore, the results obtained in this study did not show the existence of typical strain/s of this sub-zone. In the wine

Years	1		2		3		4	
Winery	**Strain**	**%**	**Strain**	**%**	**Strain**	**%**	**Strain**	**%**
A	A-I$_1$	55.2	A-I$_2$	42.9	A-IV$_3$	81.8	A-I$_4$	73.3
	A-II$_1$	24.1	A-II$_2$	21.4				
	Total	**79.3**		**64.3**		**81.8**		**73.3**
B	–	–	B-I$_2$	35.0	B-I$_3$	33.3	B-II$_4$	21.4
	–	–	B-IV$_2$	35.0	B-VIII$_3$	20.0	B-IV$_4$	42.3
	Total	–		**70.0**		53.3		**63.7**
C	–	–	C-IV$_2$	14.3	C-I$_3$	40.9	C-I$_4$	33.3
	–	–	C-V$_2$	14.3	C-IV$_3$	27.3	C-IX$_4$	16.7
			C-X$_2$	14.3				
	Total	–		**42.9**		**68.2**		**50.0**
D	–	–	D-V$_2$	35.0	D-II$_3$	63.6	D-I$_4$	35.0
	–	–	D-VI$_2$	15.0			D-III$_4$	25.0
	Total	–		**50.0**		**63.6**		**60.0**

Clones with box in gray within the same winery indicate that they are the same strain, according to their mtDNA restriction profiles

Table 2.
Major S. cerevisiae strains in the fermentations of Rioja Oriental in 4 consecutive years (1–4).

Winery A				Winery B			Winery C			Winery D		
1	**2**	**3**	**4**	**2**	**3**	**4**	**2**	**3**	**4**	**2**	**3**	**4**
A-I$_1$	A-I$_2$	A-I$_3$	A-I$_4$	**B-I$_2$**	B-I$_3$	B-I$_4$	C-I$_2$	C-I$_3$	C-I$_4$	D-I$_2$	D-I$_3$	D-I$_4$
A-II$_1$	**A-II$_2$**	A-II$_3$	**A-II$_4$**	B-II$_2$	B-II$_3$	B-II$_4$	C-II$_2$	C-II$_3$	**C-II$_4$**	D-II$_2$	D-II$_3$	D-II$_4$
A-III$_1$	A-III$_2$	A-III$_3$	A-III$_4$	B-III$_2$	B-III$_3$	B-III$_4$	C-III$_2$	C-III$_3$	C-III$_4$	D-III$_2$	D-III$_3$	D-III$_4$
A-IV$_1$	A-IV$_2$	A-IV$_3$	A-IV$_4$	B-IV$_2$	B-IV$_3$	B-IV$_4$	C-IV$_2$	C-IV$_3$	C-IV$_4$	D-IV$_2$	D-IV$_3$	D-IV$_4$
A-V$_1$	A-V$_2$		***A-V$_4$***	B-V$_2$	B-V$_3$	B-V$_4$	C-V$_2$	C-V$_3$	C-V$_4$	D-V$_2$	D-V$_3$	D-V$_4$
A-VI$_1$	A-VI$_2$		A-VI$_4$		B-VI$_3$	**B-VI$_4$**	C-VI$_2$	C-VI$_3$	C-VI$_4$	D-VI$_2$		D-VI$_4$
	A-VII$_2$				B-VII$_3$	B-VII$_4$	C-VII$_2$	C-VII$_3$	C-VII$_4$	D-VII$_2$		D-VII$_4$
					B-VIII$_3$	B-VIII$_4$	C-VIII$_2$	C-VIII$_3$	C-VIII$_4$	D-VIII$_2$		D-VIII$_4$
					B-IX$_3$		C-IX$_2$		C-IX$_4$	D-IX$_2$		
							C-X$_2$		C-X$_4$	D-X$_2$		
							C-XI$_2$		C-XI$_4$			

Clones in bold and with box in gray in different columns show the same clones in different cellars

Table 3.
S. cerevisiae clones identified in each elaboration of Rioja Oriental in 4 consecutive years (1-4).

area of Charentes in Cognac, a strain widely distributed throughout the area had been described, which was considered representative of the wine region. But unlike our data, this strain was the dominant one in all the samples where it appeared [18]. Likewise, in other ecological studies, widely disseminated strains were isolated in an area, and it was thought that they may be typical strains of that area [22].

2.2.2 Ecology of the yeast population during alcoholic fermentation in Rioja Alta sub-zone

In this sub-zone, five wineries were sampled. Vinification was by both the destemming and crushing (F and H) and the carbonic maceration methods (E, G, and I).

As happened in the Rioja Oriental, alcoholic fermentation in the Rioja Alta sub-zone was carried out by yeasts of the *S. cerevisiae* species, since the non-*Saccharomyces* group was identified mainly in the early stages of the process. However, in the third harvest of cellars E and F, three colonies belonging to non-*Saccharomyces* genera were isolated in tumultuous fermentation (data not shown). The detection of these yeasts in advanced stages of winemaking had also been reported in other works [23, 24]. In total, 450 yeasts belonging to the *S. cerevisiae* species were studied.

Clonal variability was high in this sub-zone in all the wineries (**Table 4**). From the 450 isolated colonies of *S. cerevisiae*, 177 different clones were identified, resulting in high clonal variability in Rioja Alta, which ranged between 35% in winery H and 39% in winery F, with the exception of cellar I, in which 50% was reached. It is noteworthy that clonal variability depended on the year analyzed, being generally lower the first year and higher the third. These results could be related to the climatological characteristics of the harvest.

The index of diversity was high and similar for all the wineries with a value of 0.95–0.96. These results were due to the high number of different strains that participated in each fermentation, which in the case of wineries F and I was favored by the low presence of common clones in different campaigns within the same winery (**Table 4**). In wineries E, G, and H, although there were more strains that appeared in different campaigns, they did so in a low percentage, and, therefore, the index remained high.

The population evolution in this sub-zone was similar to Rioja Oriental, since the alcoholic fermentation was carried out by different clones during the fermentation process. In most of the vinifications, 1–3 majority clones were detected, representing at least 43% of the population (**Table 5**). However, in this sub-zone certain exceptions were found. Thus, in the first year analyzed in winery F, one clone represented 65% of the total and so was therefore dominant and responsible for the fermentation. This could be due to the special characteristics of this campaign (frosts at the end of April, high rainfall and strong winds in spring, hail storms and low temperatures in summer) that negatively influenced the grape ripening and meant that few yeasts were able to adapt and develop to carry out the fermentation. These conditions were more adverse in winery F, due to its geographic situation. It is the winery located at higher altitude and closer to the mountains.

On the other hand, in wineries G and I, there were vinification processes in which no major clones were isolated, which could be related to the high number of different clones that participated in the fermentations. What these wineries have in common is that they follow the carbonic maceration method and have been making wine for more than 100 years. Santamaría et al. [15] showed that the number of different S. *cerevisiae* strains and the frequency of their appearance varied according to age.

When comparing the clones found in the five wineries (**Table 6**), only pattern I, coming from the last year studied in the four wineries, was common in all of them. When comparing the restriction profiles of the four strains, all of them showed the same electrophoretic pattern (**Figure 2**).

The presence of the same strain in four wineries could be due to the fact that the villages where wineries G, H, and I are located are geographically very close to each other (**Figure 1**). In cellar F, a little further away, a part of the grapes came from the same area as the other three. In this fourth year of study, strong winds were reported by grape growers in September and October. These winds could be responsible for transporting yeasts from one area to another, which would explain the appearance

Winery	Years	Isolates of *S. cerevisiae* analyzed	Number of different clones	Clonal variability (%)	Index of diversity	Genotypes detected in previous years	
						Number	%
E	1	30	10	33.3	0.80	–	–
	2	30	14	46.7	0.93	2	6.6
	3	12	8	66.7	0.91	3	41.6
	4	20	8	40.0	0.87	1	5.0
	Total	**92**	**34**	**37.0**	**0.96**		
F	1	20	5	25.0	0.57	–	
	2	23	10	43.5	0.91	0	0
	3	19	9	47.4	0.85	0	0
	4	20	9	45.0	0.79	1	5.0
	Total	**82**	**32**	**39.0**	**0.95**		
G	1	30	13	43.3	0.93	–	–
	2	23	11	47.8	0.92	0	0
	3	21	12	57.1	0.92	5	57.3
	4	30	12	40.0	0.85	3	40.0
	Total	**104**	**40**	**38.5**	**0.95**		
H	1	20	9	45.0	0.88	–	–
	2	30	13	43.3	0.92	3	36.6
	3	20	11	55.0	0.88	4	50.0
	4	30	13	43.3	0.80	4	23.2
	Total	**100**	**35**	**35.0**	**0.95**		
I	1	–	–	–	–	–	–
	2	22	9	40.9	0.82	–	–
	3	30	20	66.7	0.97	3	13.3
	4	20	11	55.0	0.76	1	5.0
	Total	**72**	**36**	**50.0**	**0.96**		

Table 4.
Clonal diversity of S. cerevisiae yeasts in Rioja Alta wineries in 4 consecutive years (1–4).

of the same strain in the four wineries. This yeast could come from the grapes, since in none of the three previous vintages had this strain been isolated in any of the cellars. This strain dominated the beginning of the fermentations and was later replaced by other indigenous strains, since in no case was it isolated at the end of the fermentation processes. These results would concur with those of Schütz and Gafner [19], who consider that the population of yeasts can be considered dependent on the harvest and the vineyard, and also with Le Jeune et al. [25], who indicated that the populations involved in spontaneous alcoholic fermentation result from a balance between the *S. cerevisiae* strains present in the grape and in the cellar.

The data obtained would show the existence of extensive microbiota in each winery, which, together with the microflora that accompanies the grapes, will develop according to the characteristics of the harvest. The fermentations would be the result of the sequence of different yeasts, and despite having found a widely spread yeast in the sub-zone 1 year, it did not appear in any other winery in the

Years	1		2		3		4	
Winery	**Strain**	**%**	**Strain**	**%**	**Strain**	**%**	**Strain**	**%**
E	E-I_1	40.0	E-I_2	13.3	E-I_3	25.0	E-I_4	20.0
	E-II_1	20.0	E-IV_2	16.7	E-III_3	25.0	E-IV_4	25.0
			E-XI_2	13.3			E-V_4	20.0
	Total	**60.0**		**43.3**		**50.0**		**65.0**
F	F-III_1	65.0	F-II_2	13.0	F-I_3	25.0	F-I_4	45.0
			F-III_2	21.8	F-V_3	30.0		
			F-VIII_2	17.4				
	Total	**65.0**		**52.2**		**55.0**		**45.0**
G	G-I_1	20.0	G-V_2	20.0	G-I_3	14.3	G-I_4	33.3
			G-VII_2	15.0	G-IV_3	28.6	G-III_4	20.0
			G-VII_2	15.0			G-VI_4	17.4
	Total	**20.0**		**50.0**		**42.9**		**70.7**
H	H-II_1	25	H-III_2	20.0	H-I_3	30.0	H-I_4	43.3
	H-III_1	20	H-X_2	13.3	H-II_3	20.0	H-XI_4	13.3
	Total	55		**33.3**		**50.0**		**56.6**
I	–	–	I-I_2	18.2	I-XV_3	15.0	I-I_4	50.0
			I-IV_2	36.4				
	Total	–		**54.6**		**15.0**		**50.0**

Table 5.
Major S. cerevisiae strains in the fermentations of Rioja Alta in 4 consecutive years (1–4).

other three campaigns studied. So, we do not consider that in Rioja Alta, there are representative strains from either the wineries or the sub-zone.

2.2.3 Ecology of the yeast population during alcoholic fermentation in the Rioja Alavesa sub-zone

In this sub-zone, two wineries (J and K) were sampled, where wines were produced by the carbonic maceration method. As in the other two sub-zones, the non-*Saccharomyces* yeasts participated at the beginning of the alcoholic fermentation, and later they were replaced by *S. cerevisiae* strains, which were the ones that directed and carried out the process until the end. In total, 149 colonies of these species were isolated.

The level of clonal variability was very high in both cellars, especially in winery K, where it reached 60% (**Table 7**). It is noteworthy that the highest percentage of clonal variability in both wineries was obtained in the third year, as in the vinifications in the Rioja Alta, which confirms the importance of the characteristics of the harvest in the microbiota that drives the fermentations.

As in the rest of the sub-zones, the high number of different clones identified in each fermentation and the few common yeasts between campaigns (**Table 7**) provided high indexes of diversity, which reached values of 0.95 and 0.98 for cellars J and K, respectively. These indices were of the same order as those found in Rioja Alta and higher than those of Rioja Oriental. Likewise, the number of yeasts that participated in more than one campaign was small in the two wineries, with the

Winery E				Winery F				Winery G				Winery H				Winery I		
1	2	3	4	1	2	3	4	1	2	3	4	1	2	3	4	2	3	4
E-I$_1$	E-I$_2$	E-I$_3$	E-I$_4$	F-I$_1$	F-I$_2$	F-I$_3$	F-I4	G-I$_1$	G-I$_2$	G-I$_3$	G-I$_4$	H-I$_1$	H-I$_2$	H-I$_3$	H-I$_4$	I-I$_2$	I-I$_3$	I-I$_4$
E-II$_1$	E-II$_2$	E-II$_3$	E-II$_4$	F-II$_1$	F-II$_2$	F-II$_3$	F-II$_4$	G-II$_1$	G-II$_2$	G-II$_3$	G-II$_4$	H-II$_1$	H-II$_2$	H-II$_3$	H-II$_4$	I-II$_2$	I-II$_3$	II-II$_4$
E-III$_1$	E-III$_2$	E-III$_3$	E-III$_4$	F-III$_1$	F-III$_2$	F-III$_3$	F-III$_4$	G-III$_1$	G-III$_2$	G-III$_3$	G-III$_4$	H-III$_1$	H-III$_2$	H-III$_3$	H-III$_4$	I-III$_2$	I-III$_3$	II-III$_4$
E-IV$_1$	E-IV$_2$	E-IV$_3$	E-IV$_4$	F-IV$_1$	F-IV$_2$	F-IV$_3$	F-IV$_4$	G-IV$_1$	G-IV$_2$	G-IV$_3$	G-IV$_4$	H-IV$_1$	H-IV$_2$	H-IV$_3$	H-IV$_4$	I-IV$_2$	I-IV$_3$	II-IV$_4$
E-V$_1$	E-V$_2$	E-V$_3$	E-V$_4$	F-V$_1$	F-V$_2$	F-V$_3$	F-V$_4$	G-V$_1$	G-V$_2$	G-V$_3$	G-V$_4$	H-V$_1$	H-V$_2$	H-V$_3$	H-V$_4$	I-V$_2$	I-V$_3$	II-V$_4$
E-VI$_1$	E-VI$_2$	E-VI$_3$	E-VI$_4$		F-VI$_2$	F-VI$_3$	F-VI$_4$	G-VI$_1$	G-VI$_2$	G-VI$_3$	G-VI$_4$	H-VI$_1$	H-VI$_2$	H-VI$_3$	H-VI$_4$	I-VI$_2$	I-VI$_3$	II-VI$_4$
E-VII$_1$	E-VII$_2$	E-VII$_3$	E-VII$_4$		F-VII$_{98}$	F-VII$_3$	F-VII$_4$	G-VII$_1$	G-VII$_2$	G-VII$_3$	G-VII$_4$	H-VII$_1$	H-VII$_2$	H-VII$_3$	H-VII$_4$	I-VII$_2$	I-VII$_3$	II-VII$_4$
E-VIII$_1$	E-VIII$_2$	E-VIII$_3$	E-VIII$_4$		F-VIII$_2$	F-VIII$_3$	F-VIII$_4$	G-VIII$_1$	G-VIII$_2$	G-VIII$_3$	G-VIII$_4$	H-VIII$_1$	H-VIII$_2$	H-VIII$_3$	H-VIII$_4$	I-VIII$_2$	I-VIII$_3$	II-VIII$_4$
E-IX$_1$	E-IX$_2$				F-IX$_2$	F-IX$_3$	F-IX$_4$	G-IX$_1$	G-IX$_2$	G-IX$_3$	G-IX$_4$	H-IX$_1$	H-IX$_2$	H-IX$_3$	H-IX$_4$	I-IX$_2$	I-IX$_3$	II-IX$_4$
E-X$_1$	E-X$_2$				F-X$_2$			G-X$_1$	G-X$_2$	G-X$_3$	G-X$_4$		H-X$_2$	H-X$_3$	H-X$_4$		I-X$_3$	II-X$_4$
	E-XI$_2$							G-XI$_1$	G-XI$_2$	G-XI$_3$	G-XI$_4$		H-XI$_2$	H-XI$_3$	H-XI$_4$		I-XI$_3$	II-XI$_4$
	E-XII$_2$							G-XII$_1$		G-XII$_3$	G-XII$_4$		H-XII$_2$		H-XII$_4$		I-XII$_3$	
	E-XIII$_2$							G-XIII$_1$					H-XIII$_2$		H-XIII$_4$		I-XIII$_3$	
	E-XIV$_2$																I-XIV$_3$	
																	I-XV$_3$	
																	I-XVI$_3$	
																	I-XVII$_3$	
																	I-XVIII$_3$	
																	I-XIX$_3$	
																	I-XX$_3$	

Clones with box in gray in different columns show the same clone in different cellars

Table 6.
S. cerevisiae clones identified in each elaboration of Rioja Alta in 4 consecutive years (1–4).

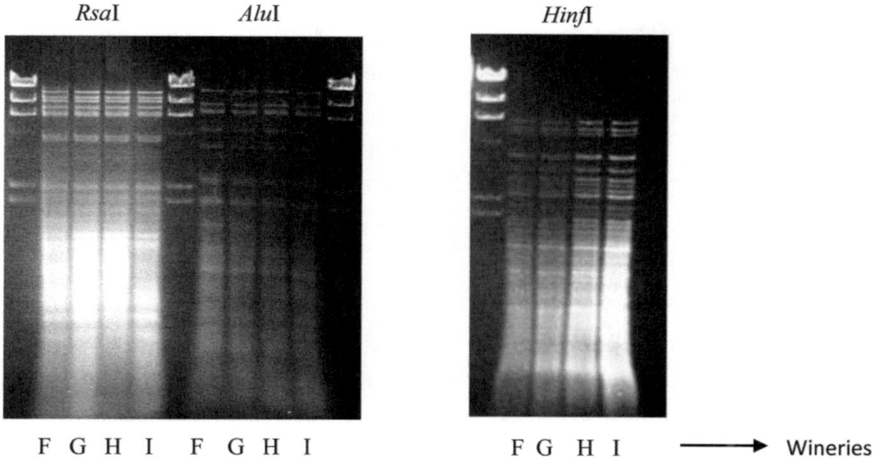

Figure 2.
Restriction patterns obtained with different restriction endonucleases, from the patterns I isolated in four Rioja Alta wineries in year 4.

Winery	Years	Isolates of *S. cerevisiae* analyzed	Number of different clones	Clonal variability (%)	Index of diversity	Genotypes detected in previous years	
						Number	%
J	1	29	15	51.7	0.93	–	–
	2	30	13	43.3	0.92	3	29.9
	3	20	14	70.0	0.94	5	50.0
	Total	**79**	**34**	**43.0**	**0.95**		
K	1	20	12	60.0	0.95	–	–
	2	30	14	46.7	0.92	0	0
	3	20	19	95.0	0.99	3	15.0
	Total	**70**	**42**	**60.0**	**0.98**		

Table 7.
Clonal diversity of S. cerevisiae yeasts in Rioja Alavesa wineries in 3 consecutive years (1–3).

exception of the third year of study in winery J. Therefore, the microbiota responsible for the fermentation was different in each harvest.

Regarding the population dynamics of the fermentation, it was observed that in the two wineries, the fermentations were carried out by different *S. cerevisiae* strains that followed each other during the different fermentative phases. In each vinification process, several major clones were detected, which represented at least 30% of the yeasts that carried out the fermentation (**Table 8**). The exception was the third year, in which most of the strains that were found in a fermentative phase were replaced by others in the following stage, particularly in winery K. This fact was also found in some of the vinifications in wineries G and I in Rioja Alta. All four wineries are over a hundred years old and conduct vinifications using the carbonic maceration method.

The comparison of the different strains found in the two cellars during the 3 years of study (**Table 9**) showed the existence of a single common clone, corresponding to the fermentations of the second year. Therefore, the data obtained would show the existence of extensive microbiota in each winery. The fermentations would be the result of the sequence of different populations of yeasts, as there are no representative strains of the sub-zone.

Years	1		2		3	
Winery	Strain	%	Strain	%	Strain	%
J	$J\text{-}IV_1$	20.7	$J\text{-}III_2$	13.3	$J\text{-}III_3$	25
	$J\text{-}I_1$	13.8	$J\text{-}IV_2$	13.3		
			$J\text{-}V_2$	13.3		
			$J\text{-}VI_2$	16.6		
	Total	**34.5**		**56.5**		**25**
K	$K\text{-}VI_1$	15.5	$K\text{-}III_2$	16.7	$K\text{-}II_3$	6.7
	$K\text{-}XI_1$	15.0	$K\text{-}IV_2$	16.7		
			$K\text{-}VI_2$	13.3		
	Total	**30.5**		**46.7**		**6.7**

Table 8.
Major S. cerevisiae strains in the fermentations of Rioja Alavesa in 3 consecutive years (1–3).

Winery J			Winery K		
1	2	3	1	2	3
$J\text{-}I_1$	**$J\text{-}I_2$**	$J\text{-}I_3$	$K\text{-}I_1$	$K\text{-}I_2$	$K\text{-}I_3$
$J\text{-}II_1$	$J\text{-}II_2$	$J\text{-}II_3$	$K\text{-}II_1$	**$K\text{-}II_2$**	$K\text{-}III_3$
$J\text{-}III_1$	$J\text{-}III_2$	$J\text{-}VI_3$	$K\text{-}III_1$	$K\text{-}III_2$	$K\text{-}IV_3$
$J\text{-}IV_1$	$J\text{-}IV_2$	$J\text{-}VII_3$	$K\text{-}IV_1$	$K\text{-}IV_2$	$K\text{-}V_3$
$J\text{-}V_1$	$J\text{-}VI_2$	$J\text{-}IX_3$	$K\text{-}V_1$	$K\text{-}V_2$	$K\text{-}VI_3$
$J\text{-}VI_1$	$J\text{-}VII_2$	$J\text{-}X_3$	$K\text{-}VI_1$	$K\text{-}VI_2$	$K\text{-}VIII_3$
$J\text{-}VII_1$	$J\text{-}VIII_2$	$J\text{-}XII_3$	$K\text{-}VII_1$	$K\text{-}VII_2$	$K\text{-}IX_3$
$J\text{-}VIII_1$	$J\text{-}IX_2$	$J\text{-}XIII_3$	$K\text{-}VIII_1$	$K\text{-}VIII_2$	$K\text{-}X_3$
$J\text{-}IX_1$	$J\text{-}X_2$	$J\text{-}XIV_3$	$K\text{-}IX_1$	$K\text{-}IX_2$	$K\text{-}XI_3$
$J\text{-}X_1$	$J\text{-}XIII_2$		$K\text{-}X_1$	$K\text{-}X_2$	$K\text{-}XII_3$
$J\text{-}XI_1$			$K\text{-}XI_1$	$K\text{-}XI_2$	$K\text{-}XIII_3$
$J\text{-}XII_1$			$K\text{-}XII_1$	$K\text{-}XII_2$	$K\text{-}XV_3$
$J\text{-}XIII_1$				$K\text{-}XIII_2$	$K\text{-}XVI_3$
$J\text{-}XIV_1$				$K\text{-}XIV_2$	$K\text{-}XVII_3$
$J\text{-}XV_1$					$K\text{-}XVIII_3$
					$K\text{-}XIX_3$
Clones In bold in different columns show the same clone in different cellars					

Table 9.
S. cerevisiae clones identified in each elaboration of Rioja Alavesa in 3 consecutive years (1–3).

3. Conclusions

As in other ecological studies of wine fermentations, spontaneous alcoholic fermentations in the Rioja qualified designation of origin are mainly conducted by yeasts of the *S. cerevisiae* species, and the non-*Saccharomyces* species have only been detected in the early stages. These fermentations have been carried out by different

S. cerevisiae strains that have appeared throughout the different stages of the process. Out of the 915 colonies of *S. cerevisiae* analyzed, 330 different clones have been identified, which means a very high clonal diversity.

Different agronomic and technological factors can influence in the diversity of the yeasts present in each vinification, such as the age of the winery, the winemaking system employed, and the climate conditions that prevailed during the ripening period of the grapes. The vinifications carried out in newly constructed wineries presented a lower clonal diversity than those which took place in older wineries. The clonal diversity was higher in vinifications conducted by carbonic maceration than in those carried out after crushing and destemming grapes. Unfavorable climatology during the vegetative period decreased the number of strains that participated in the fermentation.

There were very few common strains that participated in the fermentations carried out in successive years within the same winery, and hardly any common strains were detected in different wineries of the same sub-zone during the 4 years studied. All this allows us to affirm that there are no representative "typical" strains of the wineries, nor of the sub-zones and, therefore, of the Rioja designation of origin.

Conflict of interest

The authors declare that they have no conflict of interest.

Author details

Pilar Santamaría, Rosa López, Maria del Patrocinio Garijo, Rocío Escribano, Lucía González-Arenzana, Isabel López-Alfaro and Ana Rosa Gutiérrez*
ICVV, Institute of Vine and Wine Sciences, Government of La Rioja, University of La Rioja, CSIC, Logroño, La Rioja, Spain

*Address all correspondence to: ana-rosa.gutierrez@unirioja.es

References

[1] Varela C, Siebert T, Cozzolino D, Rose L, McLean H, Henschke PA. Discovering a chemical basis for differentiating wines made by fermentation with wild indigenous and inoculated yeasts: Role of yeasts volatile compounds. Australian Journal of Grape and Wine Research. 2009;**15**:238-248. DOI: 10/1111/j.1755-0238.2009.00054.x

[2] Santamaría P, López R, López E, Garijo P, Gutiérrez AR. Permanence of yeast inocula in the winery ecosystem and presence in spontaneous fermentations. European Food Research and Technology. 2008;**227**(5):1563-1567. DOI: 10.1007/s00217-008-0855-5

[3] Gutiérrez AR, Santamaría P, Epifanio S, Garijo P, López R. Ecology of spontaneous fermentation in one winery during five consecutive years. Letters in Applied Microbiology. 1999;**29**:411-415. DOI: 10.1046/j.1472-765X.1999.00657.x

[4] Vezinhet F, Hallet JN, Valade M, Poulard A. Ecological survey of wine strains by molecular methods of identification. American Journal of Enology and Viticulture. 1992;**43**:83-86

[5] Blanco P, Ramilo A, Cerdeira M, Orriols I. Genetic diversity of wine *Saccharomyces cerevisiae* strains in an experimental winery from Galicia (NW Spain). Antonie Van Leeuwenhoek. 2006;**89**:351-357. DOI: 10.1007/s10482-005-9038-6

[6] Torija M, Rozès N, Poblet M, Guillamón JM, Mas A. Yeast population dynamics in spontaneous fermentations: Comparison between two different wine-producing areas over a period of three years. Antonie Van Leeuwenhoek. 2001;**79**:345-352. DOI: 10.1023/A:1012027718701

[7] Knight S, Klaere S, Fedrizzi B, Goddard MR. Regional microbial signatures positively correlate with differential wine phenotypes: Evidence for a microbial aspect to terroir. Scientific Reports. 2015;**5**:14233. DOI: 10.1038/srep14233

[8] Capece A, Granchi L, Guerrini S, Mangani S, Romaniello R, Vincenzini M, et al. Diversity of *Saccharomyces cerevisiae* strains isolated from two italian wine-producing regions. Frontiers in Microbiology. 2016;**7**:1-11. DOI: 10/3389/fmicb.2016.01018

[9] Tristezza M, Fantastico L, Vetrano C, Bleve G, Corallo D, Grieco F, et al. Molecular and technological characterization of *Saccharomyces cerevisiae* strains isolated from natural fermentation of Susumaniello grape must in Apulia, southern Italy. International Journal of Microbiology. 2014:897428. DOI: 10.1155/2014/897428

[10] Schvarczová E, Stefaniková J, Jankura E, Kolek E. Selection of autochthonous *Saccharomyces cerevisiae* strains for production of typical Pinot gris wines. Journal of Food and Nutrition Research. 2017;**56**(4):389-397

[11] Orlic S, Vojvoda T, Babic KH, Arroyo-López FN, Jeromel A, Kocina B, et al. Diversity and oenological characterization of indigenous *Saccharomyces cerevisiae* associated with Zilavka grapes. World Journal of Microbiology and Biotechnology. 2010;**26**:1483-1489. DOI: 10.1007/s11274-010-0323-9

[12] Querol A, Barrio E. A rapid and simple method for the preparation of yeast mitocondrial DNA. Nucleic Acids Research. 1990;**18**:1657

[13] López I. Detección y control por técnicas de la Biología Molecular de bacterias lácticas autóctonas responsables de la fermentación maloláctica en vinos de la D.O.Ca. Rioja [thesis]. University of La Rioja; 2004

[14] Clemente-Jiménez JM, Mingorance-Cazorla L, Martínez-Rodríguez S, Las Heras-Vázquez FJ, Rodríguez-Vico F. Molecular characterization and oenological properties of wine yeasts isolated during spontaneous fermentation of six varieties of grape must. Food Microbiology. 2004;**21**:149-155. DOI: 10.1016S0740-0020(03)00063-7

[15] Santamaría P, Garijo P, López R, Tenorio C, Gutiérrez AR. Analysis of yeast population during spontaneous alcoholic fermentation: Effect of the age of the cellar and the practice of inoculation. International Journal of Food Microbiology. 2005;**103**:49-56. DOI: 10.1016/j.ijfoodmicro.2004.11.024

[16] Jemec KP, Cadez N, Zagorc T, Vuvic V. Yeast population dynamics in five spontaneous fermentations of Malvasia musts. Food Microbiology. 2001;**18**(3):247-259. DOI: 10.1016/fmic.2001.0396

[17] Sangorrín M, Zajonskovsky I, van Broock M, Caballero A. The use of killer biotyping in an ecological survey of yeast in an old Patagonian winery. World Journal of Microbiology and Biotechnology. 2002;**18**(2):115-120. DOI: 10.1023/A:101441722 2890

[18] Versavaud A, Courcoux P, Roulland C, Dulau L, Hallet JN. Genetic diversity and geographical distribution of wild *Saccharomyces cerevisiae* strains from the wine-producing area of Charentes, France. Applied Environmental Microbiology. 1995;**61**(10):3521-3529

[19] Schütz M, Gafner J. Dynamics of the yeast strain population during spontaneous alcoholic fermentation determined by CHEF gel electrophoresis. Letters in Applied Microbiology. 1994;**19**(4):253-257. DOI: 10.1111/j.1472-765X.1994.tb00957.x

[20] Granchi L, Ganucci D, Viti C, Giovannetti L, Vincenzini M. *Saccharomyces cerevisiae* biodiversity in spontaneous commercial fermentations of grape musts with adequate and inadequate assimilable-nitrogen content. Letters in Applied Microbiology. 2003;**36**:54-58. DOI: 10.1046/j.1472-765X.2003.01263.x

[21] Garijo P. Estudio del aire como vía de diseminación de microorganismos enológicos [thesis]. University of La Rioja; 2013

[22] Van der Westhuizen TJ, Augustyn OPH, Pretorius IS. Geographical distribution of indigenous *Saccharomyces cerevisiae* strains isolated from vineyard in the Coastal Regions of the Western Cape in South Africa. South African Journal of Enology and Viticulture. 2000;**21**(1):3-9. DOI: 10.21548/21-1-2179

[23] Fleet GH. Yeast interactions and wine flavour. International Journal of Food Microbiology. 2003;**86**(1-2):11-22. DOI: 10.1016/S0168-1605(03)00245-9

[24] Ocón E, Garijo P, López I, Santamaría P, Gutiérrez AR, Tenorio C, et al. Quantitative and qualitative analysis of non-*Saccharomyces* yeasts in spontaneous alcoholic fermentations. European Food Research and Technology. 2010;**230**(6):885-891. DOI: 10.1007/s00217-010-1233-7

[25] Le Jeune C, Claude E, Demuyter C, Lollier M. Evolution of the population of *Saccharomyces cerevisiae* from grape to wine in a spontaneous fermentation. Food Microbiology. 2006;**23**:709-716. DOI: 10.1016/j.fm.2006.02.007

Chapter 8

Yeast Strain Optimization for Enological Applications

David José Moreira Ferreira and Jessica Noble

Abstract

In the world of winemaking, tradition and innovation have always been side by side, on the one hand a culture of several centuries and on the other the need to constantly improve and answer new challenges. Consumers' preferences, climate changes, and fermentation efficiency are some of the modern questions that winemakers have to consider. Yeast, at the center of the fermentation, has revealed itself as the perfect platform to answer many of these challenges. By understanding the metabolism and the genetic basis that modulate specific phenotypes of yeast during fermentation, an era of yeast optimization has surfaced in the last decades and pushed research even further. In this chapter we will focus the attention on two of the most successful techniques to that end, quantitative trait locus (QTL) and evolutionary engineering. QTL relies on a highly precise identification of the genome regions that control a phenotype of interest. The transfer of these regions to selected wine yeasts is then possible by a technique called backcrossing. Evolutionary engineering induces the yeast itself to modify its genetic background to adapt to a selective pressure and improve its fitness. The right choice of pressure leads to the improvement of its performances in enological conditions.

Keywords: winemaking, yeast optimization, QTL, backcrossing, evolutionary engineering

1. Introduction

Nowadays, most of the enological fermentations are performed using selected wine yeast strains. Historically, and to some extent to this day, the selected wine yeasts have been found exploring the microbial flora present on the grapes, in the cellar, or in the vineyards. Next, a long process of characterization and selection is conducted in order to identify the yeast strain corresponding to a specific demand [1–5]. Since many years, the knowledge about wine yeasts has exponentially increased thanks to numerous scientific studies as well as the immense gain in the understanding of their metabolism. Consumption and requirements in nutrients (sugars, lipids, nitrogen, sulfur, vitamins, minerals), synthesis and production of biomass and metabolites (ethanol, glycerol, acids, alcohols, esters, sulfur compounds), resistance to stresses, and deficiencies have been well characterized. At the same time, the market trend and consumers' preference evolution results in a growing demand for new wine yeast strains combining different properties of interest or adapted to specific winemaking conditions and to global climate change. Consequently, meeting those steadily increasing requirements started to be a challenge. It becomes harder to find a strain combining all the properties

IntechOpen

of interest [3, 6]. The development of wine yeast strain optimization strategies provided then a possible way out [7–9].

Optimization strategies of wine yeasts can be divided in two categories: the first exploits the existing diversity inside the *Saccharomyces cerevisiae* genus that has been recently demonstrated to be immense [10, 11] and the second allows to go further creating new phenotypes.

The exploitation of the natural diversity can be done by breeding. Breeding strategies of wine yeasts to combine properties of the parental strains have been implemented for many years [12, 13]. This can be done by sexual breeding or protoplast fusion for strains impaired in sporulation or mating. However, breeding without prior knowledge of the genetic basis of the properties of interest may lead to aleatory results, as most of the phenotypes are governed by complex regulations and often involve interactions. Additionally, a major drawback is that wine yeast strains are particularly difficult to mate due to the low spore viability and homothallism typical of this group [7]. Nowadays, more rational and powerful methods supported by the rise of the "omics" (genomic, transcriptomic, metabolomic studies, etc.), such as directed hybridization, can be carried out. Directed hybridization takes advantage of the knowledge of gene(s) of interest to follow and direct their transfer from one wine yeast strain to another [14].

On the other hand, going beyond the existing phenotypes can be performed by inducing new mutations. Mutagenesis by chemical or physical ways can be used to induce aleatory mutations inside the genome of wine yeasts [9]. Although simple to perform, this approach delivers quite random results with potential deleterious effects and requires massive clone screening, which can be unpractical depending on the phenotype being tested [15].

Evolutionary engineering also allows to go further the common phenotypes by continuously applying specific stressful conditions to a population of yeasts and selecting natural mutants presenting a higher fitness under those conditions [16–19]. This strategy is particularly powerful when the genetic bases of the phenotypes are not known.

Finally, the GMO strategy, also called genetic engineering, can be considered. This strategy exploits a set of molecular tools in order to manipulate the genetic characteristics of yeasts. In comparison to conventional improvement strategies that can transfer a large number of both specific and nonspecific genes to the recipient or may be responsible for some nontargeted variations in the genome, genetic engineering only transfers a small block of desired genes. Thus, this strategy is less time-consuming and yields more reliable products. However, the use of GMOs in food is strictly regulated in the EU and requires a heavy declaration, traceability procedures, and mandatory labeling even if no trace of the GMO can be found in the final product [20]. Although in some countries, the use of GMOs in food applications can be more easily allowed, the lack of background and studies to assess their impact on food safety, public health, and environment led to the creation of strict regulations and legislation during the 1990s. Several European regulations (e.g., EC258/97, EC1829/2003, 65/2004) have been issued to regulate every aspect of GMO use in the EU [21]. In enology, different strains were genetically modified, for instance, to obtain better aromatic profiles [22–24] or to overproduce glycerol and reduce ethanol yield [25, 26]. However their use in the EU and other countries is far from simple: long and costly administrative procedures, international and local regulations, consumer distrust, and the desire to keep the process within traditional boundaries point to a future in the wine industry without GMO [15]. More recently, the development of a marker-free, high-throughput, and multiplexed genome editing approach, the clustered regularly interspaced short palindromic repeats and CRISPR-associated protein 9 (Cas9) (CRISPR-Cas9) immune system, an easier and traceless method of genome editing, has also been classified by the European Court of Justice as a GMO and is subject to the same controls [27]. It becomes clear that this kind of approaches cannot be reasonably developed for wine applications in the current context.

In this chapter, we will develop more in depth the two most widely used approaches for wine yeast improvement, directed hybridization through quantitative trait locus (QTL) mapping combined with backcrossing cycles and evolutionary engineering. These approaches currently provide very efficient, GMO-free strategies that have been greatly contributing for yeast optimization, particularly in winemaking.

2. QTL mapping and backcrossing

2.1 Identification of the molecular basis of technological properties: QTL mapping

Numerous properties and phenotypes of wine yeasts are quantitative traits. These present continuous variations among individuals, in opposition to qualitative ones showing discrete variations. Those quantitative traits are due to complex genetic mechanisms, often linked to interactions between several loci. It is possible to identify the genetic determinants of such phenotypes using a QTL mapping. A quantitative trait locus is defined as a region of the genome, often scattered, associated with the phenotypic variation of a quantitative trait. The first study using the principles of QTL was done almost 100 years ago [28]. At the time, Sax [28] performed a genetic analysis correlating the size of beans with the color of pigmentation. Shortly after, the concept was applied to agriculture and since then has been widely used in many different organisms such as *Drosophila melanogaster* [29] and *Arabidopsis thaliana* [30], in crops [31], and in yeast [32, 33].

Thanks to those approaches, chromosomic regions, genes, or even mutations, responsible for several wine yeast properties, have been deciphered. These include traits like acetic acid production, sporulation, ethanol tolerance, growth at high temperature, flocculation, wine aroma production, amino acid consumption, nitrogen requirement, fermentative performances, and sulfur compound production [34–45]. These studies have shown some phenotypes to be particularly complex.

The QTL mapping method is divided into three steps. First, a recombinant population is constituted, second, this population is then phenotyped and genotyped, and, lastly, a statistical analysis to link the regions of the genome to the phenotypes is performed (**Figure 1**).

The recombinant population is usually constituted from a hybrid obtained by crossing two parental strains, selected based on their phenotypic diversity. We can note that it is also possible to start directly with a highly heterozygous diploid parental strain, e.g., selected after evolutionary engineering. The hybrid is induced to sporulate to generate a population of meiotic segregants. The meiotic segregants passed through recombination so that each segregant possesses a random distribution of the alleles of the two parents. As the recombination rate is a crucial point in the accuracy of the final mapping of the QTL, it is also possible to generate an F2 segregant population to increase the allelic mixing. In that case, the initial meiotic segregant population, F1, is submitted to random crossing before a second sporulation round to constitute the F2 haploid segregant population [42, 43].

The phenotyping of the segregant population is a crucial step that can be limiting in the QTL approach. Each segregant has to be phenotyped individually for the trait of interest. The higher the number of segregants that are phenotyped, the

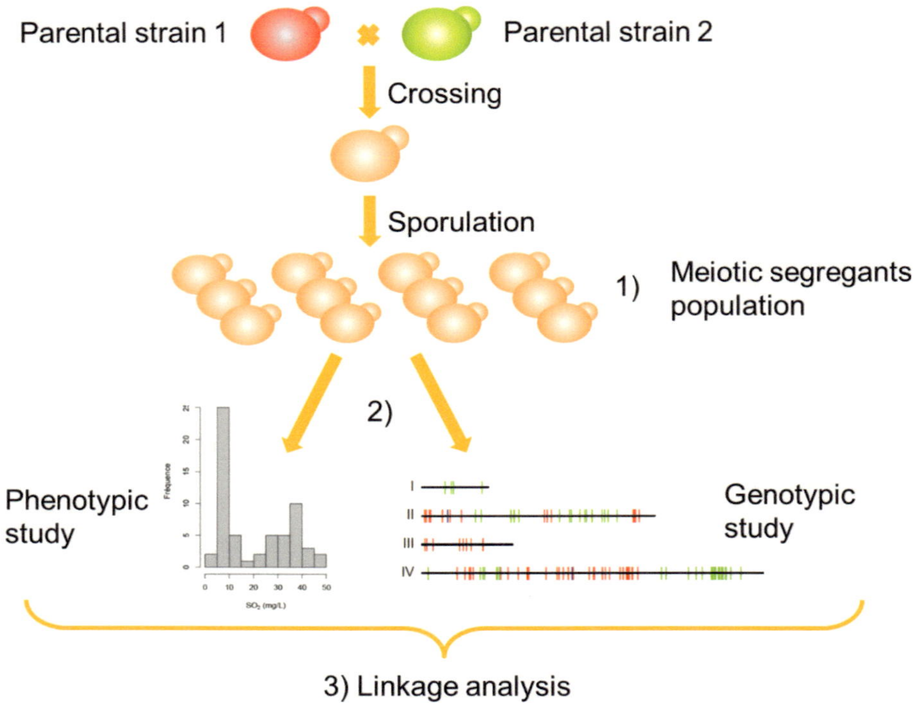

Figure 1.
Schematic representation of the QTL mapping strategy divided in three steps: (1) constitution of the meiotic segregants population, (2) phenotypic and genotypic study, and finally (3) linkage analysis.

better the precision will be in the mapping of the QTL. Some phenotypes can be measured on plates, such as ethanol tolerance; however, numerous phenotypes of interest for wine yeasts require to perform enological fermentations.

The next step is to create a genetic map constituted by molecular markers differentiating the two parental strains. The aim is to obtain the most homogeneous and dense distribution of the markers throughout the genome. The better the coverage is, the more accurate and precise the QTL mapping will be. Then, genotyping of the segregants attributes a parental origin to each marker. Nowadays, the development of sequencing approaches and the reduction of their costs allow to genotype the strains using whole-genome re-sequencing implementing next-generation sequencing technology [46]. This is done for parental strains as well as for the selected segregants.

Different approaches of QTL mapping can be carried out, using individual genotyping or bulk segregant analysis (BSA). For the individual genotyping, each segregant is genotyped, and a linkage analysis identifies the regions that are more likely to be involved in the phenotype. The powerful method of interval mapping is often used [47]. This method is based on the distances between markers. For each marker, the probability that this locus is a true QTL is calculated by a model. A significant threshold can be established by permutation testing. This approach is based on the hypothesis of a single QTL, but it is possible to identify other QTLs by a composite interval mapping that will iteratively scan the genome and add known QTL to the regression model as QTLs are identified.

In the BSA approach, the segregants that present the same phenotype are pooled together [48, 49]. The aim is to identify the regions of the genome that are common to all the segregants presenting the same phenotype. The allelic frequency between the two bulks or with the control is studied and allows detecting gene variants involved in the phenotype.

A powerful extension of BSA has also been developed: extreme QTL (X-QTL) mapping [50]. This approach is based on the generation of segregating populations of very large size. Those populations composed of large numbers of progeny with extreme trait values can be constituted using selection for drug or stress resistance or by cell sorting. Pooled allelic frequencies are then determined.

The genomic regions identified by QTL mapping strategies can vary from few to 1000 kilobases. Inside those regions, the sequences of the genes are compared, and non-synonymous mutations between the parental strains are searched in the coding region and the promotor/terminator regions. A study of the function of the genes located in this region using databases allows identifying candidate genes. A functional validation of the candidate genes can then be performed. Allelic replacement and reciprocal hemizygosity analysis (RHA) are the common ways to validate the impact of an allele on the phenotype. Allelic replacement consists in deleting the candidate gene in a parental strain and replacing by the allele of the opposite parent. Hemizygotes are constructed using the hybrid of the parental strains and deleting only one copy of the gene. The obtained strains are tested for their phenotypes. Thanks to those approaches, genes, mutations, or even translocations have been validated for diverse wine yeast properties, such as lag phase duration, fermentation capacity under nitrogen starvation, and ester production [42, 51, 52].

2.2 Transferring properties of interest from one strain to another: backcrossing

Once markers or mutations have been identified thanks to a QTL mapping strategy, it is possible to manage their transfer from one wine yeast strain to another. Introgression, also called backcrossing, or selection assisted by molecular markers, consists in recursive hybridization between a strain possessing the allele of interest and a strain to improve (**Figure 2**).

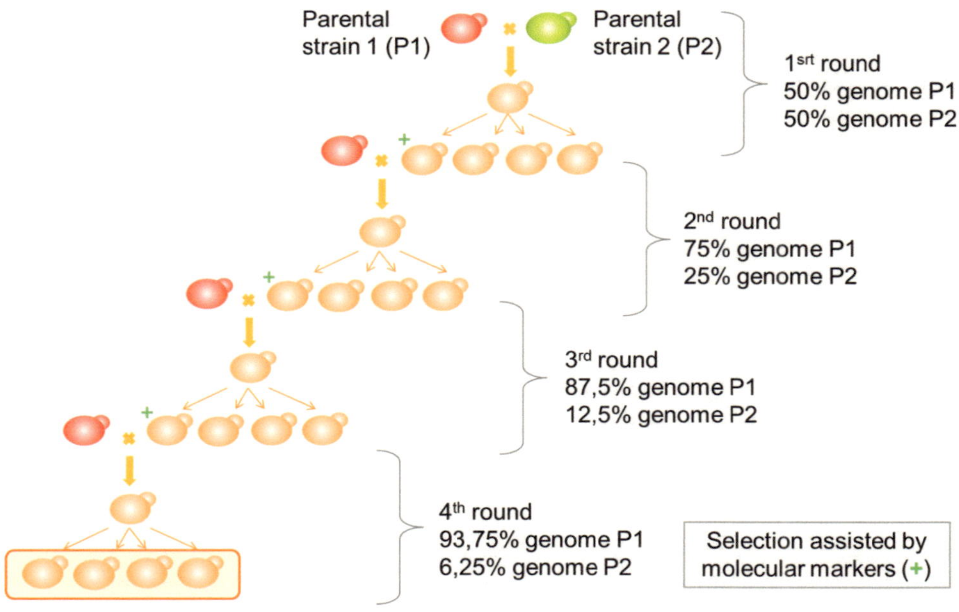

Figure 2.
Schematic representation of backcrossing cycles or recursive hybridization between the receptor strain (parental strain 1, in red) and the donor strain (parental strain 2, in green). The molecular markers (green cross) are followed at each step by PCR. The final strain (in orange) possesses a major part of its genome coming from the receptor strain and a small part transferred from the donor and containing the region of interest.

The first step is the selection of a "receptor" strain. This strain possesses a good genetic background and presents numerous properties of interest, except the one aimed to be enhanced. This strain will be crossed with a "donor" strain that possesses the property of interest. A first cross results in a hybrid possessing 50% of the genome of each parental strain. This hybrid is induced to sporulate, and a population of meiotic segregants is constituted. A segregant with the right marker or allele of interest is selected using a simple identification by PCR. This segregant is crossed again with the receptor strain. The second hybrid possesses 75% of the genome of the receptor strain and 25% of the donor strain. Several cycles of breeding/sporulation are performed to regenerate the genome of the receptor strain and to recover its good properties. Generally, four cycles are sufficient and lead to a strain possessing more than 93% of the genome of the receptor strain and less than 7% from the donor, including the genes of interest.

This approach has been implemented in plants for many years [53, 54]. Its application to the improvement of wine yeasts has started more than 10 years ago [14] and since then it has been applied to generate numerous wine yeast strains The production of H_2S, lag phase, and POF character [14], volatile thiol release [55], or SO_2, H_2S, and acetaldehyde [56] have been improved using this approach.

3. Evolutionary engineering

3.1 Evolutionary engineering as a simulation of nature

Evolution is one of the most important processes present in nature to which all living beings are submitted. After traveling around the world collecting much data, Charles Darwin published the book *On the Origin of Species* on the mid-nineteenth century explaining his theory of evolution based on natural selection. To this day, aside from minor revisions, this theory is the one broadly accepted within the scientific community to best explain evolution. In short, the theory bases itself on the fact that genetic variation occurs among individuals of the same species in a given population leading to phenotypic variations as in morphology, physiology, and behavior traits. In each specific environment, different traits confer different survival rates and different reproduction chances. Upon natural selection, advantageous traits can be passed from generation to generation in a stable heritability. By combining these principles, the progeny of the fittest (best adapted) in a given environment will gradually replace the members of a population and take over. In the case of adverse conditions or sudden environment change, this is one of the main mechanisms on which species rely to keep thriving and avoid extinction.

Evolutionary engineering, also designated as adaptive, directed, or experimental evolution, is an approach where these very same principles of evolution are applied to a selected population in a known and controlled environment [57]. The main difference from nature's evolution is the orientation of the natural selection toward specific selective pressures, the ones which best represent a given environment where we look for an evolution. Over time, individuals initially not optimally adapted may evolve and gain a higher fitness with the accumulation of natural and positive mutations for that specific environment. As the fittest, these individuals will be able to better utilize the available resources, grow faster, and multiply faster in higher number. The natural course is then for their progeny to gradually become dominant within the population, leading to the evolved individuals initially sought. Multiple evolutionary engineering experiments have been performed with different organisms such as *Drosophila* [58], domestic mouse [59, 60], the unicellular algae *Chlamydomonas* [61], *Pseudomonas fluorescens* [62], *Escherichia coli* [63], and *S. cerevisiae* [64] demonstrating how the evolution principles can be successfully applied to different living beings and contexts.

3.2 Evolutionary engineering applied to yeast: why and how?

Yeast has been the focus of many evolutionary studies due to its potential to generate academic knowledge as well as its broad range of applications. Its success in the evolutionary context is related to different advantages such as the high number of individuals that can be obtained within the same population, the easiness of maintenance/growth of populations with relatively low costs, and a fast generation time. Additionally, evolutionary engineering is a non-GMO technique. As explained before, no direct human manipulation occurs since the yeast itself improves and evolves its genetic background with natural mutations. Thus, evolved yeasts are perfectly safe and can be used in any food and/or beverage context without restrictions. Finally, another positive feature is the simplicity and empirical way on how the evolutionary engineering can be performed [21]. Contrary to other approaches, no genetic characterization or deep knowledge about the selected yeast is required. Nonetheless, when planning a yeast evolutionary engineering, key parameters need to be defined.

3.2.1 Selective pressures

During an evolutionary experiment, the mutations and consequently the diversity generated are completely random and cannot be predicted or controlled [65]. Therefore, it is crucial to identify the selective pressures that will best conduct the selection of positive mutations toward the desired phenotype. Once they are clearly identified, a proper initial experimental characterization should be performed to identify the intensity that these selective pressures should have. If too low there will be no selection, and if too high, yeast will struggle to continue in culture. Additionally, if a further industrial application is predicted, evolutionary engineering should be performed in conditions as close as possible to the actual conditions in which the yeast will later perform. By doing so, yeasts will not only face the selective pressures chosen to drive evolution but also all the other constraints naturally present [66]. For instance, in wine fermentation conditions, yeasts need to cope with stress factors as diverse as low pH, nutrient deprivation, ethanol, osmotic, and oxidative stress which are commonly present [67]. Therefore, it is preferable to use natural or synthetic must as a media that closely mimics realistic conditions while allowing the control and modification of specific parameters [68].

3.2.2 Strain choice

Depending on the final objective, the choice of a yeast strain can vary. Different laboratory strains have been used in evolutionary approaches with the main goal of generating academic knowledge. On the opposite side of the spectrum, industrial yeast strains have also attracted major interest due to the possibility of improving their efficiency and resistance [9, 66, 69]. Ploidy can also influence strain choice. Haploid strains have the advantage of evolving faster, making it easier to later on identify the mutations that lead to the evolved phenotype [70]. However, they are more sensitive to deleterious mutations that could easily become lethal, whereas diploid strains, as most wine yeast strains, have an increased ability to buffer such mutations. This way diploid strains tend to be more stable and robust when submitted to evolutionary engineering.

3.2.3 Cultivation

Microbial evolutionary engineering approaches are typically done in one of the two ways: serial cultivation (batch) and continuous cultivation (chemostat). Both are equally valid, and the choice will depend on the experimental conditions and objectives. With serial cultivation, the principle is to aliquot the culture into a new fresh medium at regular intervals (**Figure 3**). This is often used to select for microorganisms with shorter lag phase or higher growth rate, but certain regimes might also allow the selection for higher biomass formation or a better survival after nutrient depletion [18, 19, 69, 71]. Due to manipulation easiness and economic maintenance, this method allows several parallel cultures, often performed in shake flasks. On the downside, batch cultures are prone to some uncontrolled parameters and fluctuations in population density, growth rate, or dissolved oxygen [57]. In continuous cultivations bioreactor vessels are typically used. Here, all experimental parameters such as medium influx rate, temperature, oxygenation, and pH are continuously monitored leading to constant growth rates and population densities (**Figure 3**). Continuous cultures usually favor selection for higher substrate affinity [69]. The major disadvantages are the much higher costs and limitation in parallel experiments, depending on how many chemostats are available [57].

Independent of the cultivation method used, running simultaneous evolutionary engineering approaches of the same condition is advised. Woods et al. [72], using *Escherichia coli* in 12 identical and parallel evolutionary engineering experiments, showed that different random mutations can be fixed in different populations. As a consequence, the final outcome of each evolutionary process can vary. Identical phenotypes can be obtained with equivalent or different mutations; however different phenotypes can also be obtained despite the same exact conditions. Having parallel experiments increases the chances of success.

3.2.4 Duration

How long an evolutionary engineering approach lasts is highly case dependent and somehow unpredictable due to the randomness of mutations. Rather than

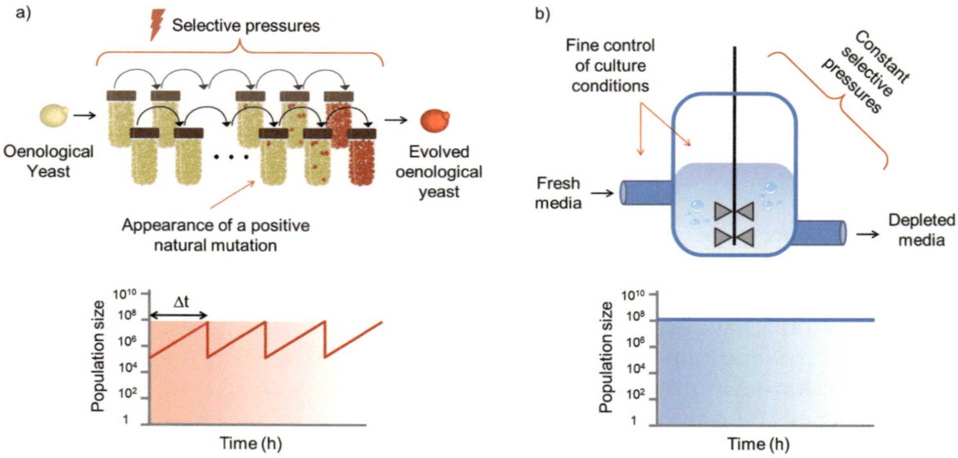

Figure 3.
Illustration of both directed evolution strategies: (a) serial transfer and (b) continuous culture. (a) Done with regular inoculations/transfers to fresh media which makes it similar to a batch. Once inoculated, populations increase over time and interact with medium with no intervention until new transfer. (b) A continuous nutrient feed that allows a constant population over time, permanently under the same conditions. Similar principles as a fed batch.

absolute time, duration is often measured by the number of generations. Natural mutations mostly occur when microorganisms divide, and since experimental conditions can modulate cell division from a few hours to several days, measuring the number of generations is a more accurate evolution timescale. If selective pressures are effective, in yeast a positive evolution is frequently observed between the 50th and the 200th generation. However, it might be the case that after many more generations, no evolution is observed. In this scenario, it might be the case that the approach setup, conditions, or parameters such as the selective pressures need to be reviewed. The best strategy to optimize duration is to regularly screen the evolving populations. By early detecting a positive evolution, the approach can either be stopped at the right moment or pursued if the evolved phenotype is still not satisfactory.

3.3 From the bench to the cellar

Once a positive evolution is detected in a wine evolutionary approach, a thorough work of validation needs to be done before an evolved wine yeast can actually be used in a cellar. The first step, often at laboratory scale, is to submit evolved yeasts to the evolutionary conditions in direct comparison with the parental strain, separately or in competition, to evaluate the relative improvement of the phenotype [73]. If acceptable, this comparison should also be validated in different realistic conditions where the yeast might perform. Typically, natural or synthetic musts are used in order to better reproduce enological fermentation conditions [68]. Aside from the characterization itself, this first screening allows for the search of possible trade-offs. A trade-off occurs when a particular phenotypic trait gets improved at the expense of one or more other phenotypic traits that get worsen. This is well illustrated in a study by Wenger et al. [74] who successfully evolved *S. cerevisiae* for a higher fitness in anaerobic glucose-limited media. Despite this, when in aerobic, carbon-rich environments, the evolved clones performed less well than their ancestor due to a trade-off. Similarly, yeast cells evolved for efficient galactose consumption which presented trade-offs when grown on glucose as a carbon source [75]. In winemaking context it is fundamental for yeast traits such as aroma production or fermentation efficiency to be kept at high standards and free of trade-offs. To note that in an evolutionary approach, the higher the number of generations occurred, the higher the chances of unrelated mutation accumulation. This reinforces the fact that the approach should be stopped as soon as a positive evolution is detected to avoid the accumulation of potential trade-offs.

Another fundamental test is to propagate and dry the yeast under industrial conditions, often the method used to produce commercialized wine yeast strains. Propagation and drying represent as the major sources of stress for yeast including oxidative, osmotic, and desiccation stresses which the evolved strains need to endure at least as well as the parental strains [21, 76–78]. The final stage of validation is the scale-up to pilot and industrial fermentation volumes, often performed by cellars with tanks of several hectoliters. If the evolved wine yeast strain performance is satisfactory both for the evolved phenotype and the remaining important traits, the evolutionary engineering process is then a success from an industrial point of view, and the yeast can be commercialized. From an academic point a view, new knowledge can also be generated by studying the new genetic profile in correlation to the evolved phenotype and how the evolved strains differ from the parental one. Approaches to conduct this characterization include genome microarray hybridization and direct DNA sequencing [75, 79, 80].

3.4 Successful evolutionary engineering in winemaking

To illustrate the potential of evolutionary engineering approaches in winemaking, few examples can be used where technical or field problematics were successfully solved by using this approach with validated evolved strains.

While using a long-term batch culturing on gluconate (a carbon source poorly assimilated by *S. cerevisiae*), Cadière et al. [19] evolved a commercial wine yeast strain obtaining interesting results. Evolved clones presented a carbon flux through the pentose phosphate pathway which increased by 6% when compared to the parental strains. This also resulted in a higher fermentation rate, lower levels of acetate production, and increased production of aroma compounds. As the process was carried out at a laboratory scale but in realistic (enological) conditions, the same phenotypic improvements were verified when the evolved strain was used in pilot-scale trials [81]. It was identified that the evolved strain produced higher levels of phenyl ethanol, isobutanol, isoamyl alcohol, ethyl acetate, isoamyl acetate, and ethyl esters [82].

Other authors were able to obtain a stable wine yeast strain with slightly enhanced glycerol production. By employing sulfite as a selective agent in an alkaline pH, Kutyna et al. [83] obtained evolved clones with an increase of 41% in glycerol production, which can have a benefic impact in wine organoleptic properties. To reduce the final ethanol content in wine, Tilloy et al. [18] submitted a wine strain to hyperosmotic stress for 200 generations, which yielded evolved clones that grew better under osmotic stress and glucose starvation and produced markedly more glycerol but also succinate and 2,3-butanediol. The approach was then complemented with an intra-strain breeding strategy that further increased the glycerol yield and reduced ethanol production in wine by up to 1.3% (v/v).

More recently, López-Malo et al. [80] performed an evolutionary process aiming for a higher performance for low-temperature fermentations (12°C). It was discovered that inositol and mannoprotein limitations were responsible for an evolution toward shorter fermentation times and higher final populations. After genome sequencing, it was discovered that an SNP in the gene *GAA1*, fundamental in inositol and mannoprotein synthesis, was at the basis of the improvement.

4. Conclusions

For a long time, innovation in yeast applications was mainly based on empirical observation and selection of natural isolates. In wine fermentation, despite the hundreds of wine yeast strains well characterized and commercially available, this diversity started to become insufficient to effectively answer all modern problematics. Consumers' preferences (e.g., specific aromas), improvements in fermentation efficiency, or counterbalance climate change are examples of key challenges that winemakers currently face and to which they require rapid and viable solutions. The emergence in the last decades of the different techniques discussed in this chapter allowed major advances in that sense. QTL mapping/backcrossing and evolutionary engineering are particularly two techniques that excel in providing solutions to specific applied issues.

QTL mapping is relevant as most of the enological traits of interest are governed by multiple loci and present a continuous variation in a population. Thanks to recently growing genetic tools, the study of the genetic determinants is becoming easier, and QTL mapping can be performed using molecular markers or whole-genome sequencing. Once the alleles of interest are known, they can be transferred from one strain to another using introgression. This constitutes a powerful natural approach to combine traits of interest of two wine yeast strains and/or to improve a strain conferring it a new property. On the other hand, some phenotypes and traits of interest can be hard to improve due to their complex regulation by different loci in the genome. If QTL mapping can precisely identify their genetic basis, evolutionary engineering is a solid alternative for a direct improvement of a trait to which low or no knowledge might be available. Often performed in the industrial context itself, this approach can provide both applied and academic outcomes with a relative simple and cost-effective methodology. Using this technique, most of the wine yeast traits of interest can be improved which leaves the future of winemaking with an immense potential in terms of innovation.

By combining relatively simple principles with high precision in addressing the problematics at their basis, QTL mapping and evolutionary engineering offer high rates of success. This justifies their initial success within the academia. In combination with their non-GMO status, this was quickly transferred to application and industry such as winemaking. Despite the precision that these techniques already offer, it is very likely that in the coming years their efficiency will continue to increase, while their cost will be reduced. Sequencing and whole-genome sequencing are following this exact trend and becoming more and more current. Identifying mutations or DNA regions responsible for specific phenotypic traits will then be more accessible with even more accurate results. In addition to other techniques that may emerge in the meantime, this suggests a bright future for wine yeast optimization and a continuous progress in winemaking.

Acknowledgements

The authors thank Anne Ortiz-Julien and Jose-Maria Heras for their support in writing this manuscript.

Conflict of interest

The authors declare no conflict of interest.

Author details

David José Moreira Ferreira and Jessica Noble*
Lallemand SAS, Blagnac, France

*Address all correspondence to: jnoble@lallemand.com

References

[1] Degre R. Selection and commercial cultivation of wine yeast and bacteria. In: Fleet GH, editor. Wine Microbiology and Biotechnology. Harwood Academic: Reading; 1993. pp. 421-447

[2] Zagorc T, Maráz A, Cadez N, Jemec KP, Péter G, Resnik M, et al. Indigenous wine killer yeasts and their application as a starter culture in wine fermentation. Food Microbiology. 2001;**18**(4):441-451. DOI: 10.1006/fmic.2001.0422

[3] Mannazzu I, Clementi F, Ciani M. Strategies and criteria for the isolation and selection of autochthonous starters. In: Ciani M, editor. Biodiversity and Biotechnology of Wine Yeasts. Trivandrum, India: Research Signpost; 2002. pp. 19-35

[4] Lopes CA, Rodríguez ME, Sangorrín M, Querol A, Caballero AC. Patagonian wines: The selection of an indigenous yeast starter. Journal of Industrial Microbiology & Biotechnology. 2007;**34**(8):539-546. DOI: 10.1007/s10295-007-0227-3

[5] Capece A, Romaniello R, Siesto G, Pietrafesa R, Massari C, Poeta C, et al. Selection of indigenous *Saccharomyces cerevisiae* strains for Nero d'Avola wine and evaluation of selected starter implantation in pilot fermentation. International Journal of Food Microbiology. 2010;**144**(1):187-192. DOI: 10.1016/j.ijfoodmicro.2010.09.009

[6] Romano P, Fiore C, Paraggio M, Caruso M, Capece A. Function of yeast species and strains in wine flavour. International Journal of Food Microbiology. 2003;**86**:169-180

[7] Pretorius IS. Tailoring wine yeast for the new millennium: Novel approaches to the ancient art of winemaking. Yeast. 2000;**16**(8):675-729

[8] Pretorius IS, Bauer FF. Meeting the consumer challenge through genetically customized wine-yeast strains. Trends in Biotechnology. 2002;**20**:426-432

[9] Steensels J, Snoek T, Meersman E, Picca Nicolino M, Voordeckers K, Verstrepen KJ. Improving industrial yeast strains: Exploiting natural and artificial diversity. FEMS Microbiology Reviews. 2014;**38**(5):947-995. DOI: 10.1111/1574-6976.12073

[10] Liti G, Carter DM, Moses AM, et al. Population genomics of domestic and wild yeasts. Nature. 2009;**458**:337-341. DOI: 10.1038/nature07743

[11] Peter J, De Chiara M, Friedrich A, et al. Genome evolution across 1,011 *Saccharomyces cerevisiae* isolates. Nature. 2018;**556**(7701):339-344. DOI: 10.1038/s41586-018-0030-5

[12] Thornton RJ. New yeast strains from old—The application of genetics to wine yeasts. Food Technology in Australia. 1983;**35**:46-50

[13] Romano P, Soli MG, Suzzi G, Grazia L, Zambonelli C. Improvement of a wine *Saccharomyces cerevisiae* strain by a breeding program. Applied and Environmental Microbiology. 1985;**50**(4):1064-1067

[14] Marullo P, Yvert G, Bely M, Aigle M, Dubourdieu D. Efficient use of DNA molecular markers to construct industrial yeast strains. FEMS Yeast Research. 2007;**7**:1295

[15] Cebollero E, Gonzalez-Ramos D, Tabera L, et al. Transgenic wine yeast technology comes of age: Is it time for transgenic wine? Biotechnology Letters. 2007;**29**:191-200

[16] Sauer U. Evolutionary engineering of industrially important microbial

phenotypes. Advances in Biochemical Engineering/Biotechnology. 2001;**73**: 129-169

[17] Portnoy VA, Bezdan D, Zengler K. Adaptive laboratory evolution—Harnessing the power of biology for metabolic engineering. Current Opinion in Biotechnology. 2011;**22**:590-594

[18] Tilloy V, Ortiz-Julien A, Dequin S. Reduction of ethanol yield and improvement of glycerol formation by adaptive evolution of the wine yeast *Saccharomyces cerevisiae* under hyperosmotic conditions. Applied and Environmental Microbiology. 2014;**80**(8):2623. DOI: 10.1128/AEM.03710-13

[19] Cadière A, Ortiz-Julien A, Camarasa C, Dequin S. Evolutionary engineered *Saccharomyces cerevisiae* wine yeast strains with increased in vivo flux through the pentose phosphate pathway. Metabolic Engineering. 2011;**13**:263-271

[20] Schuller D, Casal M. The use of genetically modified *Saccharomyces cerevisiae* strains in the wine industry. Applied Microbiology and Biotechnology. 2005;**68**:292-304. DOI: 10.1007/s00253-005-1994-2

[21] Pérez-Torrado R, Querol A, Guillamón JM. Genetic improvement of non-GMO wine yeasts: Strategies, advantages and safety. Trends in Food Science and Technology. 2015;**45**:1-11

[22] Pérez-González JA, González R, Querol A, et al. Construction of a recombinant wine yeast strain expressing beta-(1,4)-endoglucanase and its use in microvinification processes. Applied and Environmental Microbiology. 1993;**59**:2801-2806

[23] Ganga MA, Piñaga F, Vallés S, et al. Aroma improving in microvinification processes by the use of a recombinant wine yeast strain expressing the *Aspergillus nidulans* xlnA gene. International Journal of Food Microbiology. 1999;**47**:171-178

[24] Manzanares P, Orejas M, Vicente Gil J, et al. Construction of a genetically modified wine yeast strain expressing the *Aspergillus aculueatus* rhaA gene, encoding an a-L-Rhamnosidase of Enological interest. Microbiology. 2003;**69**:7558-7562

[25] Michnick S, Roustan J, Remize F, et al. Modulation of glycerol and ethanol yields during alcoholic fermentation in *Saccharomyces cerevisiae* strains overexpressed or disrupted for GPD1 encoding glycerol 3-phosphate dehydrogenase. Yeast. 1997;**793**:783-793

[26] Remize F, Roustan JL, Sablayrolles JM, et al. Glycerol overproduction by engineered *Saccharomyces cerevisiae* wine yeast strains leads to substantial changes in by-product formation and to a stimulation of fermentation rate in stationary phase. Applied and Environmental Microbiology. 1999;**65**:143-149

[27] Callaway E. CRISPR plants now subject to tough GM laws in European Union. Nature. 2018;**560**(7716):16. DOI: 10.1038/d41586-018-05814-6

[28] Sax K. The Association of Size Differences with seed-coat pattern and pigmentation in *Phaseolus vulgaris*. Genetics. 1923;**8**:552-560

[29] Mackay TFC. The genetic architecture of quantitative traits: Lessons from *Drosophila*. Current Opinion in Genetics & Development. 2004;**14**:253-257

[30] Shindo C, Bernasconi G, Hardtke CS. Natural genetic variation in *Arabidopsis*: Tools, traits and prospects for evolutionary ecology. Annals of Botany. 2007;**99**:1043-1054

[31] Würschum T. Mapping QTL for agronomic traits in breeding populations. Theoretical and Applied Genetics. 2012;**125**:201-210

[32] Steinmetz LM, Sinha H, Richards D, Spiegelman JI, Oefner PJ, McCusker JH, et al. Dissecting the architecture of a quantitative trait locus in yeast. Nature. 2002;**416**(6878):326-330. DOI: 10.1038/416326a

[33] Brem RB, Yvert G, Clinton R, Kruglyak L. Genetic dissection of transcriptional regulation in budding yeast. Science. 2002;**296**(5568):752-755. DOI: 10.1126/science.1069516

[34] Ben-Ari G, Zenvirth D, Sherman A, David L, Klutstein M, Lavi U, et al. Four linked genes participate in controlling sporulation efficiency in budding yeast. PLoS Genetics. 2006;**2**(11):e195. DOI: 10.1371/journal.pgen.0020195

[35] Hu XH, Wang MH, Tan T, Li JR, Yang H, Leach L, et al. Genetic dissection of ethanol tolerance in the budding yeast *Saccharomyces cerevisiae*. Genetics. 2007;**175**(3):1479-1487. DOI: 10.1534/genetics.106.065292

[36] Marullo P, Bely M, Masneuf-Pomarede I, Aigle M, Dubourdieu D. Inheritable nature of enological quantitative traits is demonstrated by meiotic segregation of industrial wine yeast strains. FEMS Yeast Research. 2004;**4**(7):711-719

[37] Marullo P, Yvert G, Bely M, Masneuf-Pomare I, Durrens P, Dubourdieu D, et al. Single nucleotide resolution QTL mapping of an enological trait from derived wild *Saccharomyces cerevisiae* strains. FEMS Yeast Research. 2007;7:941-952

[38] Sinha H, Nicholson BP, Steinmetz LM, McCusker JH. Complex genetic interactions in a quantitative trait locus. PLoS Genetics. 2006;**2**(2):e13. DOI: 10.1371/journal.pgen.0020013

[39] Sinha H, David L, Pascon RC, Clauder-Münster S, Krishnakumar S, Nguyen M, et al. Sequential elimination of major-effect contributors identifies additional quantitative trait loci conditioning high-temperature growth in yeast. Genetics. 2008;**180**(3):1661-1670. DOI: 10.1534/genetics.108.092932

[40] Steyer D, Ambroset C, Brion C, Claudel P, Delobel P, Sanchez I, et al. QTL mapping of the production of wine aroma compounds by yeast. BMC Genomics. 2012;**13**(1):573. DOI: 10.1186/1471-2164-13-573

[41] Ambroset C, Petit M, Brion C, Sanchez I, Delobel P, Guérin C, et al. Deciphering the molecular basis of wine yeast fermentation traits using a combined genetic and genomic approach. G3. 2011;**1**:263-281. DOI: 10.1534/g3.111.000422

[42] Brice C, Sanchez I, Bigey F, Legras JL, Blondin B. A genetic approach of wine yeast fermentation capacity in nitrogen-starvation reveals the key role of nitrogen signaling. BMC Genomics. 2014;**15**:495. DOI: 10.1186/1471-2164-15-495

[43] Eder M, Sanchez I, Brice C, Camarasa C, Legras JL, Dequin S. QTL mapping of volatile compound production in *Saccharomyces cerevisiae* during alcoholic fermentation. BMC Genomics. 2018;**19**(1):166. DOI: 10.1186/s12864-018-4562-8

[44] Noble J, Sanchez I, Blondin B. Identification of new *Saccharomyces cerevisiae* variants of the MET2 and SKP2 genes controlling the sulfur assimilation pathway and the production of undesirable sulfur compounds during alcoholic fermentation. Microbial Cell Factories. 2015;**14**:68. DOI: 10.1186/s12934-015-0245-1

[45] Trindade de Carvalho BT, Holt S, Souffriau B, Lopes Brandão R,

Foulquié-Moreno MR, Thevelein JM. Identification of novel alleles conferring superior production of rose flavor phenylethyl acetate using polygenic analysis in yeast. Microbiology 2017;**8**(6):1-21. pii: e01173-17. DOI: 10.1128/mBio.01173-17

[46] Metzker ML. Sequencing technologies—The next generation. Nature Reviews Genetics. 2010;**11**:31-46. DOI: 10.1038/nrg2626

[47] Lander E, Botstein D. Mapping Mendelian factors underlying quantitative traits using RFLP linkage maps. Genetics. 1989;**121**:185-199

[48] Michelmore RW, Paran I, Kesseli RV. Identification of markers linked to disease-resistance genes by bulked segregant analysis: A rapid method to detect markers in specific genomic regions by using segregating populations. Proceedings of the National Academy of Sciences of the United States of America. 1991;**88**:9828-9832

[49] Pais TM, Foulquié-Moreno MR, Thevelein JM. QTL mapping by pooled-segregant whole-genome sequencing in yeast. Methods in Molecular Biology. 2014;**1152**:251-266. DOI: 10.1007/978-1-4939-0563-8_15

[50] Ehrenreich IM, Torabi N, Jia Y, Kent J, Martis S, Shapiro JA, et al. Dissection of genetically complex traits with extremely large pools of yeast segregants. Nature. 2010;**464**(7291):1039-1042. DOI: 10.1038/nature08923

[51] Zimmer A, Durand C, Loira N, Durrens P, Sherman DJ, et al. QTL dissection of lag phase in wine fermentation reveals a new translocation responsible for *Saccharomyces cerevisiae* adaptation to sulfite. PLoS One. 2014;**9**(1):e86298. DOI: 10.1371/journal.pone.0086298

[52] Holt S, Trindade de Carvalho B, Foulquié-Moreno MR, Thevelein JM. Polygenic analysis in absence of major effector ATF1 unveils novel components in yeast flavor ester biosynthesis. Microbiology. 2018;**9**(4):e01279-e01218. DOI: 10.1128/mBio.01279-18

[53] Tanksley SD, Nelson JC. Advanced backcross QTL analysis: A method for the simultaneous discovery and transfer of valuable QTLs from unadapted germplasm into elite breeding lines. Theoretical and Applied Genetics. 1996;**92**:191-203

[54] Hospital F, Charcosset A. Marker-assisted introgression of quantitative trait loci. Genetics. 1997;**147**:1469-1485

[55] Dufour M, Zimmer A, Thibon C, Marullo P. Enhancement of volatile thiol release of *Saccharomyces cerevisiae* strains using molecular breeding. Applied Microbiology and Biotechnology. 2013;**97**:5893-5905. DOI: 10.1007/s00253-013-4739-7

[56] Berlese-Noble J, Raynal C, Silvano A, Granes D, Bonnefond C, Ortiz-Julien A. New research, new yeast, new selection tool. Australian and New Zealand Grapegrower and Winemaker. 2014;**602**:48-53

[57] Dragosits M, Mattanovich D. Adaptive laboratory evolution—Principles and applications for biotechnology TL-12. Microbial Cell Factories. 2013;(12):VN-r:64

[58] Rose MR, Vu LN, Park SU, et al. Selection on stress resistance increases longevity in *Drosophila melanogaster*. Experimental Gerontology. 1992;**27**:241-250

[59] Dohm MR, Richardson CS, Garland T. Exercise physiology of wild and random-bred laboratory house mice and their reciprocal hybrids. The American Journal of Physiology. 1994;**267**:1098-1108

[60] Dumke CL, Rhodes JS, Garland T Jr, et al. Genetic selection of mice for high voluntary wheel running: Effect on skeletal muscle glucose uptake. Journal of Applied Physiology. 2001;**91**:1289-1297

[61] Colegrave N, Kaltz O, Bell G. The ecology and genetics of fitness in *Chlamydomonas*. Viii. The dynamics of adaptation to novel environments after a single episode of sex. Evolution. 2002;**56**:14-21

[62] Rainey PB, Travisano M. Adaptive radiation in a heterogeneous environment. Nature. 1998;**394**:69-72

[63] Lenski RE, Travisano M. Dynamics of adaptation and diversification: A 10,000-generation experiment with bacterial populations. Proceedings of the National Academy of Sciences. 1994;**91**:6808-6814

[64] Paquin C, Adams J. Frequency of fixation of adaptive mutations is higher in evolving diploid than haploid yeast populations. Nature. 1983;**302**:495-500

[65] Luria S, Delbrück M. Mutations of bacteria from virus sensitivity to virus resistance. Genetics. 1943;**28**:491-511

[66] Çakar ZP, Turanli-Yildiz B, Alkim C, et al. Evolutionary engineering of *Saccharomyces cerevisiae* for improved industrially important properties. FEMS Yeast Research. 2012;**12**:171-182

[67] Querol A, Fernández-Espinar MT, Del Olmo M, et al. Adaptive evolution of wine yeast. International Journal of Food Microbiology. 2003;**86**:3-10

[68] Bely M, Sablayrolles JM, Barre P. Automatic detection of assimilable nitrogen deficiencies during alcoholic fermentation in oenological conditions. Journal of Fermentation and Bioengineering. 1990;**70**:246-252

[69] Bachmann H, Pronk JT, Kleerebezem M, et al. Evolutionary engineering to enhance starter culture performance in food fermentations. Current Opinion in Biotechnology. 2015;**32**:1-7

[70] Gerstein AC, Cleathero LA, Mandegar MA, et al. Haploids adapt faster than diploids across a range of environments. Journal of Evolutionary Biology. 2011;**24**:531-540

[71] Rabbers I, van Heerden JH, Nordholt N, et al. Metabolism at evolutionary optimal states. Metabolites. 2015;**5**:311-343

[72] Woods R, Schneider D, Winkworth CL, et al. Tests of parallel molecular evolution in a long-term experiment with *Escherichia coli*. Proceedings of the National Academy of Sciences. 2006;**103**:9107-9112

[73] Elena SF, Lenski RE. Microbial genetics: Evolution experiments with microorganisms: The dynamics and genetic bases of adaptation. Nature Reviews. Genetics. 2003;**4**:457-469

[74] Wenger JW, Piotrowski J, Nagarajan S, et al. Hunger artists: Yeast adapted to carbon limitation show trade-offs under carbon sufficiency. PLoS Genetics. 2011;**7**:1-17. DOI: 10.1371/journal.pgen.1002202

[75] Hong K-K, Vongsangnak W, Vemuri GN, et al. Unravelling evolutionary strategies of yeast for improving galactose utilization through integrated systems level analysis. Proceedings of the National Academy of Sciences. 2011;**108**:12179-12184

[76] França MB, Panek AD, Eleutherio ECA. The role of cytoplasmic catalase in dehydration tolerance of *Saccharomyces cerevisiae*. Cell Stress & Chaperones. 2005;**10**:167-170

[77] Singh J, Kumar D, Ramakrishnan N, et al. Transcriptional response

of *Saccharomyces cerevisiae* to desiccation and rehydration. Society. 2005;**71**:8752-8763

[78] Gómez-Pastor R, Pérez-Torrado R, Cabiscol E, et al. Transcriptomic and proteomic insights of the wine yeast biomass propagation process. FEMS Yeast Research. 2010;**10**:870-884

[79] Dunham MJ, Badrane H, Ferea T, et al. Characteristic genome rearrangements in experimental evolution of *Saccharomyces cerevisiae*. Proceedings of the National Academy of Sciences of the United States of America. 2002;**99**:16144-16149

[80] López-Malo M, García-Rios E, Melgar B, et al. Evolutionary engineering of a wine yeast strain revealed a key role of inositol and mannoprotein metabolism during low-temperature fermentation. BMC Genomics. 2015;**16**:537

[81] Cadière A, Aguera E, Caillé S, et al. Pilot-scale evaluation the enological traits of a novel, aromatic wine yeast strain obtained by adaptive evolution. Food Microbiology. 2012;**32**:332-337

[82] Mouret JR, Cadiere A, Aguera E, et al. Dynamics and quantitative analysis of the synthesis of fermentative aromas by an evolved wine strain of *Saccharomyces cerevisiae*. Yeast. 2014;**32**:259-267

[83] Kutyna DR, Varela C, Stanley GA, et al. Adaptive evolution of *Saccharomyces cerevisiae* to generate strains with enhanced glycerol production. Applied Microbiology and Biotechnology. 2012;**93**:1175-1184

Chapter 9

Strategies to Improve the Freshness in Wines from Warm Areas

Antonio Morata, Iris Loira, Juan Manuel del Fresno, Carlos Escott, María Antonia Bañuelos, Wendu Tesfaye, Carmen González, Felipe Palomero and Jose Antonio Suárez Lepe

Abstract

Trends in wine consumption are continuously changing. The latest in style is fresh wine with moderate alcohol content, high acidity, and primary aromas reminiscent of grapes, whereas certain fermentative volatiles may also influence the freshness of the wine. In addition, the effects of climate change on the composition of the grapes (high sugar content and low acidity) are adverse for the quality of the wine, also considering the microbiological stability. Herein, different strategies aiming at improving wine freshness are presented, and their performance in winemaking is discussed: among them, the addition of organic acids able to inhibit malolactic fermentation such as fumaric acid; the use of acidifying yeasts for alcoholic fermentation, such as *Lachancea thermotolerans*; and the selection of non-*Saccharomyces* yeasts with β-glucosidase activity in order to release terpene glycosides present in the must.

Keywords: wine freshness, organic acids, *Lachancea thermotolerans*, high acidity, climate change

1. Wine freshness

Wine freshness is an unspecific concept which includes parameters concerning acidity, aroma, alcohol content, and even color. It is also strongly correlated with fruit maturity, but the grapes from warm areas frequently have excessive sugar content that produces high alcoholic degree (>13%v/v) and low acidity (pH > 3.8). Wines produced with these grapes are normally winey, with unpleasant taste, scarce aromaticity mainly supported by higher alcohols with low levels of fruity esters, and a lack of sourness being usually less appreciated by the consumers. Moreover, these wines have a complex management during production and storage, because the low acidity produces higher sensibility to microbial spoilage and also because of the oxidation due to the low contents of molecular and free SO_2. For a better management and preservation, these wines are frequently dosed with tartaric acid, thus favoring a more suitable management which counteracts both oxidative and spoilage processes but at the same time produces a typical excessive and over-perceived sourness.

2. Wine acidity

Types of acidity in wine: wine acidity is due to the organic acids from grapes, mainly tartaric, malic, and citric acids. There are also other acids that are formed during alcoholic and malolactic fermentations (e.g., acetic, fumaric, succinic, and lactic acids) [1]. Among the grape acids, the most stable and with higher repercussion in pH is the tartaric acid. Malic acid is metabolized by lactic acid bacteria (LAB) during malolactic fermentation (MLF), and its influence in pH is not too relevant. Moreover, potassium contents in soil affect the levels of tartaric acid in grape and must, forming potassium tartrates that are highly insoluble, especially in a polar condition. The precipitation of these salts, especially when ethanol level increases during the alcoholic fermentation, produces the reduction of tartaric acid contents with a subsequent pH augmentation.

Harvesting time is another strongly influential parameter; the sooner the grape is harvested, the higher the acidity. However, acidity decreases significantly when the collection is retarded beyond the normal harvesting conditions because the enologist looks out for the optimum skin phenolic ripeness and also a good seed maturity especially in red varieties. Some alternatives have been proposed to keep acidity using non-matured grapes; one interesting proposal is the use of unripe bunches coming from cluster thinning. These grapes are pressed obtaining a high-acidity must which later is cleaned of astringency and excessive vegetal taints by using adsorbents, such as activated charcoal or other products. The juice is mixed with the matured and well-balanced grape to both reduce the pH and improve the acidity [2].

3. Wine aroma: influence of both winemaking practices and biotechnologies in freshness

The lack of freshness in the aroma fraction is produced by a relative excess of higher alcohols regarding the fruity esters (especially acetate esters) and varietal aromatic compounds (terpenes, thiols, etc.). It makes the smell simple, warm, and flat. The approach to improve this shortcoming in wines is variable according to the type of wine. Wines made with terpenic varieties can be improved by physical techniques such as cryomacerations, to enhance the extraction of varietal aromatic compounds; however, significant differences in aroma cannot always be perceived when cold soak is used to make prefermentative macerations in red wines [3, 4]. Conversely, color extraction is usually increased when cold soak is used [4, 5]. On the other hand, the use of cold soak can influence the yeast populations that can be developed in wine. It has been observed that macerations at 14°C favor the development and growth of *Hanseniaspora uvarum* and *Candida zemplinina*, but when temperature is kept at 8°C, the predominant yeast specie is *Saccharomyces cerevisiae* (Sc) (**Figure 1a**) [6]. In addition, fermentation at low temperature, 15°C instead of 28°C, has also proven the formation of higher flowery aroma [7], thus enhancing the freshness. Finally, the optimization of harvesting time, delaying or alternatively advancing the time window to collect the grapes, can help to optimize the concentration of aromatic compounds.

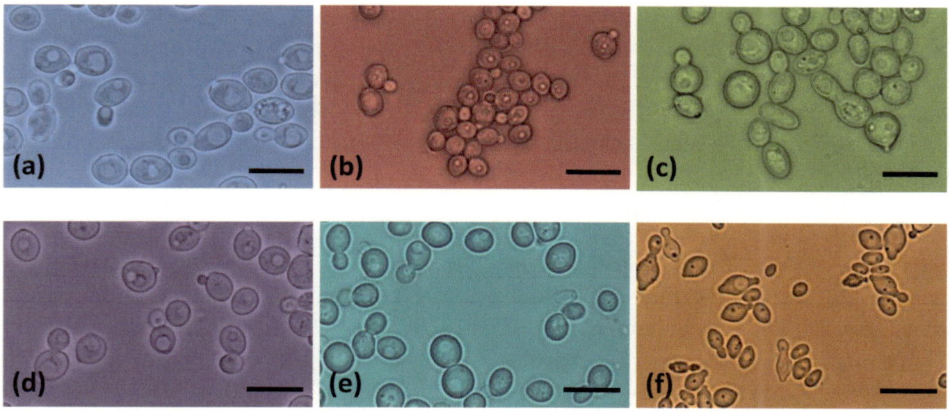

Figure 1.
Yeast morphology and asexual reproduction by budding. (a) Saccharomyces cerevisiae, (b) Torulaspora delbrueckii, (c) Wickerhamomyces anomalus. (d) Lachancea thermotolerans, (e) Metschnikowia pulcherrima, and (f) Kloeckera apiculata. Scale = 10 μm.

High contents of aldehydes have been related to oxidative off-flavors and reduced freshness in wines [8, 9]. Methional is an especially defective compound with a typical smell of boiled potato [9]. Moreover, other compounds like phenylacetaldehyde, with a typical honey smell, may increase the heaviness and sweetness, thereby reducing the wine freshness.

Conversely, several aromatic compounds have been described as enhancers of freshness; among them furaneol together with homofuraneol enhance red wine quality and fruitiness [10, 11] and ethyl 2-hydroxy-4-methylpentanoate contributes with the smell of fresh blackberries [12]. High contents of ethyl propanoate, ethyl 2-methylpropanoate, and ethyl 2-methylbutanoate have also been correlated with blackberry aromas, and ethyl butanoate, ethyl hexanoate, ethyl octanoate, and ethyl 3-hydroxybutanoate conferred redberry aromas [13]. Moreover, the formation of fruity (isoamyl acetate, ethyl butyrate, etc.) or floral esters (2-phenylethyl acetate) increases the sensation of fresh complexity in white wines, especially when accompanied by suitable acidity.

In the last years, the use of non-*Saccharomyces* yeasts has been described as an efficient tool to promote the formation of esters during fermentation. Species such as *Torulaspora delbrueckii* (**Figure 1b**) in sequential and mixed fermentations have been used extensively to promote the formation of fruity esters like isoamyl and isobutyl acetate [14] and floral esters such as 2-phenylethyl acetate [15]. Moreover, 3-ethoxy propanol is formed during the fermentation with *T. delbrueckii*, and it is not found in *S. cerevisiae* single fermentations [15]. The presence of this later compound is correlated with blackcurrant nuances in red wines [16].

Wickerhamomyces anomalus (formerly *Pichia anomala*, **Figure 1c**) has also been described as a good producer of isoamyl acetate and, in general, several acetate and ethyl esters [17–21]. Sequential fermentations in which *W. anomalus* is involved have a more complex aroma and an increased fruitiness that can help to improve the freshness of wines from warm areas. Concerning terpenic varieties, the expression of several enzymes, β-D-glucosidase, α-L-arabinofuranosidase, α-L-rhamnosidase, and β-D-xylosidase, can help to hydrolyze bonded terpenes to free aglycones enhancing varietal aroma [21, 22]. Nevertheless, β-glucosidase activity can be detrimental for the processing of red grape varieties since this enzyme may degrade anthocyanins, affecting their stability and causing an unwanted color loss in red wines [23].

Fermentation of Syrah and Sauvignon blanc musts by *Lachancea thermotolerans* (Lt) increased the formation of 2-phenylethanol, phenethyl propionate, ethyl salicylate, methyl salicylate, and 3-methylthio-1-propanol [24]. The release of varietal terpenes and volatile thiols can be promoted by Lt because the β-D-glucosidase [25] and carbon-sulfur lyase [26] enzymatic activities have been described in some strains.

Metschnikowia pulcherrima (Mp) in single fermentations has shown an excessive production of ethyl acetate with negative sensory repercussion [27]. However, the mixed use of *M. pulcherrima* with *S. uvarum* diminishes the production of ethyl acetate simultaneously increasing the formation of 2-phenyl ethanol and 2-phenylethyl acetate [27]. Furthermore, the use of mixed fermentations Mp/Sc produces high content of acetate esters and β-damascenone with reduced levels of C6 alcohols in ice wines made from Vidal blanc grape variety [28]. The β-glucosidase and β-lyase enzymatic activities have also been described in Mp [29, 30].

Most of the acetate esters can be enhanced by using *Hanseniaspora*/*Kloeckera* (**Figure 1f**) species [31, 32]. Several works with *H. vineae* in lab assays, but also industrial wines made in sequential fermentations with *S. cerevisiae*, have demonstrated a fruitier aroma with increased concentrations of both 2-phenylethyl acetate and ethyl acetate [31–33]. Moreover, the de novo formation of several aromatic compounds such as benzyl alcohol, benzaldehyde, *p*-hydroxybenzaldehyde, and *p*-hydroxybenzyl alcohol in the absence of precursors has been verified during the fermentation with *H. vineae* [34, 35]. Concerning enzymes, it has been observed that β-glucosidase activity, which facilitates the release of free terpenes increasing the varietal aroma, can be 6.6-fold higher in *H. vineae* than *S. cerevisiae* [36].

4. Yeast to improve acidity

The fermentation with *Saccharomyces cerevisiae* (*S. cerevisiae*) strains usually does not affect significantly the pH values. Some strains are able to degrade (more commonly) or produce malic acid. However, concerning malic acid production, even when the amount can reach up to 1 g/L [37], this happens in musts with low acidity, where this amount is inefficient to produce a suitable pH reduction. Under enological conditions, most of the malic acid producing *S. cerevisiae* strains (4%) are able to release 0.3–1 g/L of malic acid. It should also be considered that in red wines and some white and rose wines, malic acid is usually transformed into lactic acid during the MLF. It makes the effect of this natural acidification under enological conditions even lower.

Acidification by the use of non-*Saccharomyces* yeasts: In the last years, the species *Lachancea thermotolerans* (formerly *Kluyveromyces thermotolerans*) has been used for acidification purposes in several beverages as wine [38, 39] and beer [40–42]. The maximum alcoholic degree reached by *L. thermotolerans* ranges 5–9% v/v during fermentation [38, 43, 44], so it must be used mixed or sequentially with *S. cerevisiae* or *S. pombe* to completely ferment the sugars [45]. *L. thermotolerans* has shown the ability to modify significantly the pH in grape musts even at industrial level in crushed red grape [39], decreasing the initial value in 0.5 pH units. Indeed, a higher decrease in pH may be obtained (up to 1 pH unit) when Lt is used for the malt fermentation in beer production, due to the lower buffer effect of this matrix [46]. The acidification produced by *L. thermotolerans* is a consequence of the metabolization of sugars to lactic acid. Moreover, metabolic properties, physiology, nutritional requirements, and enological applications of this yeast have been recently reviewed [45]. Some strains can produce extremely high concentrations of lactic acid, higher than 16 g/L [47]. This acidification is produced not only with some sugar degradation and a slight effect in the alcoholic degree [39] but also with a low production of volatile acidity [38, 48]. What is especially interesting is that lactic acid is stable under enological conditions; it does not degrade during processing or storage, so it can affect permanently the pH values. Moreover, in some situations, a synergistic effect in the production of lactic acid when *L. thermotolerans* is used in co-inoculation together with *Oenococcus oeni* has been observed [39].

Most of the acidification occurs at the beginning, during the first 3–4 days of fermentation. This facilitates the production of lactic acid even under enological conditions because it is just at the beginning of the fermentation when the wild population is lower and the implantation of *L. thermotolerans* can succeed (**Figure 2**). The typical industrial acidification with *L. thermotolerans* includes a subsequent inoculation with *S. cerevisiae* to completely ferment the sugars in a sequential fermentation (**Figure 2**). This is necessary because the fermentative power of *L. thermotolerans* is always lower than 9% v/v.

In warm areas, the acidification by *L. thermotolerans* may increase the microbial stability of wines, especially during barrel aging, and it also increases the effectivity of sulfur dioxide because the contents of free and molecular SO_2 are much higher at pH 3.5 than at 3.9. This pH reduction is feasible under enological conditions as it was previously seen.

Yeasts can influence wine color by affecting the production of stable pigments, such as pyranoanthocyanins or polymeric pigments. In addition, yeast strains with low ability to adsorb grape anthocyanins in their cell walls are suitable to decrease color loss during fermentation, and, finally, yeasts can affect color stability and intensity by pH reduction [49]. The effect of *L. thermotolerans* on color stability and the formation of stable pigments have been studied recently [50]. However, this study revealed that a low effect in the formation of these pigments can be promoted with the *S. cerevisiae* when it is used in either mixed or sequential fermentation to completely ferment the sugars. Concerning color stability, acidity is a main parameter to protect anthocyanins in wine and to increase color intensity by a

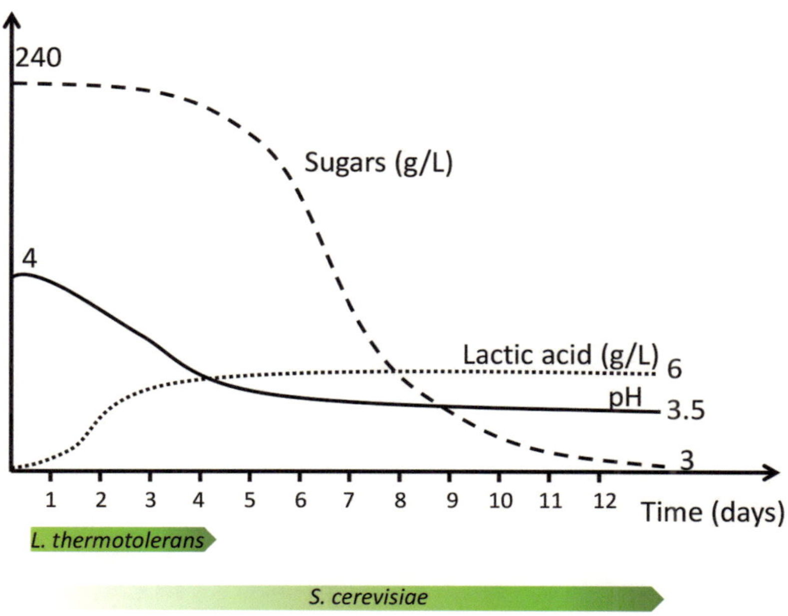

Figure 2.
Evolution of the pH, lactic acid level, and sugar content during the sequential fermentation with L. thermotolerans and S. cerevisiae.

hyperchromic effect. Indirectly, as pH affects the levels of both molecular and free sulfur dioxide, it may also promote a protective effect on color.

From a sensory perspective, the biological acidification with *L. thermotolerans* produces a good and perceptible sourness, thus increasing wine freshness [39]. Usually, no unpleasant nuances of dairy foods are found, even when higher levels of ethyl lactate are produced, but the levels of acetoin and diacetyl in the sequential fermentations with *S. cerevisiae* are quite controlled and similar to single *S. cerevisiae* fermentations [39].

5. Yeast selection to improve acidity, aromatic profile, or color

The selection of yeast strains to obtain non-*Saccharomyces* able to improve the wine freshness in terms of acidity, aromatic profile, or color starts with the isolation of a yeast collection from a vine environment, mainly grapes, and also leaves, wood, or soil. After that, the yeast can be initially classified by using both selective and differential agar media. Later, the pre-identified yeasts can be confirmed by PCR amplification of the ribosomal region spanning the internal transcribed spacers (ITS1 and ITS2) and the 5.8S rRNA gene using as primers the ITS1 and ITS4 [51], the subsequent sequencing and the comparison of the sequence in a genomic database that facilitates the proper identification of genus and species [45]. Microfermentations in triplicate can be performed in order to select specific yeast strains with improved properties, e.g., a *L. thermotolerans* strain with suitable production of lactic acid, during spontaneous fresh must fermentation. Later, the production of lactic acid and whatever other metabolites with repercussion in wine sensory quality can be evaluated by instrumental analysis (**Figure 3**).

Yeast selection can be focused on the identification of strains with specific properties of technological, fermentative, or sensory repercussion during wine

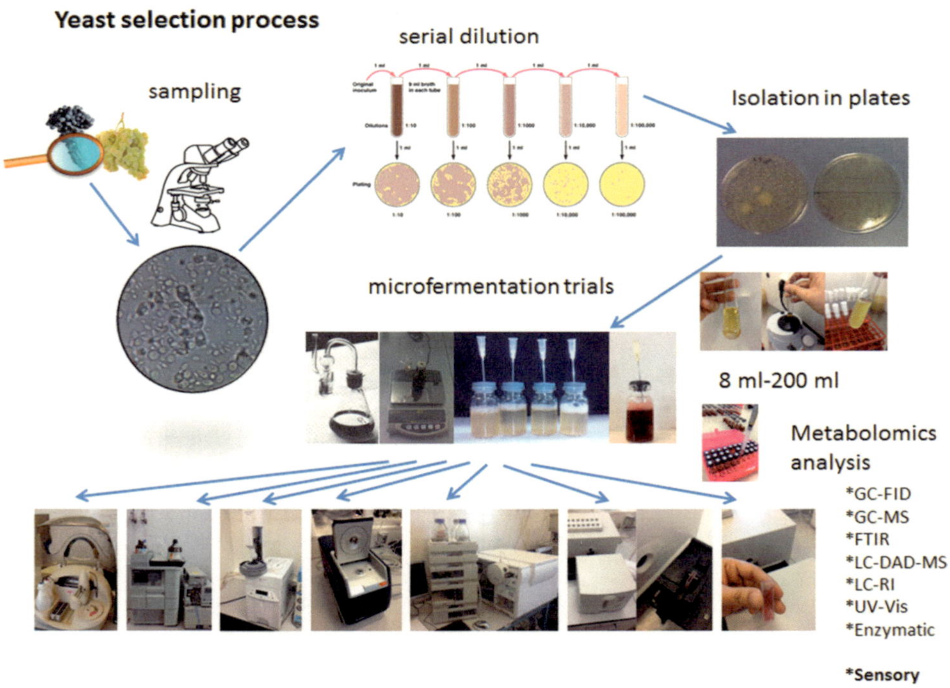

Figure 3.
Isolation of wild yeast and selection protocol under a metabolic approach.

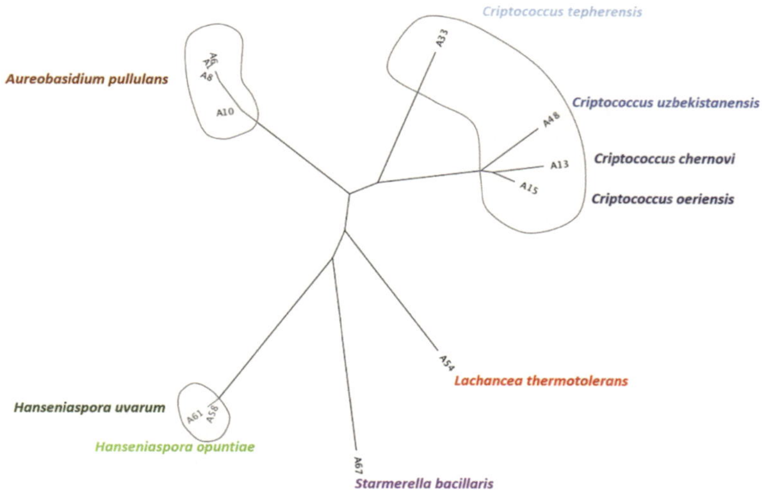

Figure 4.
*Phylogenetic tree of the wild non-*Saccharomyces *yeast species that were found in the grapes of a vineyard from a warm region.*

fermentation [52–54]. These properties can be targeted to improve color by the formation of stable pigments as vitisins [55, 56], vinylphenolic pyranoanthocyanins [57], and polymeric pigments [50, 58], the enhancement of aroma by the production of esters or enzymatic activities able to release varietal aroma [59, 60], or the improvement of the mouthfeel and flavor by the production/release of polyalcohols, polysaccharides [61, 62], acids [39, 45], etc.

The isolation of wild yeasts and the subsequent sequencing and comparison of the rDNA can help to elucidate the yeast microbioma from a vineyard (**Figure 4**). Normally, when the wild yeast populations are evaluated at the grape maturity stage, several mold species are frequently found together with apiculate yeasts such as those which belong to the genus *Kloeckera* or *Hanseniaspora*, making difficult to isolate and identify *S. cerevisiae* strains. Apiculate yeast can reach populations of 2–4 log CFU/mL.

6. Ternary sequential inoculations in warm areas: biotechnological approach to improve freshness

The use of sequential fermentations with non-*Saccharomyces* species has been used to improve wine acidity, aromatic and flavor complexity, and freshness. As reviewed in Section 3, non-*Saccharomyces* yeasts such as *H. vineae*, *T. delbrueckii*, *W. anomalus*, *M. pulcherrima*, *K. apiculata*, *S. bombicola*, and *C. stellata* improve aroma by either the increased production of acetate esters or the development of enzymatic activities that enhance the varietal aroma. Some of them can also increase sweetness and body by the production of polyalcohols such as glycerol or 2,3-butanediol. Moreover, it is currently possible to control pH in fermentation by the formation of suitable amounts of lactic acid with *L. thermotolerans*. The use of sequential combinations of two yeasts is already used at industrial level, but the combination of three yeast species (**Table 1**), namely, ternary inoculations, is less explored as a biotechnology to improve freshness in warm areas. In this case, it is more similar to what happens in a spontaneous fermentation according to the principle of succession: the fermentation is started by an apiculate yeast, followed by a medium fermentative power yeast like *T. delbrueckii*, *L. thermotolerans*, or

Aroma and flavor improvement	pH and acidity	To completely deplete sugars
Hanseniaspora vineae *Torulaspora delbrueckii* *Wickerhamomyces anomalus* *Metschnikowia pulcherrima* *Kloeckera apiculata* *Starmerella bombicola* *Candida stellata*	*Lachancea thermotolerans*	*Saccharomyces cerevisiae* *Schizosaccharomyces pombe*

Table 1.
Potential combinations of three yeasts to improve freshness.

M. pulcherrima, and finally the sugars are completely depleted by *S. cerevisiae* to obtain a dry wine. In ternary fermentations, the use of several non-*Saccharomyces* species to improve aroma and flavor must be completed with *L. thermotolerans* to decrease pH, improve the acidity, and, therefore, enhance the wine freshness. Lastly, the sugars are finished by *S. cerevisiae* or alternatively *S. pombe*. Using the latter species, it would be possible to make interesting wines in the absence of *S. cerevisiae*.

7. Conclusions

The use of fermentation biotechnologies such as sequential ternary fermentations with non-*Saccharomyces* emerges as a natural and useful bio-tool to improve freshness in warm areas. The use of *L. thermotolerans* favors a powerful pH modulation by the production of a stable acid without the production of off-flavors. Yeast selection to obtain appropriate non-*Saccharomyces* strains facilitates the development of safer and sensory-improved fermentation, with the added advantage of protecting the wine typicity, compared to the traditional fermentation driven by a single yeast, especially when only *S. cerevisiae* is used.

Acknowledgements

The authors would like to thank the European Regional Development Fund (ERDF), through the National Smart Growth Operational Programme FEDER INTERCONECTA EXP-00111498/ITC-20181125, project: FRESHWINES.

Author details

Antonio Morata*, Iris Loira, Juan Manuel del Fresno, Carlos Escott,
María Antonia Bañuelos, Wendu Tesfaye, Carmen González, Felipe Palomero
and Jose Antonio Suárez Lepe
enotecUPM, Universidad Politécnica de Madrid, Madrid, Spain

*Address all correspondence to: antonio.morata@upm.es

References

[1] Comuzzo P, Battistutta F. Acidification and pH control in red wines. In: Red Wine Technology. London, UK: Elsevier; 2019. pp. 17-34

[2] Kontoudakis N, Esteruelas M, Fort F, Canals J, Zamora F. Use of unripe grapes harvested during cluster thinning as a method for reducing alcohol content and pH of wine. Australian Journal of Grape and Wine Research. 2011;**17**(2):230-238

[3] Gardner DM, Zoecklein BW, Mallikarjunan K. Electronic nose analysis of cabernet sauvignon (*Vitis vinifera* L.) grape and wine volatile differences during cold soak and postfermentation. American Journal of Enology and Viticulture. 2011;**62**(1):81-90

[4] Casassa LF, Bolcato EA, Sari SE. Chemical, chromatic, and sensory attributes of 6 red wines produced with prefermentative cold soak. Food Chemistry. 2015;**174**:110-118

[5] Gómez-Míguez M, González-Miret ML, Heredia FJ. Evolution of colour and anthocyanin composition of Syrah wines elaborated with pre-fermentative cold maceration. Journal of Food Engineering. 2007;**79**(1):271-278

[6] Maturano YP, Mestre MV, Esteve-Zarzoso B, Nally MC, Lerena MC, Toro ME, et al. Yeast population dynamics during prefermentative cold soak of cabernet sauvignon and Malbec wines. International Journal of Food Microbiology. 2015;**199**:23-32

[7] Molina AM, Swiegers JH, Varela C, Pretorius IS, Agosin E. Influence of wine fermentation temperature on the synthesis of yeast-derived volatile aroma compounds. Applied Microbiology and Biotechnology. 2007;**77**(3):675-687

[8] San-Juan F, Ferreira V, Cacho J, Escudero A. Quality and aromatic sensory descriptors (mainly fresh and dry fruit character) of Spanish red wines can be predicted from their aroma-active chemical composition. Journal of Agricultural and Food Chemistry. 2011;**59**(14):7916-7924

[9] Culleré L, López R, Ferreira V. The instrumental analysis of aroma-active compounds for explaining the flavor of red wines. In: Red Wine Technology. London, UK: Elsevier; 2019. pp. 283-307

[10] Ferreira V, San Juan F, Escudero A, Cullere L, Fernandez-Zurbano P, Saenz-Navajas MP, et al. Modeling quality of premium Spanish red wines from gas chromatography-olfactometry data. Journal of Agricultural and Food Chemistry. 2009;**57**(16):7490-7498

[11] Ferreira V, Sáenz-Navajas M, Campo E, Herrero P, de la Fuente A, Fernández-Zurbano P. Sensory interactions between six common aroma vectors explain four main red wine aroma nuances. Food Chemistry. 2016;**199**:447-456

[12] Falcao L, Lytra G, Darriet P, Barbe J. Identification of ethyl 2-hydroxy-4-methylpentanoate in red wines, a compound involved in blackberry aroma. Food Chemistry. 2012;**132**(1):230-236

[13] Pineau B, Barbe J, Van Leeuwen C, Dubourdieu D. Examples of perceptive interactions involved in specific "red-" and "black-berry" aromas in red wines. Journal of Agricultural and Food Chemistry. 2009;**57**(9):3702-3708

[14] Renault P, Coulon J, de Revel G, Barbe J, Bely M. Increase of fruity aroma during mixed *T. delbrueckii*/*S. cerevisiae* wine fermentation is linked to specific esters enhancement. International

Journal of Food Microbiology. 2015;**207**:40-48

[15] Loira I, Vejarano R, Bañuelos M, Morata A, Tesfaye W, Uthurry C, et al. Influence of sequential fermentation with *Torulaspora delbrueckii* and *Saccharomyces cerevisiae* on wine quality. LWT-Food Science and Technology. 2014;**59**(2):915-922

[16] Tao Y, Zhang L. Intensity prediction of typical aroma characters of cabernet sauvignon wine in Changli County (China). LWT-Food Science and Technology. 2010;**43**(10):1550-1556

[17] Rojas V, Gil JV, Piñaga F, Manzanares P. Acetate ester formation in wine by mixed cultures in laboratory fermentations. International Journal of Food Microbiology. 2003;**86**(1-2):181-188

[18] Kurita O. Increase of acetate ester-hydrolysing esterase activity in mixed cultures of *Saccharomyces cerevisiae* and *Pichia anomala*. Journal of Applied Microbiology. 2008;**104**(4):1051-1058

[19] Domizio P, Romani C, Lencioni L, Comitini F, Gobbi M, Mannazzu I, et al. Outlining a future for non-*Saccharomyces* yeasts: Selection of putative spoilage wine strains to be used in association with *Saccharomyces cerevisiae* for grape juice fermentation. International Journal of Food Microbiology. 2011;**147**(3):170-180

[20] Izquierdo Cañas PM, García-Romero E, Manso JMH, Fernández-González M. Influence of sequential inoculation of *Wickerhamomyces anomalus* and *Saccharomyces cerevisiae* in the quality of red wines. European Food Research and Technology. 2014;**239**(2):279-286

[21] Padilla B, Gil J, Manzanares P. Challenges of the non-conventional yeast *Wickerhamomyces anomalus* in winemaking. Fermentation. 2018;**4**(3):68

[22] Padilla B, Gil JV, Manzanares P. Past and future of non-*Saccharomyces* yeasts: From spoilage microorganisms to biotechnological tools for improving wine aroma complexity. Frontiers in Microbiology. 2016;**7**:411

[23] Romero-Cascales I, Fernández-Fernández JI, Ros-García JM, López-Roca JM, Gómez-Plaza E. Characterisation of the main enzymatic activities present in six commercial macerating enzymes and their effects on extracting colour during winemaking of Monastrell grapes. International Journal of Food Science and Technology. 2008;**43**(7):1295-1305

[24] Whitener MEB, Carlin S, Jacobson D, Weighill D, Divol B, Conterno L, et al. Early fermentation volatile metabolite profile of non-*Saccharomyces* yeasts in red and white grape must: A targeted approach. LWT-Food Science and Technology. 2015;**64**(1):412-422

[25] Rosi I, Vinella M, Domizio P. Characterization of β-glucosidase activity in yeasts of oenological origin. The Journal of Applied Bacteriology. 1994;**77**(5):519-527

[26] Zott K, Thibon C, Bely M, Lonvaud-Funel A, Dubourdieu D, Masneuf-Pomarede I. The grape must non-*Saccharomyces* microbial community: Impact on volatile thiol release. International Journal of Food Microbiology. 2011;**151**(2):210-215

[27] Varela C, Sengler F, Solomon M, Curtin C. Volatile flavour profile of reduced alcohol wines fermented with the non-conventional yeast species *Metschnikowia pulcherrima* and *Saccharomyces uvarum*. Food Chemistry. 2016;**209**:57-64

[28] Zhang B, Shen J, Duan C, Yan G. Use of indigenous *Hanseniaspora vineae* and *Metschnikowia pulcherrima* co-fermentation with *Saccharomyces cerevisiae* to improve the aroma diversity of Vidal blanc icewine. Frontiers in Microbiology. 2018;**9**:2303

[29] Ganga MA, Carriles P, Raynal C, Heras JM, Ortiz-Julien A, Dumont A. Vincular la *Metschnikowia Pulcherrima* y la *Saccharomyces cerevisiae* Para Una Máxima Revelación del Aroma en Vinos Blancos; 2014. Available from: http://www.lallemandwine.com/wp-content/uploads/2014/10/Flavia-Lee-el-documento.pdf

[30] Barbosa C, Lage P, Esteves M, Chambel L, Mendes-Faia A, Mendes-Ferreira A. Molecular and phenotypic characterization of *Metschnikowia pulcherrima* strains from Douro wine region. Fermentation. 2018;**4**(1):8

[31] Viana F, Belloch C, Vallés S, Manzanares P. Monitoring a mixed starter of *Hanseniaspora vineae*-*Saccharomyces cerevisiae* in natural must: Impact on 2-phenylethyl acetate production. International Journal of Food Microbiology. 2011;**151**(2):235-240

[32] Martin V, Valera M, Medina K, Boido E, Carrau F. Oenological impact of the *Hanseniaspora/Kloeckera* yeast genus on wines—A review. Fermentation. 2018;**4**(3):76

[33] Medina K, Boido E, Fariña L, Gioia O, Gomez M, Barquet M, et al. Increased flavour diversity of chardonnay wines by spontaneous fermentation and co-fermentation with *Hanseniaspora vineae*. Food Chemistry. 2013;**141**(3):2513-2521

[34] Martin V, Boido E, Giorello F, Mas A, Dellacassa E, Carrau F. Effect of yeast assimilable nitrogen on the synthesis of phenolic aroma compounds by *Hanseniaspora vineae* strains. Yeast. 2016;**33**(7):323-328

[35] Martin V, Giorello F, Fariña L, Minteguiaga M, Salzman V, Boido E, et al. De novo synthesis of benzenoid compounds by the yeast *Hanseniaspora vineae* increases the flavor diversity of wines. Journal of Agricultural and Food Chemistry. 2016;**64**(22):4574-4583

[36] Hu K, Jin G, Xu Y, Tao Y. Wine aroma response to different participation of selected *Hanseniaspora uvarum* in mixed fermentation with *Saccharomyces cerevisiae*. Food Research International. 2018;**108**:119-127

[37] Yéramian N, Chaya C, Suárez Lepe JA. L-(−)-malic acid production by *Saccharomyces* spp. during the alcoholic fermentation of wine. Journal of Agricultural and Food Chemistry. 2007;**55**(3):912-919

[38] Comitini F, Gobbi M, Domizio P, Romani C, Lencioni L, Mannazzu I, et al. Selected non-*Saccharomyces* wine yeasts in controlled multistarter fermentations with *Saccharomyces cerevisiae*. Food Microbiology. 2011;**28**(5):873-882

[39] Morata A, Bañuelos MA, Vaquero C, Loira I, Cuerda R, Palomero F, et al. *Lachancea thermotolerans* as a tool to improve pH in red wines from warm regions. European Food Research and Technology. 2019;**245**(4):885-894

[40] Domizio P, House JF, Joseph CML, Bisson LF, Bamforth CW. *Lachancea thermotolerans* as an alternative yeast for the production of beer. Journal of the Institute of Brewing. 2016;**122**(4):599-604

[41] Callejo MJ, González C, Morata A. Use of non-*Saccharomyces* yeasts in bottle fermentation of aged beers. In: Kanauchi M, editor. Brewing Technology. Rijeka, Croatia: IntechOpen; 2017

[42] Callejo MJ, Navas JG, Alba R, Escott C, Loira I, González MC, et al. Wort

fermentation and beer conditioning with selected non-*Saccharomyces* yeasts in craft beers. European Food Research and Technology. 2019;**245**:1229-1238

[43] Kapsopoulou K, Kapaklis A, Spyropoulos H. Growth and fermentation characteristics of a strain of the wine yeast *Kluyveromyces thermotolerans* isolated in Greece. World Journal of Microbiology and Biotechnology. 2005;**21**(8-9):1599-1602

[44] Gobbi M, Comitini F, Domizio P, Romani C, Lencioni L, Mannazzu I, et al. *Lachancea thermotolerans* and *Saccharomyces cerevisiae* in simultaneous and sequential co-fermentation: A strategy to enhance acidity and improve the overall quality of wine. Food Microbiology. 2013;**33**(2):271-281

[45] Morata A, Loira I, Tesfaye W, Bañuelos M, González C, Suárez Lepe J. *Lachancea thermotolerans* applications in wine technology. Fermentation. 2018;**4**(3):53

[46] Vanooteghem M. Use of Non-*Saccharomyces* Yeasts in the Brewing of Craft Beers; 2019

[47] Banilas G, Sgouros G, Nisiotou A. Development of microsatellite markers for *Lachancea thermotolerans* typing and population structure of wine-associated isolates. Microbiological Research. 2016;**193**:1-10

[48] Kapsopoulou K, Mourtzini A, Anthoulas M, Nerantzis E. Biological acidification during grape must fermentation using mixed cultures of *Kluyveromyces thermotolerans* and *Saccharomyces cerevisiae*. World Journal of Microbiology and Biotechnology. 2007;**23**(5):735-739

[49] Morata A, Loira I, Suárez-Lepe JA. Influence of yeasts in wine colour. In: Morata A, Loira I, editors. Grape and Wine Biotechnology. Rijeka, Croatia: IntechOpen; 2016

[50] Escott C, Morata A, Ricardo-da-Silva J, Callejo M, González M, Suarez-Lepe JA. Effect of *Lachancea thermotolerans* on the formation of polymeric pigments during sequential fermentation with *Schizosaccharomyces pombe* and *Saccharomyces cerevisiae*. Molecules. 2018;**23**(9):2353

[51] White TJ, Bruns TD, Lee SB, Taylor JL. Amplification and direct sequencing of fungal ribosomal RNA genes for phylogenetics. In: PCR-Protocols and Applications—A Laboratory Manual. San Diego, CA, USA: Academic Press; 1990

[52] Fleet GH. Wine yeasts for the future. FEMS Yeast Research. 2008;**8**(7):979-995

[53] Suárez-Lepe JA, Morata A. New trends in yeast selection for winemaking. Trends in Food Science & Technology. 2012;**23**(1):39-50

[54] Suárez-Lepe JA, Morata A. Levaduras Para Vinificación en Tinto. Madrid, Spain: AMV Ediciones; 2015

[55] Morata A, Gómez-Cordovés MC, Colomo B, Suárez JA. Pyruvic acid and acetaldehyde production by different strains of *Saccharomyces cerevisiae*: Relationship with vitisin A and B formation in red wines. Journal of Agricultural and Food Chemistry. 2003;**51**(25):7402-7409

[56] Morata A, Gómez-Cordovés MC, Calderón F, Suárez JA. Effects of pH, temperature and SO_2 on the formation of pyranoanthocyanins during red wine fermentation with two species of *Saccharomyces*. International Journal of Food Microbiology. 2006;**106**:123-129

[57] Morata A, González C, Suárez-Lepe JA. Formation of vinylphenolic pyranoanthocyanins by selected yeasts fermenting red grape musts supplemented with hydroxycinnamic acids. International Journal of Food Microbiology. 2007;**116**(1):144-152

[58] Escott C, Del Fresno JM, Loira I, Morata A, Tesfaye W, González MC, et al. Formation of polymeric pigments in red wines through sequential fermentation of flavanol-enriched musts with non-*Saccharomyces* yeasts. Food Chemistry. 2018;**239**:975-983

[59] Lambrechts MG, Pretorius IS. Yeast and its importance to wine aroma—A review. South African Journal of Enology and Viticulture. 2000;**21**:97-129

[60] Swiegers JH, Bartowsky EJ, Henschke PA, Pretorius I. Yeast and bacterial modulation of wine aroma and flavour. Australian Journal of Grape and Wine Research. 2005;**11**(2):139-173

[61] Palomero F, Morata A, Benito S, Calderón F, Suárez-Lepe JA. New genera of yeasts for over-lees aging of red wine. Food Chemistry. 2009;**112**(2):432-441

[62] Loira I, Vejarano R, Morata A, Ricardo-da-Silva JM, Laureano O, González MC, et al. Effect of *Saccharomyces* strains on the quality of red wines aged on lees. Food Chemistry. 2013;**139**(1-4):1044-1051

Chapter 10

Yeast from Distillery Plants: A New Approach

Beatriz García-Béjar, Pilar Fernández-Pacheco, Ana Briones and María Arévalo-Villena

Abstract

Nowadays, there is more and more interest in the microbiological resources from different ecosystems, not only because this would allow knowing more about the microbial biodiversity related with these substrata but also because it provides an opportunity to study their characteristics and technological properties which may be of potential interest. This knowledge may allow finding future biotechnological applications for these microorganisms on bio-conservation and reuse of agricultural by-products and may also lead to studies on the improvement of raw material processing. Some raw materials and processing plants in wine and related industries constitute a suitable place for yeast growth; for example, musts, wines in cellars, piquettes, bagasse, pomace, grape skins and yeast lees in the ethanol industry all provide an inexhaustible supply of yeasts. Few microbiological studies have been published so far about the biodiversity of the yeast population in distillery plants. For that reason, the aim of this research was to determine yeast biodiversity and their distribution in different distillery plants in the La Mancha region which are at least 100 years old.

Keywords: distilleries, non-*Saccharomyces*, *Saccharomyces* spp., wine by-products, cell vitality, biocontrol

1. Introduction

Agricultural residues from food industries are an important raw material involved in bioethanol production. Traditionally, residual juice, molasses and pomace from sugarcane, agave and sugar beet have been widely used in South America for obtaining distilled beverages such us cachaça, tequila and rum. The distillation process is used to isolate, select and concentrate pleasant volatile compounds from the previously fermented liquids and concentrate the alcohol content. Additionally, certain long esters from yeast cells are extracted by distillation and transferred to the final product [1].

Microbial communities from these raw materials and their fermented and distillate beverages not only are interesting due to their role in the aroma production, but their biodiversity and other biotechnological properties are also important. Yeast populations from these ecosystems have not been studied very much, and any studies on them have normally been focused on tequila [2–5], rum [6] and cachaça beverages [7–10].

IntechOpen

Yeasts are able to spread from diverse niches to many environments, especially in the vegetable world [11]. Crops and processing plants provide a good niche for yeast growth. In fact, grape crops, musts and wines have been thoroughly studied [12–14], although distillate products and their industry have not been analyzed in Spain in spite of the fact that it is believed to be a new environment for yeast biodiversity study and its biotechnological applications.

In recent years, Spain has been established as the vineyard of the world, presenting the largest surface area (13%) dedicated to this crop [15]. The wine industry is an important sector in Spain which grew considerably throughout 2018. The number of cellars has increased by 6.8% with wine production also increasing (26%) and current production being 40.9 million hectolitres.

Castilla-La Mancha is the world's largest vine-growing region with an annual wine production of around 17 million hectolitres in the 2017–2018 vintage, which accounted for nearly 50% of the total Spanish production. Part of this large production is derived from the distillery industry; in the last year, nearly 250,000 hectolitres were transformed into alcoholic derivatives (16). There is a total of 33 authorized distilleries for wine by-product distillation, 13 of which are located in the La Mancha region. These industries process not only wine but also sweet grape pomaces and its fermented products, obtaining around 4–4.5 million hectolitres [16].

Wine production generates around 600,000 tons of grape derivatives annually such as fermented red skins, which still contain reducing sugars and ethanol, and sweet pomace (from white wine vinification). These by-products, as well as yeast lees and flocculated yeasts, are transported to distilleries where the ethanol is extracted. As **Figure 1** shows, sweet pomaces are mixed and stored for 10–15 days, starting a spontaneous fermentation process. Then, pomace and grape skins are washed with water at 50°C in a heat diffusion system in order to extract the residual sugars and ethanol. After that, a liquid is obtained which is a mixture of alcohol (3–4%) (V/V), water and sugar and is called fermented or sweet piquette. On the other hand, a liquid is drained during the storage of solid organic waste which is mixed with the piquette and fermented for 2–3 days in a stainless-steel container, obtaining a higher alcohol

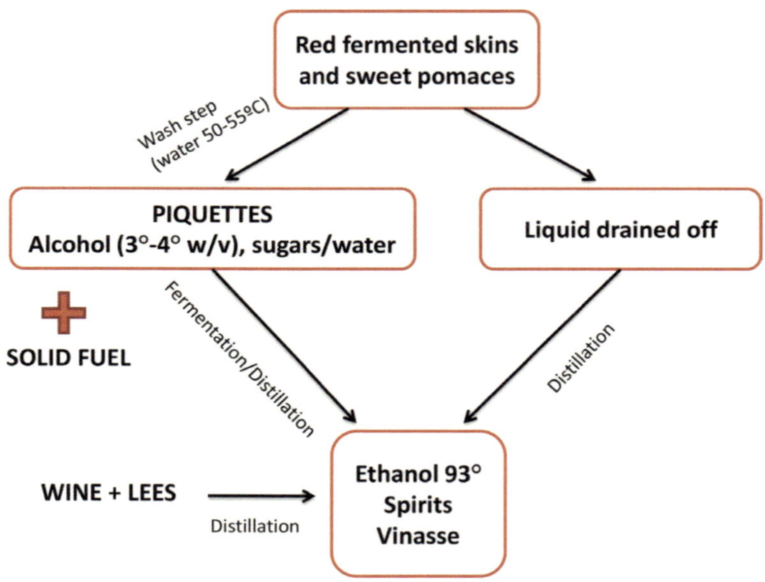

Figure 1.
Flowchart that shows all the steps involved in alcohol production from skin, lees and pomace.

content (4–5% V/V). Finally, red fermented skins are washed at a lower temperature with the aim of extracting the residual ethanol [17, 18].

The fermented piquettes and the drained liquid are distilled, producing a 93% (V/V) alcohol content product. Then, a dehydration process is carried out until the ethanol concentration of 99.9% is reached. This is mainly used in gasoline as an anti-detonating additive. Residues from distillation can be used as solid fuel (solid residue or "bagasse") or as fertilizer (liquid residue or "vinasse") [17].

Spontaneous fermentations during this last process are produced by non-*Saccharomyces* and *Saccharomyces* biota present in the environment whose biodiversity has not been widely studied [19].

The lack of information about yeast ecology in this habitat and, more specifically, in this territory has prompted the aim of this research.

2. Sample collection

Six of the largest distilleries in Europe, which are at least 100 years old and are found in the towns of Argamasilla de Alba (A), Campo de Criptana (B), Madridejos (C), Villarrobledo (D), Daimiel (E) and Tomelloso (F) in the La Mancha region (**Figure 2**), were selected to carry out the study. La Mancha is the principal area for the production of bioethanol and distillates in Spain.

A total of 47 samples were randomly collected from sweet piquettes [20], fermented piquettes [19], flocculated lees [7] and plant oil [1] throughout the pomace-based ethanol production process, and they were transported to the laboratory under aseptic and refrigerated conditions.

Samples and/or their dilutions were spread on YPD agar plates (10 g/L yeast extract, 20 g/L peptone, 20 g/L glucose and 20 g/L agar); chloramphenicol and sodium propionate were added to inhibit bacteria and mold growth, respectively. Plates were incubated at 28°C/72 hours. Then, samples displaying fewer than 30 colonies were centrifuged to concentrate the cells, and the pellet was directly spread

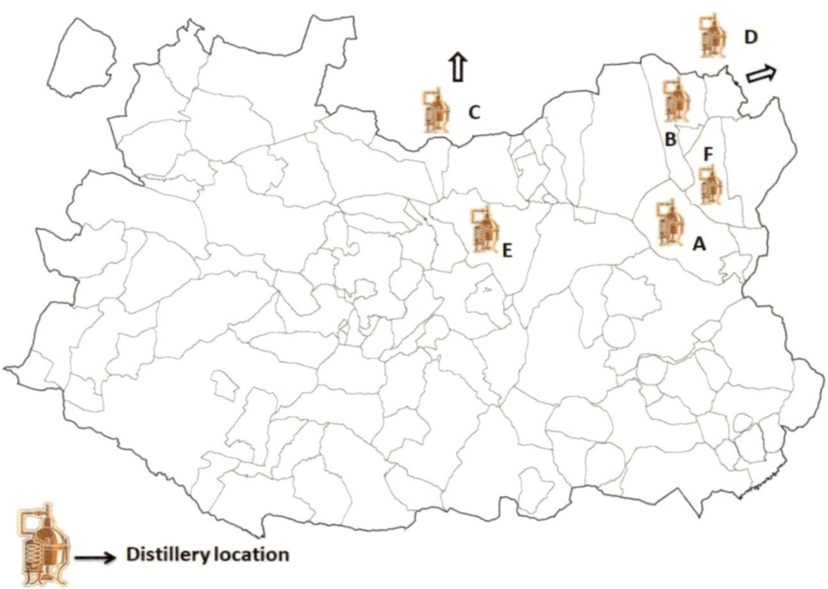

Figure 2.
Location of the distilleries included in this research in the La Mancha region (Spain).

on YPD agar. Plates with sufficiently separated colonies were replicated onto lysine agar medium (Oxoid, Basingstoke, UK) to distinguish between *Saccharomyces* sp. and non-*Saccharomyces* sp.

The isolates were obtained from 19 samples. A sample was not taken from distillery F, which is possibly due to the hot washing of the skins which would drastically decrease the number of cells.

A total of 210 purified isolates were obtained, 144 *Saccharomyces* and 66 non-*Saccharomyces*, and were stored in 15% glycerol at −80°C until they were studied.

3. Yeast classification by genetic identification

Saccharomyces spp. yeasts were the predominant profile in all distillery plants. However, the number of non-*Saccharomyces* species varied between distilleries.

3.1 Non-*Saccharomyces* yeasts

Genetic species identification was done using the polymerase chain reaction/restriction fragment length polymorphism (PCR-RFLP) technique, by amplifying the 5.8S rRNA gene using ITS1 and ITS4 [20]. Amplified products were digested (37°C for 7 h) with the three restriction endonucleases *Hinf* I, *Hae* III and *Cfo* I.

Both PCR products and their restriction fragments were separated on agarose gel with GelGreen™ (Biotium), and the results were visualized using a GeneFlash documentation system. For those isolates that could not be identified by PCR-RFLP analysis, the region D1/D2 from the domain 26D rRNA gene was sequenced using NL1 and NL4 primers. If any variation existed due to the action of the NL4 primer, LR6, NL3A and NL2A primers were used as alternatives. Finally, for those samples in which the percentage of identity at species level was less than 99%, the ITS region was sequenced using ITS1 and ITS4 primers [17]. In **Table 1** all isolates are shown, classified at the species level with 99% similarity and the NCBI accession number obtained. A percentage of similarity lower than 99% was obtained with isolates 23, 33, 48 and 62 using the primers NL1/NL4. Sequencing of the 5.8S rRNA + ITS region confirmed this with a similarity of 99%.

Non-*Saccharomyces* yeasts were mainly distributed in sweet piquettes (45.5%) without ethanol, 43.3% were found in fermented piquettes, where the ethanol concentration varied between 4% and 5% (v/v). Finally, 18.2% and 3% were isolated from plant soil and sedimented yeast lees, respectively (**Figure 3**).

As can be observed in **Figure 4**, non-*Saccharomyces* yeasts were more present in plant C (14%) and in plant D (47%) due to the difference in the age of the distilleries and the specific elaboration process followed.

The 66 isolates were cataloged as 8 genera and 20 species, which belonged mainly to the genera *Pichia* (38.0%), *Candida* (22.7%), *Hanseniaspora* (18.2%) and *Torulaspora* (10.6%). The remaining 10% belonged to *Zygosaccharomyces*, *Lachancea*, *Ogataea* and *Saccharomycodes*.

There were four predominant species that were identified as *Pichia galeiformis*, *Torulaspora delbrueckii*, *Hanseniaspora osmophila* and *Candida lactis-condensi*. All these results showed that a considerable diversity exists in this environment, unlike in grape must fermentations [21].

With regard to the substrata of isolation (**Figure 3**), *T. delbrueckii*, *H. osmophila*, *P. kudriavzevii*, *C. lactis-condensi* and *P. anomala* were isolated from sweet piquettes, while *P. galeiformis* and *C. ethanolica* were found in fermented piquettes, from which other species, such us *S'codes ludwigii*, *P. bimundalis*, *Zygosaccharomyces bailii* and *C. sake*, were also isolated but at a very low percentage. Only two species, *L. thermotolerans* and

Species	Isolates Nº	Accession number (NCBI)
Candida ethanolica	35a, 36c, 40b 41a, 48a,	35/JX880409
		40/JQ073769
		41/JX880400
		48/JQ410478
Candida lactis-condensi	50c, 51c, 52c 53c, 54c, 55a 56a, 57a	55/JN248610
		56/JN248614
		57/JN248611
Candida sake	44a	JX880410
Candida viswanathii	39a	JQ512833
Hanseniaspora meyeri	7a	JN248602
Hanseniaspora osmophila	4b, 26a, 58a 59a, 62 a, d, 65a, 66a	4/JQ073772
		26/JQ512831
		59/JQ512840
		58/JQ512840
		62a/JQ410479
		62d/JQ410479
		65/JQ512841
		66/JQ780464
Hanseniaspora uvarum	11a, 28 a, d	11/JN248600
		28/JN512834-9
Hanseniaspora valbyensis	5a	JN248613
Hanseniaspora vineae	2a	JN248606
Lachancea thermotolerans	25 a, b, 46a	25/JQ073770
		46/JN248601
Ogataea polymorpha	19a	JN248599
Pichia anomala	10c, 20a, 21a, 22a, 27a, 32a	20/JX880399
		21/JX880404
		22/JN248608
		27/JX880405
		32/JX880406
Pichia bimundalis	43b	JQ073768
Pichia galeiformis	9c, 37a, 38a, 45b, 47c, 49c, 68c, 69c, 70c, 71c, 74b, 76c	37/JX880397
		38/JX880398
		45/JQ073767
		74/JQ073765
Pichia kudriavzevii	3 a, b, 8a, 13a, 14b, 24b	3/JN248607
		8/JN248609
		13/JX880402
		14/JQ073771
		24/JQ073766
Pichia membranaefaciens	23 a, d	23a/JQ410476
		23d/JQ410476

Species	Isolates N°	Accession number (NCBI)
S'codes ludwigii	72a, 77a	72/JX880401
		77/JQ512842
Torulaspora delbrueckii	1a, 6a, 60a, 61a, 64a, 67a, 75a	1/JN248605
		6/JQ780463
		60/JX880407
		61/JQ512830
		64/JX880408
		67/JQ512843
		75/JQ780465
Zygosaccharomyces bailii	34a	JN248597
Zygosaccharomyces fermentati	15a	JX880403

Technique that allowed identification: (a) NL1/NL4 primers; (b) NL2A/LR6 and NL2A/NL3A primers; (c) PCR-RFLP; (d) 1.8S–5.8S rRNA region sequence (ITS1/ITS4 primers).

Table 1.
Yeast isolates identified in the different distilleries studied and accession number (NCBI).

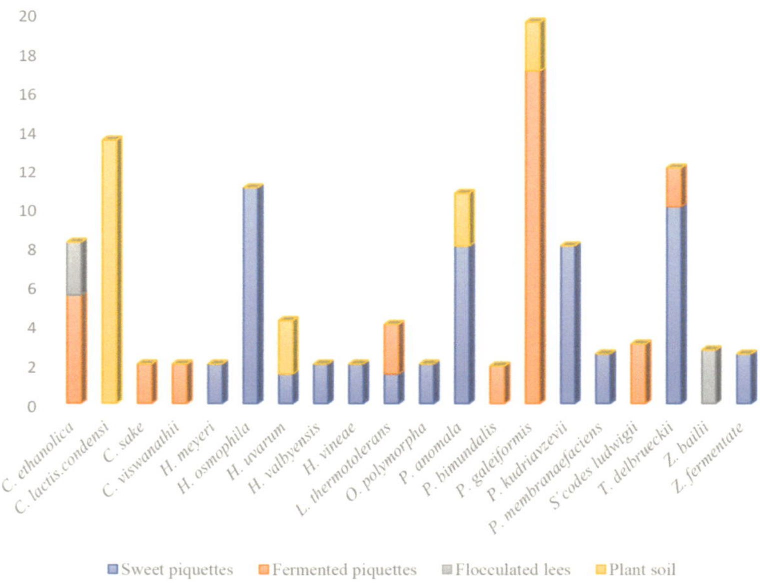

Figure 3.
Percentage of yeast species isolated in sweet and fermented piquettes, lees and plant soil.

T. delbrueckii, were found equally frequently in both sweet and fermented piquettes. Having analysed all these results, a large biodiversity of yeasts was found in the studied substrata, as was documented for grape marc by Bovo et al. [22, 23].

On the other hand, the distribution of genera (**Figure 4a**) and species (**Figure 4b**) in the studied distilleries was also analysed. *Candida* and *Pichia* genera were found in almost all of them, and *Torulaspora* and *Hanseniaspora* were found in three of the five plants in which yeasts were isolated. *P. galeiformis*, *P. kudriavzevii*, *T. delbrueckii* and *H. osmophila* were the species identified in most of the ethanol plants, with plant A being the only one where no major species were found, which contrasts with the results for the other plants (**Figure 4b**).

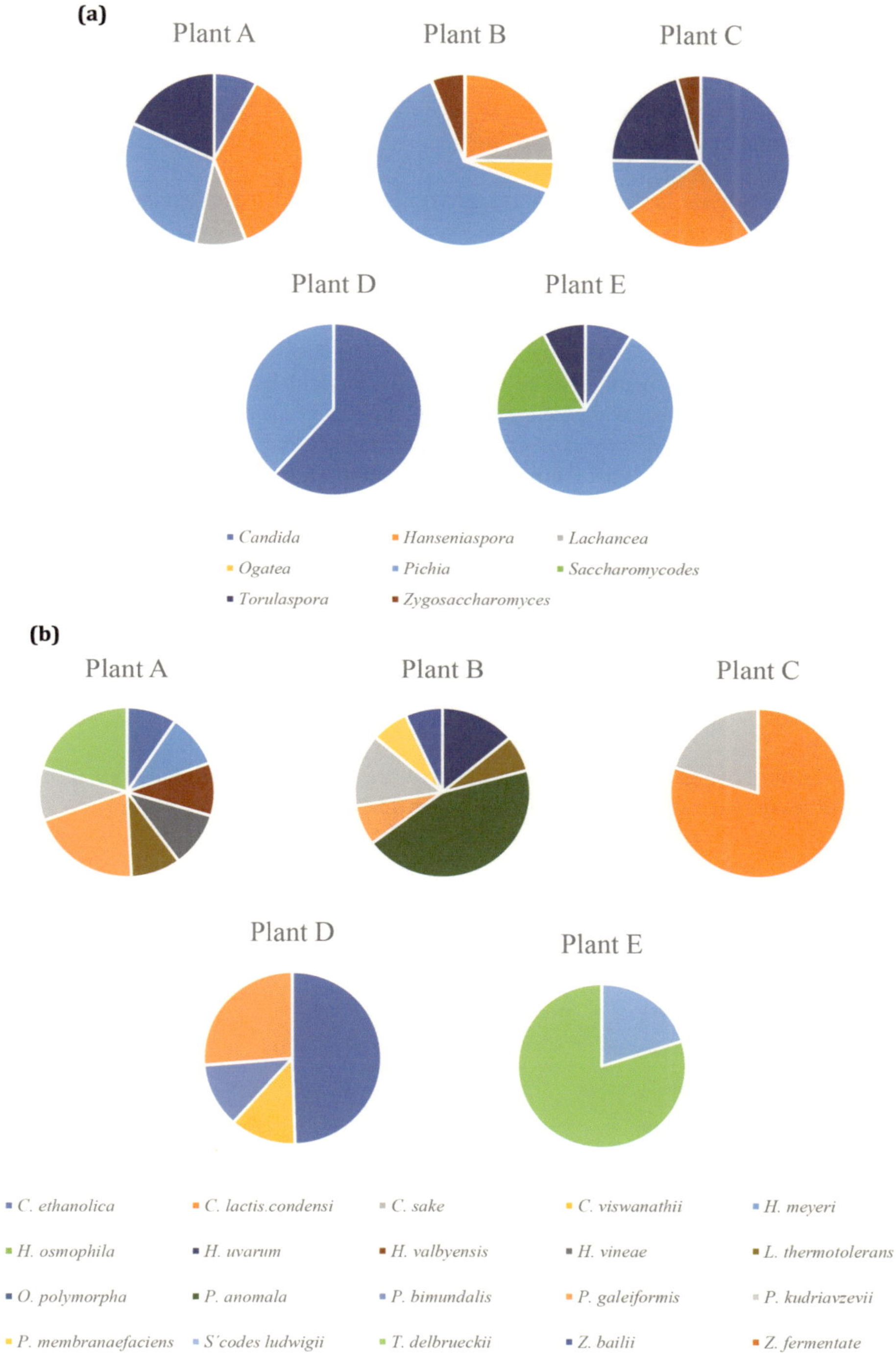

Figure 4.
Distribution of genus (4a) and species (4b) in distilleries studied.

The presence of *Candida* species (*C. sake*, *C. sorbosa*, *C. stellata*, *C. guilliermondii*, *C. karawaiewii* and *C. citrea*), *P. membranaefaciens*, *P. guilliermondii*, *K. marxianus* and large *Saccharomyces* spp. populations has been previously documented in Brazilian distilleries [7, 8]. These results confirmed that the yeast profiles in the distilleries of the two regions are very different and it is evident that the

Spanish industry is an interesting yeast niche. Additionally, some of these genera and species were also found by Amaya-Delgado et al. [5] and Lappe-Oliveras et al. [4] in tequila and agave beverages.

3.2 *Saccharomyces* yeasts

For *Saccharomyces* isolate characterization, a PCR-RFLP analysis was done, and the results showed that 95% of the isolates belonged to *Saccharomyces cerevisiae*, while only 3% and 2%, respectively, were identified as *S. paradoxus* and S. *bayanus.*

However, to discriminate isolate samples within the *Saccharomyces* sensu stricto group, a mitochondrial DNA restriction analysis [13] was carried out by digestion with the restriction endonuclease enzyme *Hinf* I. Restriction fragments were separated by electrophoresis on agarose gel with GelGreen™ (Biotium), and the results were visualized using a GeneFlash documentation system.

The *Saccharomyces* isolates were clustered in 105 different mtDNA patterns (**Table 2**), reflecting a variability of nearly 73% which is very high if it is compared to their variety in cellars [24–26].

Genetic patterns which involved at least 20% of the isolates were named as the "majority profile". At plants A and C, two majority profiles were characterized; at B and D, there was only one; and none was found at plant E. In addition, sweet and fermented piquettes were the substrata from which the most profiles were identified. Although patterns tended to be typical of each plant, the majority profiles accounted for 57% of the isolates at plant B and 33% and 30% at plants C and A, respectively.

Fermented piquettes presented the greatest degree of *Saccharomyces* variability, although several strains coexisted in both lees and sweet piquettes.

Plants	Sample	Isolates	Strains	Variability	Majority profile
	Fresh piquette	28	16	57	—
A	Fermented piquette	11	5	45	27%
	Lees	10	5	50	30%
	Fresh piquette	9	8	89	—
B	Fermented piquette	13	12	92	—
	Lees	7	4	57	57%
	Fresh piquette	27	22	81	22%
C	Fermented piquette	—	—	—	—
	Lees	9	7	78	33%
	Fresh piquette	—	—	—	—
D	Fermented piquette	8	7	88	27%
	Lees	—	—	—	—
	Fresh piquette	—	—	—	—
E	Fermented piquette	22	19	86	—
	Lees	—	—	—	—

Table 2.
Distribution of Saccharomyces isolates and strains in sweet and fermented piquettes and lees at the ethanol plants studied.

4. Biotechnological properties of non-*Saccharomyces*: fermentation and assimilation of carbon compounds

Fermentation of carbon compounds is particularly useful for identifying isolates with new fermentation profiles for potential applications in various fields. The carbon compounds assayed were D-glucose, D-galactose, L-arabinose, L-rhamnose, melibiose, lactose, raffinose, xylose, maltose, mannose, saccharose and cellobiose. The tests were carried out on a 96-well microtiter plate. Each well was filled with sugar solution, bromocresol green and cell suspensions (exhausting the endogenous carbon compound reserves). Finally, the wells were sealed with sterile vaseline, and the plates were incubated at 28°C/5 days. Depending on the time of the change and the intensity of colouration (from blue to yellow or yellow green), a classification system was established [27, 28].

The majority of the isolates (*Torulaspora*, *Lachancea* and *Saccharomycodes* species and *C. lactis-condensi*) fermented D-glucose either in the first 12 h or on the 5th day. D-mannose and saccharose were fermented to a lesser extent.

None of the isolates fermented xylose, lactose, arabinose, melibiose and rhamnose, and some only weakly fermented galactose, maltose and raffinose.

C. lactis-condensi fermented the majority of the sugars at a major or minor intensity. On the other hand, for galactose, raffinose and saccharose fermentation, variability was observed in species such as T. *delbrueckii*, *C. lactis-condensi*, *P. galeiformis* and *C. ethanolica*.

Only one *H. uvarum* isolate and one *H. vinae* isolate weakly fermented cellobiose, which is a sugar of great biotechnological interest in the production of bioethanol from agricultural and forest by-products.

The compounds used for the assimilation assay were mono- and disaccharides (D-glucose, maltose, lactose, L-rhamnose, xylose and cellobiose), polysaccharides (starch, carboxymethylcellulose and lignin) and alcohols (ethanol and methanol).

The tests were carried out in agar plates containing the carbon source and YNB without amino acids (Difco™). The assimilation profile was noticed as (++) abundant growth, (+) normal growth and (−) absence of growth.

Assimilation of carbon compounds, glucose and maltose were the most commonly used and, to a lesser extent, xylose and methanol. Three species of *Candida*, *C. viswanathii, C. ethanolica* and *C. sake*, and one *P. galeiformis* isolate assimilated carboxymethyl cellulose, while three *Pichia* isolates used starch. The majority of *Torulaspora* isolates and a few isolates of *P. kudriavzevii*, *P. galeiformis* and *H. osmophila* assimilated xylose. All of the *H. osmophila, H. uvarum* and *S'codes ludwigii* isolates effectively assimilated cellobiose. Ethanol was assimilated by a few *P. galeiformis* and *P. anomala* isolates. Finally, only some *L. thermotolerans, P. kudriavzevii*, *C. sake* and *C. viswanathii* isolates assimilated methanol. Thus, differences between isolates of the same species were observed, as can be seen in the fermentation tests.

5. Biotechnological properties of *Saccharomyces*: cell vitality and growth rate at different temperatures

Cell vitality and growth rates at different temperatures were carried out with the 105 strains. These properties were selected because they are considered a relevant characteristic in a fermentation process.

Cell vitality was evaluated as a measure of fermentative activity by an indirect electrical method [29].

Detection time (DT), expressed in hours, was obtained by impedance measured. It was considered that strains with lower DT presented high vitality.

DT results were clustered in five groups, as can be observed in **Figure 5**. In the interval 0.61–0.95 h, 10% of the strains studied were included, suggesting the highest vitality. Other yeasts (27%) were involved in the range between 0.96 and 1.29 h, indicating a fast cell vitality. Nevertheless, most stains (40%) were comprised between 1.30 and 1.64, and only 3% showed a low cell vitality (1.93–2.33 hours). These results indicate that yeasts from distillery plants have adequate vitality and probably they can displace the slower strains. In studies carried out by Ortíz et al. [13] and Barrajón et al. [29], it was noticed that DT of *Saccharomyces* wine strains oscillated from 0.67 to 1.80 h, although most of the strains showed a DT higher than 1.5 h.

The kinetic parameters (the maximum growth rate, generation time and maximum optical density) were studied at different temperatures (18, 24, 28, 38, 40 and 42°C) using a hurdle selection criteria. All strains were evaluated at 28°C and, depending on their specific growth rate (h^{-1}), were distributed into three groups: higher rate values correspond to the first group and the lowest to the third group.

Strains in the top range were assayed at 38 and 24°C. Likewise, strains with the best rate at 38°C were then tested at 40°C, and those which showed the best rate were again tested at 42°C. Similarly, the best strains at 24°C were also tested at 18°C.

At 28°C, 41 of the 105 evaluated strains were in the first group with the best-performing growth rates (0.25–0.32 h^{-1}), and 46 and 21 strains, respectively, were categorized in the second and third groups.

At 38°C, the groups presented 14, 13 and 14 strains, respectively, with homogenous results. Afterwards, assays were performed at 40 and 42°C, based on the same criteria. It is remarkable that at 42°C the duration of lag phase was higher than 45.5 hours. Nevertheless, at 40°C, 13 strains from the 25 studied gave the worst growth rate, which constitutes an expected result since this temperature is suboptimal.

In **Figure 6** the percentage of strains in every group was showed. It can be observed that at 38°C, strains were dispersed among three groups. Nevertheless at 18 and 40°C, most strains were included in the worst group.

In the majority of the cases, growth rates at low temperatures ($\leq$0.2 h^{-1}) were worse than those gotten at 40°C ($\geq$0.25 h^{-1}); this fact confirms that the microbial growth (outside the optimal temperature interval) is better at higher temperatures (**Table 3**). This is a logical outcome, because in distillery plants the substrate is

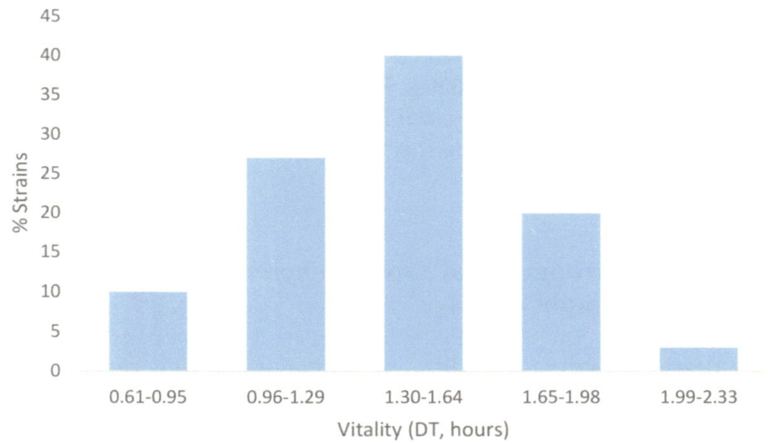

Figure 5.
Saccharomyces spp. strains grouped by their vitality according to the measurement of impedance expressed as detection time (DT, hours).

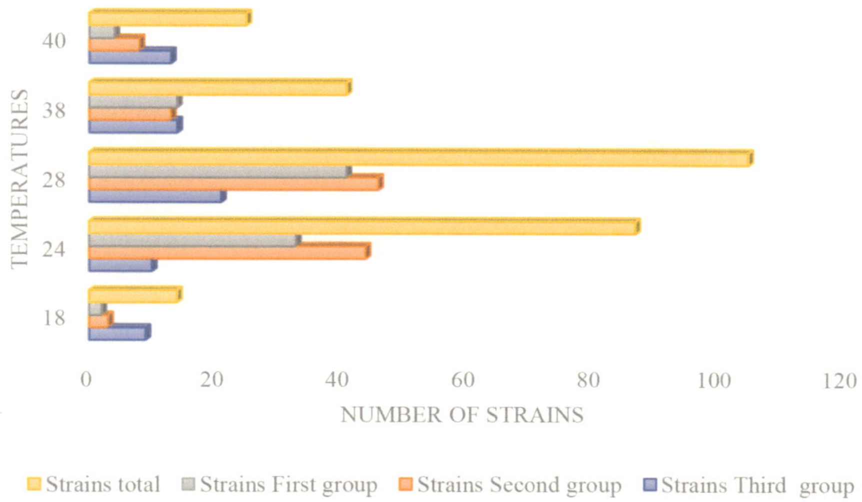

Figure 6.
Number of strains presented in each temperature range (18–40°C) based on their growth rate value. Values are means of n = 3.

	Temperatures (°C)				
Groups	**18**	**24**	**28**	**38**	**40**
First	0.17 ± 0.14	0.10 ± 0.02	0.32 ± 0.06	0.26 ± 0.06	0.32 ± 0.07
Second	0.13 ± 0.03	0.064 ± 0.01	0.21 ± 0.05	0.20 ± 0.04	0.23 ± 0.07
Third	0.09 ± 0.03	0.06 ± 0.01	0.14 ± 0.04	0.13 ± 0.07	0.13 ± 0.09

Table 3.
Distribution of Saccharomyces spp. strains based on their maximum growth rate (h^{-1}) at each temperature.

washed with warm water, and the yeasts isolated from there will grow better at higher temperatures.

The thermal washing process for the extraction of alcohol contributes to the presence of *Saccharomyces* strains with technologically interesting properties, especially in terms of vitality and resistance to high temperatures.

6. Biocontrol activity of yeast against epiphytic molds

The molds were provided from the culture collection of the University of Castilla-La Mancha (UCLM) and IVICAM (Grapevine and Wine Institute of Castilla-La Mancha). They were *Phaeomoniella (Pa.) chlamydospora*, *Neofusicoccum parvum*, *Diplodia seriata*, *Phaeoacremonium (Pm.) aleophilum* and *Aspergillus niger*.

Fungi were grown in YM agar, and pieces of agar with fungal mycelium were inserted in wells excavated in the YM agar which had been previously inoculated with yeast strains.

The results showed that there were both inter- and intraspecific variabilities.

H. meyeri, *H. uvarum*, *H. vineae* and *H. valbyensis* scarcely controlled fungal growth, and mycelium grew as in the control except for six *H. osmophila* which showed a good action against them.

However, *P. anomalous*, *P. galeiformis* and *P. kudriavzevii* effectively controlled all fungal strains including *A. niger*. Also, all *S. cerevisiae* strains except one presented good fungal growth control behaviour towards all the molds, and *A. niger* was inhibited effectively by only one of these strains. Additionally, the different *C. ethanolica* and *C. sake* have an effective action on the fungal growth, except in the case of *C. lactis-condensi*.

Finally, *T. delbrueckii* and *S'codes ludwigii* strains proved to have a large biocontrol effect not only because of their action against the growth but also because they affected every mold.

Most of the yeasts grew rapidly, forming a very dense lawn after 2 days of growth, suggesting that the mechanism of control might be based on a competition for space and nutrients. To qualitatively analyse the degree of competition between yeast and mold, the 0-day test was carried out afterwards. The assay was carried out with the yeast species which presented the best result in the previous experiment (**Figure 7**), allowing the detection of a high degree of competition between the two microorganisms.

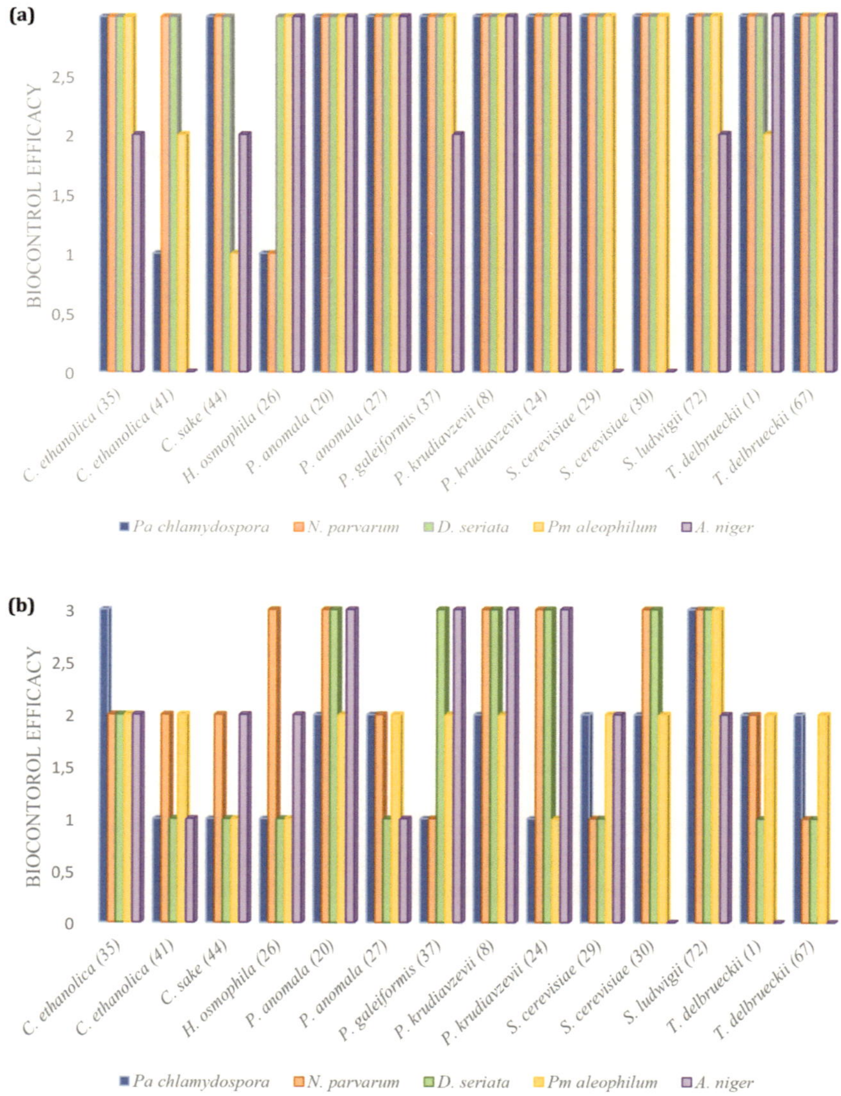

Figure 7.
Biocontrol efficacy of yeast species with 2 days (a) and 0 days (b) of preincubation time. 3: Very effective control, 2: Effective control (fungal mycelium growing slightly beyond the plug), 1: Slight control (fungal mycelium spreading in an evident form), 0: With fungal mycelium spreading similarly to the control.

The Pichia species and the only *S'codes ludwigii* assayed offered a high degree of control. One of the conclusions given by these trials is that the competition between yeast and mold for nutrients and space appeared from the first moment of contact, probably due to the very different growth rates, i.e. the yeasts have a high rate and rapidly colonize the medium preventing the development of molds. However, the inhibition mechanism may be associated with other antagonistic or enzymatic activities occurring via the production of some active compounds.

With the aim of verifying if the inhibition mechanism was produced by cell metabolites or cell wall components, the biocontrol assays were carried out with viable yeast cells, cell extract and filtered supernatant. To carry out the experiment, four wells were excavated at different points on growth fungal plates and were filled with each faction and a negative control (lysis buffer). All of them were incubated at 30°C for a maximum of 5 days in a wet chamber [30].

In most of the tests, an inhibition halo was observed with cell extracts, but when compared to the control (lysis buffer), it was difficult to identify a clear discrimination. Nevertheless, with some cell extracts, an inhibition halo slightly larger than that of the control was observed but only related to *Pm. aleophilum*. No supernatant showed antifungal activity except *H. uvarum* against *A. niger* (**Figure 8**). Finally, whole cells inhibited the molds in most cases, which is consistent with previous results except for *A. niger* which was tested with *H. osmophila*.

On the other hand, enzymatic activity such as in pectinolytic enzymes and chitinase was studied. The tests were carried out to know if the yeasts were able to degrade polygalacturonic acid and chitin. For both activities, the presence of a hydrolysis halo around the colony was considered a positive result; nevertheless, chitinolytic and pectinolytic activities were not observed in the yeasts assayed in the conditions tested.

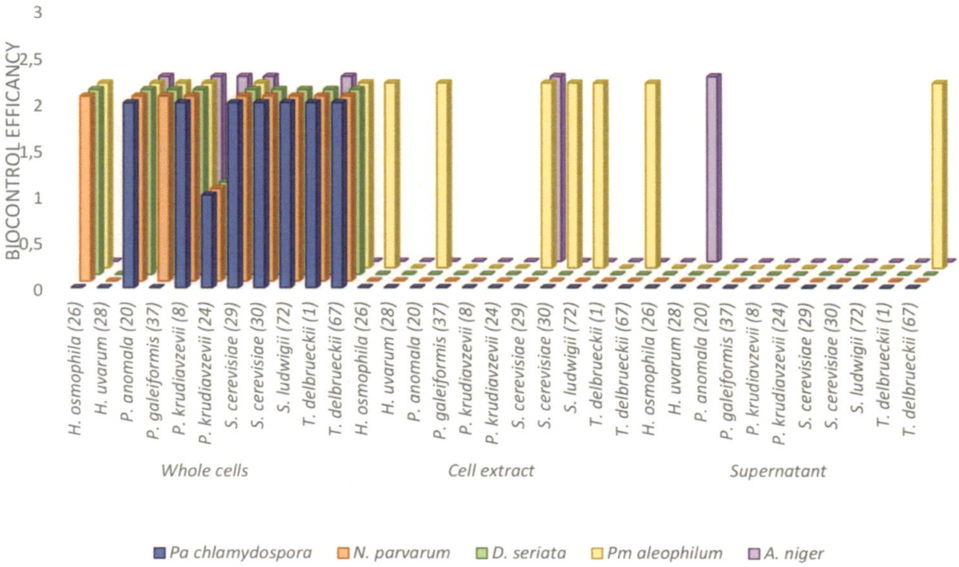

Figure 8.
Biocontrol efficacy of whole cells, cell extracts and supernatants from yeast species. 3: Very effective control, 2: Effective control (fungal mycelium growing slightly beyond the plug), 1: Slight control (fungal mycelium spreading in an evident form), 0: With fungal mycelium spreading similarly to the control.

7. Bioaccumulation of heavy metals

For bioremediation proposals, a selective elimination of metals using yeasts combined with other processes could be a feasible strategy.

Different metallic ions were tested [Cr (VI), Pb (II), Cd (II)]. Metal solutions added to inactivate biomass (obtained by thermal treatment, 5 min/121°C) were incubated at 20°C with horizontal shaking (150 rpm). Aliquots before inoculation and at time 0, 0.2, 3, 6, 24 and 48 hours were taken.

Metallic ion determination was performed by means of an inductively coupled plasma optical emission spectrometer (ICP-OES: Varian Vista-Pro, Mulgrave, VIC, Australia). Tests were semiquantitative.

Very different results were obtained depending on the yeast species as well as the metal tested for the bioaccumulation experiment (**Table 4**). The greatest metal elimination took place for Pb (II) with *H. meyeri*, *Z. bailii*, *P. membranaefaciens, P. kudriavzevii* and *S'codes ludwigii*, which presented an elimination range of around 20%, reaching 30% in some cases.

This percentage diminished by nearly half for Cd (II), with *P. kudriavzevii* having produced the highest elimination, followed by *Z. fermentati*.

Cr (VI) was eliminated in a much lower proportion, highlighting only *P. membranaefaciens* with 10% elimination, followed by the majority of the yeasts in which adsorption was not detected or was very low.

In general, the metal removal was instantaneous, and during the first 10 min of contact, no additional adsorption was observed. However, in some cases, *S'codes ludwigii* for Pb (II) and *H. uvarum* for Cd (II), the adsorption was progressive, possibly due to the different compositions of polysaccharides and proteins in the cell wall [31]. Unfortunately, *S. cerevisiae*, a by-product of the wine industry and suitable for this type of process, offered a low percentage of elimination for Pb (II) and a medium percentage for the other two metals compared with the rest of the yeasts of the same group. Appreciable desorption processes were not observed, although *P. kudriavzevii* released Cr (VI) into the media after 6 h of contact.

8. Conclusions

This initial study of yeast populations isolated from very old distilleries reflects the great existing biodiversity of this valuable yeast niche. This contrasts with what occurs in wine cellars, where the intra and interspecific variability of yeasts have been reduced drastically due to the starter use. *Saccharomyces*, *Pichia* and *Candida* are the genera found in large proportions. Some species were only isolated for certain substrates, like *T. delbrueckii* in sweet piquettes and *P. galeiformis* in fermented piquettes.

The yeast biota of these environments is varied, so these ecological niches are microbial reserves of undoubted biotechnological interest.

In fact, a great number of thermophilic *Saccharomyces* strains with a great cell vitality were found to have potential use as starters in distillery plants.

On the other hand, yeasts coming from very old distilleries might be used as biocontrol and bioremediation agents. *Pichia* sp. inhibited all molds effectively and might be produced in an aerated fermentation process and used as an antifungal postharvest treatment of fruits. In the case of *S'codes ludwigii*, *P. membranaefaciens* and *P. kudriavzevii*, the elimination of Pb (II) was achieved, with the adsorption being almost instantaneous.

P. kudriavzevii is a good candidate for both biocontrol and bioremediation because it efficiently inhibited molds and had the highest accumulation average of the tested metals.

Yeast species	Pb (II)					Cd (II)					Cr (VI)				
	Time (h)														
	0.2	3	6	24	48	0.2	3	6	24	48	0.2	3	6	24	48
C. ethanolica	2.0	1.2	2.2	2.9	4.0	2.6	0.9	2.7	1.8	2.9	-	-	-	0.6	0.2
C. lactis-condensi	10.9	10.6	10.6	9.6	9.8	0.7	0.4	1.5	-0.2	5.0	-	-	-	-	-
C. sake	4.7	10.1	8.0	10.5	10.2	2.4	2.4	0.9	2.3	-0.1	-	-	-	-	-
H. meyeri	16.8	20.7	20.5	21.4	14.2	-	-	-	-	-	-	0.6	1.2	0.5	0.9
H. osmophila	5.6	6.6	7.4	5.0	5.8	6.4	4.9	5.4	6.1	5.3	-	-	-	0.9	-
H. uvarum	5.3	5.8	9.4	9.6	10.4	3.9	6.8	6.4	8.2	8.7	4.6	4.8	6.3	4.9	1.4
H. valbyensis	3.8	9.2	5.7	6.4	5.0	0.3	0.6	0.9	-	-	3.3	3.4	2.5	3.8	1.3
H. vineae	9.9	9.8	9.5	8.6	9.4	5.2	5.6	7.4	5.8	4.5	1.9	2.6	2.4	3.0	3.3
L. thermotolerans	1.3	1.3	2.3	0.4	18.2	2.7	4.0	3.1	1.6	1.9	-	2.3	1.4	2.2	3.7
O. polymorpha	10.8	10.0	9.9	11.1	10.5	4.5	4.3	5.0	4.4	4.2	-	-	-	-	-
P. anomala	5.1	5.7	9.6	10.3	10.7	5.1	4.9	5.1	6.0	5.8	-	2.2	2.7	3.4	3.4
P. galeiformis	0.9	3.0	2.3	2.0	1.7	3.3	6.8	6.7	7.1	7.0	-	-	-	-	1.8
P. kudriavzevii	18.5	19.6	21.5	19.2	19.3	10.5	11.3	11.2	12.3	12.8	7.1	7.4	2.1	0.8	0.2
P. membranaefaciens	20.7	20.4	20.9	20.1	20.2	2.8	3.2	3.3	2.6	2.3	9.5	9.7	11.	8.9	8.2
S. cerevisiae	6.0	6.6	9.0	8.1	10.7	5.2	6.1	6.1	6.1	7.6	2.0	3.8	5.7	4.4	4.8
S. ludwigii	19.7	22.7	28.1	27.8	30.1	1.0	2.6	3.3	2.3	0.2	4.1	5.6	4.5	7.5	5.7
T. delbrueckii	3.2	4.2	5.0	7.7	8.6	2.2	2.2	3.3	3.1	3.4	0.9	3.5	3.8	5.3	4.2
Z. bailii	19.4	19.1	19.5	13.5	17.0	0.4	0.8	0.5	2.0	2.8	-	0.2	-	1.1	1.2
Z. fermentati	6.9	7.8	9.9	7.9	7.6	7.6	8.2	10.4	11.4	13.7	3.5	3.2	3.1	2.4	3.3

Table 4.
Percentage elimination of Pb (II), Cd (II) and Cr (VI) by different yeast species compared to the control

Conflict of interest

None of the authors have any conflict of interest with respect to the material contained in this manuscript.

Author details

Beatriz García-Béjar*, Pilar Fernández-Pacheco, Ana Briones and María Arévalo-Villena
Food Science and Technology, University of Castilla La Mancha, Ciudad Real, Spain

*Address all correspondence to: beatriz.gbermejo@uclm.es

References

[1] Feire-Bruno SN. Distillation of Brazilian sugar cane spirits (Cachaças). In: Zereshiki S, editor. Distillation: Advances from Modelling to Applications. 1st ed. Croatia: InTech Open; 2012. pp. 159-182. DOI: 10.5772/36718

[2] Lachance MA. Yeast communities in a natural tequila fermentation. Antonie Van Leeuwenhoek. 1995;**68**:151-160. DOI: 10.1007/BF00873100

[3] Gutierrez-Coronado ML, Acedo-Felix E, Valenzuela-Quintanar AI. Bacanora industry and its process of production. Ciencia y Tecnologia Alimentaria. 2007;**5**:394-404. DOI: 10.1080/11358120709487718

[4] Lappe-Oliveras P, Moreno-Terrazas R, Arrizon-Gavino J, Herrera-Suarez T, García-Mendoza A, Gschaedler-Mathis A. Yeasts associated with the production of Mexican alcoholic non-distilled and distilled agave beverages. FEMS Yeast Research. 2008;**8**:1037-1052. DOI: 10.1111/j.1567-1364.2008.00430. x

[5] Amaya-Delgado L, Herrera-Lopez EJ, Arrizon J, Arellano-Plaza M, Gschaedler A. Performance evaluation of *Pichia kluyveri*, *Kluyveromyces marxianus* and *Saccharomyces cerevisiae* in industrial tequila fermentation. World Journal of Microbiology and Biotechnology. 2013;**29**:875-881. DOI: 10.1007/s11274-012-1242-8

[6] Fernandez T, Marcet M, Olivera W, Martin C. Isolation and evaluation of Thermotolerant strains of *Saccharomyces cerevisiae* for Aguardiente and rum production. Ciencia y Tecnologia Alimentaria. 2008;**6**:64-70. DOI: 10.1080/11358120809487629

[7] Morais PB, Rosa CA, Linardi VR, Pataro C, Maia ABRA. Characterisation and succession of yeast populations associated with spontaneous fermentations during the production of Brazilian sugar-cane Aguardiente. World Journal of Microbiology and Biotechnology. 1997;**13**:241-243. DOI: 10.1023/A:1018558302062

[8] Pataro C, Santos A, Correa SR, Morais PB, Linardi VR, Rosa CA. Physiological characterization of yeasts isolated from artisanal fermentation in an Aguardiente distillery. Revista de Microbiologia. 1998;**29**:104-108

[9] Basso LC, De Amorim HV, De Oliveira AJ. Yeast selection for fuel ethanol production in Brazil. FEMS Yeast Research. 2008;**8**: 1155-1163. DOI: 10.1111/j.1567-1364.2008. 00428.x

[10] Vianna CR, Silva CLC, Neves MJ, Rosa CA. *Saccharomyces cerevisiae* strains from traditional fermentations of Brazilian Cachaca: Trehalose metabolism, heat and ethanol resistance. Antonie Van Leeuwenhoek. 2008;**93**:205-217. DOI: 10.1007/s10482-007-9194-y

[11] Woody ST, Ives AR, Nordheim EV, Andrews JH. Dispersal, density dependence, and population dynamics of a fungal microbe on leaf surfaces. Ecology. 2007;**88**:1513-1524. DOI: 10.1890/05-2026

[12] Barrajón N, Arévalo-Villena M, Rodríguez-Aragón LJ, Briones A. Ecological study of wine yeast in inoculated vats from La Mancha region. Food Control. 2009;**20**:778-783. DOI: 10.1016/j.foodcont.2008.10.002

[13] Ortíz M, Barrajón N, Alves-Baffi M, Arévalo Villena M, Briones A. Spontaneous must fermentation: Identification and biotechnological properties of wine yeasts. LWT- Food Science and Technology. 2013;**50**: 371-377. DOI: 10.1016/j.lwt.2012.09.019

[14] Peréz-Martín F, Seseña S, Fernández-González M, Arévalo-Villena M, Palop ML. Microbial communities in air and wine of a winery at two consecutive vintages. International Journal of Food Microbiology. 2014;**190**:44-53. DOI: 10.1016/j.ijfoodmicro.2014.08.020

[15] OIV. Organización Internacional de la Viña y el Vino [Internet]. 2019. Available from: http://www.oiv.int/es/bases-de-datos-y-estadisticas [Accessed: 18-12-2018]

[16] Mapa. Ministerio Agricultura, Pesca y Alimentación [Internet]. 2018. Available from: https://www.mapa.gob.es/es/agricultura/temas/producciones-agricolas/vitivinicultura/#para5 [Accessed: 20-12-2018]

[17] Úbeda JF, Maldonado Gil M, Chiva R, Guillamón JM, Briones A. Biodiversity of non-*Saccharomyces* yeasts in distilleries of the La Mancha region (Spain). FEMS Yeast Research. 2014;**14**:663-673. DOI: 10.1111/1567-1364.12152

[18] Úbeda J, Chacón M, Díaz-Hellín P, Ramírez-Pérez H, Briones A. Genetic and phenotypic characterisation of *Saccharomyces* spp. strains isolated in distillery plants. FEMS Yeast Research. 2016;**16**:fow035. DOI: 10.1093/femsyr/fow035

[19] Iacumin L, Manzano M, Cecchini F, Orlic S, Zironi R, Comi G. Influence of specific fermentation on natural microflora of Pomace in "grappa" production. World Journal of Microbiology and Biotechnology. 2012;**28**:1747-1759. DOI: 10.1007/s11274-011-0989-7

[20] White TJ, Bruns T, Lee S, Amplification TJ. Direct sequencing of fungal ribosomal RNA genes for Phylogenetics. In: Innis MA, Gelfand DH, Sninsky JJ, White TJ, editors. PCR Protocols: A Guide to Methods and Applications. 1st ed. London: Academic Press; 1990. pp. 315-322. DOI: 10.1016/B978-0-12-372180-8.50042-1

[21] Jolly NP, Varela C, Pretorius IS. Not your ordinary yeast: Non-*Saccharomyces* yeasts in wine production uncovered. FEMS Yeast Research. 2013;**14**:2105-2037. DOI: 10.1111/1567-1364.12111

[22] Bovo B, Andrighetto C, Carlot M, Corich V, Lonbardi A, Giacomini A. Yeast population dynamics during pilot-scale storage of grape marcs for the production of grappa, a traditional Italian alcoholic beverage. International Journal of Food Microbiology. 2009;**129**:221-228. DOI: 10.1016/j.ijfoodmicro.2008.11.025

[23] Bovo B, Giacomini A, Corich V. Effects of grape marcs acidification treatment on the evolution of indigenous yeast populations during the production of grappa. Journal of Applied Microbiology. 2011;**111**:382-388. DOI: 10.1111/j.1365-2672.2011.05060. x

[24] Ganga MA, Martınez C. Effect of wine yeast monoculture practice on the biodiversity of non-*Saccharomyces* yeasts. Journal of Applied Microbiology. 2004;**96**:76-83. DOI: 10.1046/j.1365-2672.2003. 02080.x

[25] Valero E, Cambon B, Schuller D, Casa M, Dequin S. Biodiversity of *Saccharomyces* yeast strains from grape berries of wine-producing areas using starter commercial Yeasts. FEMS Yeast Research. 2007;**7**:317-332. DOI: 10.1111/j.1567-1364.2006. 00161.x

[26] Capece A, Romaniello R, Siesto G, Romano P. Diversity of *Saccharomyces cerevisiae* yeasts associated to spontaneously fermenting grapes from an Italian heroic vine-growing area. Food Microbiology. 2012;**3**:159-166. DOI: 10.1016/j.fm.2012.03.010

[27] Barnett JA, Payne RW, Yeasts YD. Characteristics and Identification. 3rd

ed. Cambridge: Cambridge University Press; 2000. ISBN: 0521573963

[28] Kurtzman CP, Fell JW, Boekhout T. The Yeasts: A Taxonomic Study. 5th ed. Amsterdam: Elsevier Science Publisher; 2011. ISBN: 9780444521491

[29] Barrajón N, Arévalo-Villena M, Úbeda JF, Briones A. Enological properties in wild and commercial *Saccharomyces cerevisiae* yeasts: Relationship with competition during alcoholic fermentation. World Journal of Microbiology and Biotechnology. 2011;**27**:2703-2710. DOI: 10.1007/s11274-011-0744-0

[30] Úbeda JF, Maldonado Gil M, Briones A, Fernández FJ. Bio-prospecting of distillery yeasts as bio-control and bio-remediation agents. Current Microbiology. 2014;**68**:594-602. DOI: 10.1007/s00284-013-0514-3

[31] Lin Z, Ye Y, Li Q, Xu Z, Wang M. A further insight into the biosorption mechanism of Au (III) by infrared spectrometry. BMC Biotechnology. 2011;**11**:98. DOI: 10.1186/1472-6750-11-98

Section 4

Wine Technology

Chapter 11

Improvement of the Bioactive Profile in Wines and Its Incidence on Human Health: Technological Strategies

Ricardo Vejarano, Angie Gil-Calderón, Valeria Díaz-Silva and Jackeline León-Vargas

Abstract

The current lifestyle and the greater awareness of the health benefits of wine are causing an increase in demand for wines with higher levels of bioactive compounds, principally red wine. Scientific evidence supports the benefits of wine, mainly related to their antioxidant and anti-inflammatory activities. This chapter, in its first section, reviews previous studies aiming to elucidate the action mechanisms through which the bioactive compounds act on the human organism in the prevention of diseases. According to the existing literature, studies dealing with specific procedures to enhance the bioactive profile of wines are scarce. Therefore, in the second section, we pay attention to some aspects related with applicable technological strategies during the winemaking process and its incidence in the extraction and stability of bioactive compounds. Furthermore, we discuss some applicable strategies in (i) the vineyard during the vine cultivation and (ii) the raw material level in pre-fermentative stage within winery, as well as, biotechnological strategies during the fermentation and aging. All these are directed to improve the content of bioactive compounds in the wine and, thus, transmit its benefits to the consumer's health.

Keywords: wine, bioactive compounds, bioactive compounds extraction, disease prevention

1. Introduction

It is known that the content of bioactive compounds is greater in red wines, so that more health benefits can be expected by its consumption. This is the reason most studies are conducted on these wines.

Among the most studied compounds are the anthocyanins, which can be found in the skin and represent between 50 and 60% of the phenolic fraction in the red grapes (dry weight basis) [1]. For its part, the flavanols are mainly found in grape seeds with predominance of catechin over its isomer epicatechin [2], while the tannins are mostly grouped in procyanidins (catechin and epicatechin derivatives) and prodelphinidins (derived from gallocatechin and epigallocatechin) [3]. Other important groups are the stilbenes, mainly resveratrol, to which much of the protective effects of wine are attributed. Also, flavonols such as quercetin, myricetin, and kaempferol, predominant in *Vitis vinifera*, are also worth mentioning.

2. Health benefits of wine

2.1 Antioxidant activity

This activity is perhaps the most important concerning the prevention of diseases, due to the presence of phenolic compounds. Among the most important action mechanisms, the prevention of oxidative damage caused by free radicals stands out. This mechanism relies on the capture of unpaired electrons and generation of less reactive species, as well as well as the chelation of metal-ions such as Fe or Cu, to avoid the production of new free radicals [4, 5]. Other mechanisms include the interruption of self-oxidation chain reactions, deactivation of singlet oxygen, suppression of nitrosative stress, synergy with other antioxidants, activation of antioxidant enzymes, and inhibition of oxidant enzymes [6], among others.

The antioxidant efficacy would be determined by the chemical nature. For instance, the anthocyanin B-ring substitution rate is crucial due to its potential to neutralize free radicals [7], mostly in the malvidin, since it contains two methoxyl groups (-OCH_3) and one hydroxyl (-OH) group in the B-ring.

Similar behavior has been observed in gallotannins (epicatechin gallate and epigallocatechin gallate) arising from high concentration of OH groups with higher antioxidant activity than the non-gallates (catechin and epicatechin) [8]. Moreover, the antioxidant activity might improve with the synergistic tannin-tannin interaction [8] or between tannins and other compounds such as quercetin and resveratrol, reducing the lipid peroxidation caused by physical activity, for instance, in athletes [9].

The resveratrol is one of the compounds with the most antioxidant activity as it shows anti-aging activity due to its stimulant action on sirtuins [10]. Also, it is able to suppress free radical production, regulate the antioxidant enzymes activity, and induce endogenous antioxidant defenses such as Nrf2 [nuclear factor (erythroid-derived 2)-like 2] pathway [11], which regulates the expression of inflammatory markers, protecting against diseases such as Parkinson's [12].

The quercetin also contributes to reduce oxidative stress acting on the anion O_2- and over the enzymes that produce it [13].

Also, the benefits of alcohol-free red wine have been observed, which include activity increase of SOD, catalase, and glutathione reductase enzymes [14] and the production of nitric oxide (NO) [15]. The latter is closely related to a lower cardiovascular risk [16].

2.2 Anti-inflammatory activity

Inflammation is a natural bodily response against the presence of injuries or harmful agents. Among these agents, free radicals can activate the production of pro-inflammatory mediators such as tumor necrosis factor alpha (TNF-α) [17], which in turn can lead to increased oxidative stress in a cycle that contributes to the progression of many diseases.

Anti-inflammatory compounds, such as resveratrol, have been proven to be effective against cyclooxygenase (COX) enzyme, which is involved in the production of prostaglandins that stimulate the growth of tumor cells [18]; in addition, resveratrol enhances the insulin sensitivity in diabetic patients by the activation of sirtuins, which are responsible for inhibiting inflammatory processes and the secretion of TNFα factor [19, 20]. Also, resveratrol acts on microglia, involved in the defense of an injury or disease of central nervous system (CNS) [21]. Thus, the inhibition of microglial activation may help prevent several disorders. Besides, resveratrol also presents protective activity against cardiovascular diseases (CVD), by inhibiting TNFα and interleukin 6 (IL-6) [22].

Specific cases related to some pathologies are discussed in detail below.

2.3 Protection against cardiovascular diseases

There is vast evidence linking the moderate consumption of wine to lower CVD predominance, with the reports by Renaud and de Logeril [23] and St. Leger et al. [24] being pioneers in the study of the known French paradox. These studies explained the lower incidence of CVD in France despite the high consumption of saturated fats. Later studies have shown the benefits for cardiovascular risk biomarkers (**Figure 1**), which are mainly attributed to phenolic compounds.

Also, the presence of ethanol has been associated with low-density lipoprotein (LDL) and triglycerides level reduction and with the increase of high-density lipoprotein (HDL) at doses of 15–30 grams of ethanol per day [26]. Later studies suggest that moderate ethanol ingestion can increase HDL levels, apolipoprotein A1 (ApoA1) and adiponectin, in addition to lowering fibrinogen levels [27]. Nonetheless, such results suggest the need for further studies due to negative effects of excessive ingestion of ethanol.

Other compounds coming from grapes, such as melatonin and phytosterols (β-sitosterol, stigmasterol, and campesterol), have also shown protective effects against CVD either individually or in synergy with phenols [28]. Melatonin has shown effects against clinic indicators such as blood pressure, NO metabolism, and endothelial functions [29, 30] in addition to the effects on free radicals [31].

Moreover, β-sitosterol, stigmasterol, and campesterol have shown hypocholesterolemic effects by reducing the plasmatic levels of LDL (up to 10%), LDL/HDL ratio (up to 11.5%), and intestinal absorption of cholesterol (30–40%) [32–34].

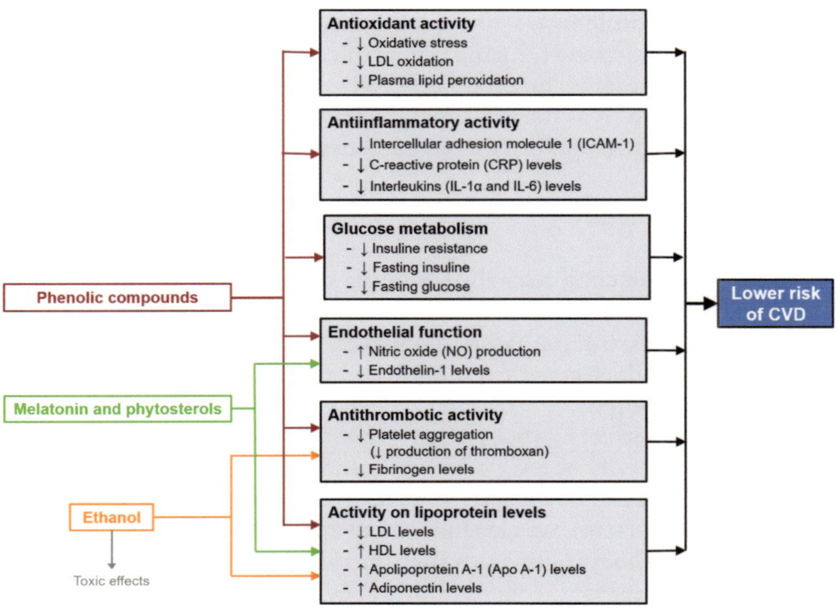

Figure 1.
Effects of wine components for cardiovascular risk factors. Adapted from Ref. [25].

2.4 Neuroprotective effects

2.4.1 Prevention of memory loss

Wine consumption could reduce the memory loss caused by cerebral circulatory insufficiency by increasing the acetylcholine levels, proteins responsible for the organization of brain cells [36], and the prevention of platelet aggregation by ethanol [37]. Other mechanisms include the resveratrol action on the telomerase enzyme, involved in preventing cell senescence and delayed cognitive impairment [38], or the action of the quercetin against cell aging by means of the activation of proteasome complex [39].

2.4.2 Action against cerebrovascular infarctions

In the Copenhagen City Heart Study, it was observed that participants who consumed wine moderately had 50% less risk of dying from cerebral infarction [40] due to the enhancement of the cerebral blood flow, the effect mainly attributed to resveratrol.

In addition, resveratrol interacts with estrogen receptors α and β, reducing cholesterol levels and the formation of atherosclerotic plaque and therefore the risk of stroke due to circulatory failure, for example, in postmenopausal women [41]. Resveratrol has also been shown neuroprotective activity against inflammatory mediators, such as interleukin 1β (IL-1β) and TNF-α, as well as keeping the levels of proteins occludin and claudin-5, of vital importance for the permeability and tissue integrity [42], and to attenuate the cellular apoptosis in ischemia-reperfusion injuries [43], which diminish cell death and the development of diseases such as Alzheimer's.

2.4.3 Antidepressant effect

This effect has been studied in rodents by administration of resveratrol, which can regulate the monoaminergic system, increasing the levels of serotonin, noradrenaline, and dopamine [44]. Also, resveratrol, quercetin, ferulic acid, ellagic acid, and proanthocyanidins can modulate the hypothalamic-pituitary-adrenal (HPA) axis activity as well as the serotonergic neurotransmission [45, 46], which are important mechanisms against anxiety and depression.

2.5 Anticarcinogenic activity

Cancer development comprises the following stages: initiation, promotion, progression, invasion, and metastasis (**Figure 2**). Initiation corresponds to DNA damage by free radicals, inflammatory mediators, cigarette smoke, radiation, etc. [47–49], which may induce genetic mutation and reproduction of mutated cells giving rise to carcinogenesis.

Greater protective effect has been observed with phenolic compounds, for example, apoptotic activity of ellagic acid [50] and delphinidin [51] in colon cancer cells. Delphinidin has also shown activity in leukemia, liver [52], and prostate cancer cells [53]. Resveratrol can also induce cell apoptosis [54].

For its part, proanthocyanidins can alter the migration and invasion processes in human pancreatic cancer [55]. Delphinidin and cyanidin has proven their antimetastatic activity in human colon cancer cells [56], while resveratrol has the same effect on lung cancer cells [57]. More specific mechanisms are shown in **Figure 2**.

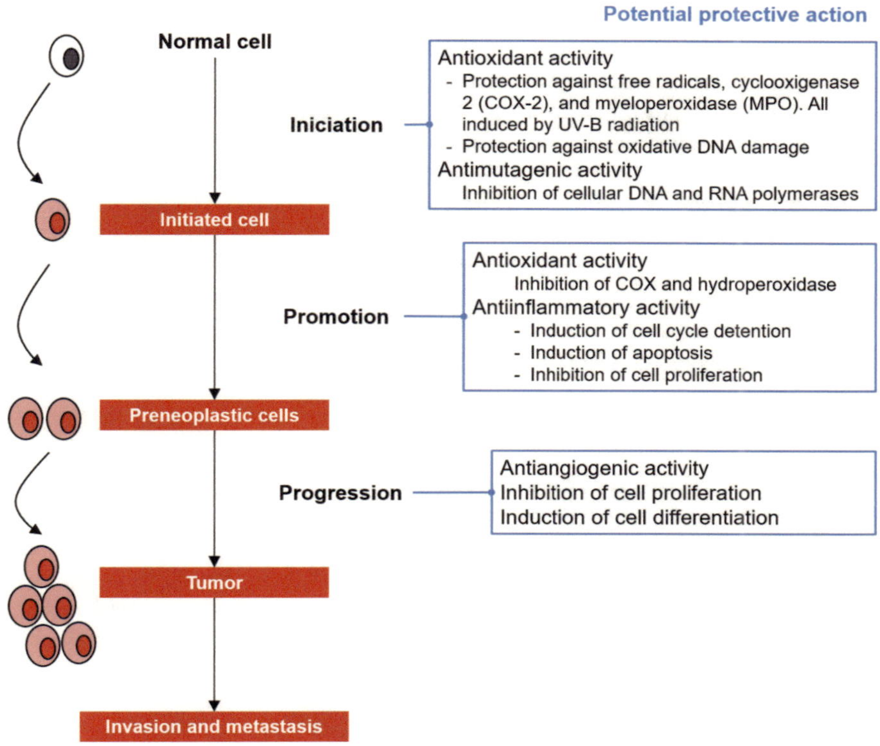

Figure 2.
Potential protective mechanisms of the phenolic compounds at different cancer stages. Adapted from Ref. [35].

2.6 Antimicrobial and antiviral activities

Red wine presents activity against *Streptococcus mutans*, *Streptococcus oralis*, *Fusobacterium nucleatum*, and *Actinomyces oris* implicated in the formation of dental cavities and periodontitis [58], in addition to *Clostridium* [59], *Candida albicans*, and *Botrytis cinerea* [60], among other microorganisms.

White wine also presents activity against *Salmonella* [61]. However, the authors argued that the effect may be associated with the presence of malic acid, since the white wine is not subjected to malolactic fermentation.

Besides, wine's activity is also effective against some viruses, which include human immunodeficiency virus (HIV) [62], hepatitis virus and adenovirus (respiratory infections), cytomegalovirus (chickenpox and infectious mononucleosis), and norovirus and rotavirus (gastroenteritis) [60].

Nonetheless, it is worth mentioning that the antimicrobial and antiviral activities showed by the wine and/or their components cannot be compared to the one attributed to antibiotics. Therefore, wine should not be used for such purposes.

3. Enhancement of bioactive compounds content

3.1 Vineyard: synthesis of bioactive compounds

The wine composition is closely related with the grape composition that mainly depends on its variety. Some compounds, such as resveratrol can reach concentrations of up to 6 mg L^{-1} in wines made of Pinot noir grapes [63], quercetin, concentrations of up to 13 mg L^{-1} in wines made of Shiraz grapes [64], or β-sitosterol, up to 106 mg/100 g of dry skin in Groppello grapes [65].

Other factors which may also induce a better synthesis of bioactive compounds at the vineyard stage are the cultivation conditions and viticulture practices (**Figure 3**). Some examples include the increase in anthocyanin and tannin levels by exposing grape bunches to sunlight and UV radiation [66], which resembles the effect observed in quercetin [67] and resveratrol [68]. In addition, agrochemical elicitation may induce the synthesis of resveratrol [69], melatonin [70], β-sitosterol, and other sterols [65].

However, conditions, such as high temperatures, can slow down the synthesis of phenolic compound, mainly anthocyanins, promoting the synthesis and accumulation of sugars in berries [71] and affecting the levels of extractable bioactive compounds during winemaking process.

3.2 Pre-fermentation treatments

Although most of the procedures are intended to enhance the physicochemical stability and sensory profile, these can be advantageous to improve the bioactive profile of wine, considering that 50% of these compounds are extracted during the winemaking process [64].

The contact time between skins and grape-must/wine can affect the content of compounds such as resveratrol, whose maximum extraction can be realized after 10 days of contact [72]. Also, the use of pre-fermentation enzymes and cold maceration can assist in the extraction of anthocyanins and tannins [73].

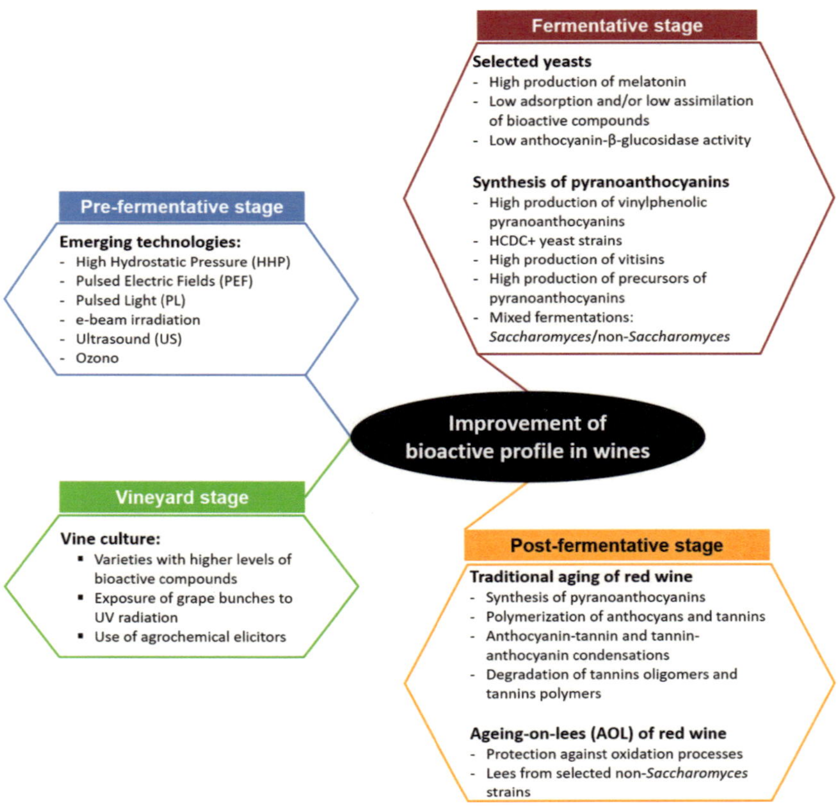

Figure 3.
Technological strategies to improve the content of bioactive compounds in red wines.

Furthermore, the emerging technologies could also be useful. Traditionally, these technologies have been studied to control microbial load of food. However, they can also be useful to improve the extraction of phenolic compounds and other molecules with positive effects on the properties of the wine. Other benefits include aroma preservation and phenolic compound protection against oxidation, since the temperature of the treated product does not change [74] and reduce SO_2 doses, an additive that can cause problems on the consumer's health [75].

These technologies can also help improve the extraction in grapes with low phenolic content, as an alternative to conventional treatments such as the use of pectolytic enzymes or the "blended" with varieties of grapes with higher phenolic content [76]. It also allows to produce wines with greater varietal character, which is preferred in the markets.

3.2.1 High hydrostatic pressure

The high hydrostatic pressure (HHP) technique can improve the extraction and protect the phenolic compounds against oxidation, given that at pressures of 600–700 MPa partial inactivation of the polyphenol oxidase enzyme is achieved [77], which enables the enhancement of the antioxidant properties of wine and, consequently, reduces the SO_2 doses [75]. HHP also allows for the maintenance of the integrity of the berry [74], facilitating the manipulation of the grape, without losses of raw material or risks of microbial contamination.

Pressures of 200 MPa have allowed the enhanced extraction of anthocyanin in red grapes, improving color intensity (26% higher) and total polyphenol index (TPI, 43% higher), with respect to the control [78]. Besides, HHP increases the selective extraction of acylated anthocyanins (up to 68% of *p*-coumarylated anthocyanins), since the HHP reduces the polarity of the grape-must due to the decrease of the water dielectric constant and the pH (molecular deprotonation at high pressures). Thus, the solubility of these anthocyanins is improved.

Higher pressures (600 MPa) were applied by Corrales et al. [79], increasing the acylated anthocyanin extraction by nine times with respect to the control at 70°C. In addition, pulsed electric field (PEF, at 3 kV cm^{-1}) technique was applied, improving the antioxidant capacity by up to three times with HHP and four times with PEF. The latter may be associated with the inactivation of oxidant enzymes.

On the other hand, the HHP favors the formation of pyranoanthocyanins, mainly derived from vitisin A at 600 MPa and 70°C [80]. Nonetheless, the anthocyanin content, like the cyanidin, can be reduced as it occurs with pulsed light (PL) and e-beam irradiation [81, 82].

3.2.2 Pulsed electric fields

The pulsed electric fields (PEF) are efficient in the extraction of phenolic compounds due to its action over the skin cell walls, reaching rates of up to 50% or higher [83], in addition to reducing the maceration time by up to 50% at a dose of 5–10 kV cm^{-1} [84].

Like HHP, the selective extraction of acylated anthocyanins can be increased by more than six times with respect to the control at 3 kV cm^{-1} [79]. Also, a higher degree of polymerization of the skin tannins can be achieved due to the greater permeability and diffusion through the fractured cell walls [85], which reduce the sensation of astringency and bitterness in the produced wines.

Also, the content of flavanols, flavonols, and hydroxycinnamic acids and derivatives can be improved after 12 months of aging in wines obtained from grapes treated with PEF, as obtained by Puértolas et al. [86] when treating Cabernet Sauvignon grapes with doses of 50 a 122 Hz, 5 kV cm^{-1} y, and 3.67 kJ kg^{-1}.

At the level of grape-musts treated with PEF, adverse effects have not been observed at doses of up to 29 kV cm^{-1} [87].

3.2.3 Ultrasound

The ultrasound (US) treatment of red grape-musts is an effective alternative to keep the level of anthocyanins up as high as 97% [88]. This fact clearly shows that the US preserves the chemical stability of these pigments. Combinations of US with heat and ethanol can also be exploited to increase the extraction of total phenols and anthocyanins and to increase the antioxidant capacity [79, 89].

3.2.4 Pulsed light

Pulsed light (PL) is a low-cost technological alternative with higher possibilities of being scaled to an industrial level than HHP, PEF, or e-beam irradiation [81]. Its efficacy varies as a function of the applied light's features. Thus, better performance is achieved with PL than with UV-C, since the former, in addition to its intensity, includes the infrared component [90].

The UV-C light (254 nm, 8.4 kJ m^{-2}, 15 min, 27°C) continuously applied produces micro-cracks in the skin of red grapes [90], inducing a high anthocyanin migration, although it is performed with lesser intensity than with HHP [74] or e-beam irradiation [82] and without affecting the external appearance of the treated berries, which facilitates their subsequent handling.

However, in wines obtained from red grapes treated with PL (12% UV-C, 10% UV-B, and 8% UV-A), a slight reduction of anthocyanins at doses of 10 pulses at 600 J has been noted. This may be associated with the oxidative degradation of these compounds by radiation [82]. Interestingly, vinylphenolic pyranoanthocyanins and vitisins have exhibited higher stability [81].

3.2.5 e-Beam irradiation

Electron beam (e-beam) irradiation can enhance the extraction of anthocyanins by up to 70% at 10 kGy [82], without affecting the external appearance of treated berries. Lower doses (0.5–3.0 kGy) have also shown improvements during extraction of anthocyanins from grape marc [91].

One disadvantage of this technology is the lowering of anthocyanin contents in the produced wines, as consequence of the induced oxidation by radiation [82]. Nonetheless, the content of vinylphenolic pyranoanthocyanins and vitisins is not affected due to the robustness of double bond in heteroaromatic ring under the induced oxidation by e-beam irradiation [82].

3.2.6 Ozone

Grapes exposed to ozone have shown greater contents of flavanols and resveratrol [92, 93]. However, the continuous exposure of berries to this gas (30 μL L^{-1}, 24 h) may produce skin hardening, causing slower extractions without affecting the final content of anthocyanins and flavanols [94].

On the other hand, the efficacy of phenolic extraction has been related with the grape variety. Wines fabricated with grapes containing high level of flavanols (as Nebbiolo) improved their color stability during winemaking procedure, especially with short expositions to ozone (<72 h, 30 μL L^{-1}) [95]. Accordingly, the anthocyanin extraction can be as high as 19% in Petit Verdot grapes treated with ozone, in addition to reduce the fermentation time [96].

3.3 Fermentation level strategies

3.3.1 Selected yeasts

The melatonin content can be increased by using *Saccharomyces* and *non-Saccharomyces* strains with high production of this compound [97], as an additional source to the melatonin coming from grapes [28]. However, some compounds like the phytosterols may be reduced during the winemaking process, since some *Saccharomyces* strains might be able to use them as nutrients [65]. Besides, contents of anthocyanins [98] and resveratrol [99] can diminish, as a result of being adsorbed by the yeast cell walls during the fermentation process.

Another issue to be aware during the winemaking process is the use of yeast with lower expression of anthocyanin-β-glucosidase activity, which is responsible for hydrolysis of anthocyanins [100].

3.3.2 Pyranoanthocyanins synthesis

The most important are vinylphenolic pyranoanthocyanins and vitisins. They present high chemical stability due to the presence of a heteroaromatic fourth ring in their structure, formed by the integration of vinylphenols, pyruvate, or acetaldehyde in the structure of the anthocyanin precursor [101], which provides resistance against oxidation and discoloration in the presence of SO_2 and/or increase of wine pH [102]. Moreover, pyranoanthocyanins possess microbiological stability, for instance, against *Dekkera/Brettanomyces*, since this yeast is not able to hydrolyze these pigments [103].

Fermentations with yeasts with hydroxycinnamate decarboxylase (HCDC+) activity have been studied as a strategy to improve the synthesis of vinylphenolic pyranoanthocyanins, from the condensation of anthocyanins with vinylphenols [101]. The vinylphenols are molecules released from hydroxycinnamic acids in grapes by the HCDC+ activity, which later on can serve as substrate to the synthesis of 4-ethylphenol by *Dekkera/Brettanomyces* [103]. By reducing the content of hydroxycinnamic acids, it is possible to prevent the synthesis of 4-ethylphenol and, in turn, the content of vinylphenolic pyranoanthocyanins can be increased.

Other interesting pyranoanthocyanin groups are the vitisins A and B, which arise from the condensation of pyruvic acid and acetaldehyde, respectively, together with the malvidin during or after the fermentation process [102].

Also, it is possible to increase vitisin A levels with *Schizosaccharomyces pombe* [104], of vinylphenolic pyranoanthocyanins in mixed fermentations of *S. cerevisiae* with *Pichia guilliermondii* [105] or by using species with high production of acetaldehyde, such as *Saccharomycodes ludwigii* [106], to improve the synthesis of vitisin B and other molecules with positive impact on the wine.

On the other hand, it is possible to enlarge the production of acetaldehyde by *S. cerevisiae* in the presence of metabolic inhibitors [71, 107], due to their effect on the alcohol dehydrogenase, which might enhance the synthesis of vitisin B.

3.4 Post-fermentation strategies

3.4.1 Traditional aging of red wine

The aging has direct effects on wine composition, since chemical and/or enzymatic oxidation processes, degradation of phenols on the presence of SO_2, and condensation and polymerization reactions [108], among others, take place at this stage, contributing to modify the content of bioactive compounds.

In general, anthocyanin, resveratrol, and flavonol levels tend to diminish with aging process [1, 108, 109]. So that, more benefits to health are attributed to young red wines. Regarding the resveratrol, hydrolysis of the glycosidic form and cis/trans isomerization take place [108], affecting its availability and activity.

At the same time, the content of pyranoanthocyanins increases through anthocyanin condensation with other molecules [101, 102]. Besides, the anthocyanic polymerization or anthocyanin-tannin condensation can be potentially increased.

Likewise, it can augment the content of monomeric flavanols from the hydrolysis of oligomeric and polymeric forms [1]. In fact, monomeric tannins possess high antioxidant capacity to act against free radicals and chelate metals [4, 5, 8], inhibit oxidative stress in cardiac hypertrophy cases, and inhibit cardiomyocyte apoptosis [110] as well as provide antimicrobial activity against oral pathogens [58].

3.4.2 Aging on lees (AOL)

In the last years, this aging technique has gained relevance in the production of red wine [109]. It consists of the release of polysaccharides from cell walls of selected yeasts lees toward the wine during its stay in barrel [111]. These released polysaccharides can enhance, among other attributes, the protection of phenolic compounds against oxidation, due to the lees that have higher oxygen affinity [112].

Nonetheless, it has been noted that anthocyanin contents can be reduced during AOL [111], especially within the first months of aging. This is a consequence of the adsorbent capacity of lees, particularly, cinnamic anthocyanins [109]. Although the loss of anthocyanins can be reduced with lees of species like *S'codes ludwigii* or *S. pombe* [111].

4. Additional considerations and future perspectives

The protective effect ascribed to bioactive compounds from wine is not only related to only one compound but also to a combined effect of several of these compounds and to their interactions with other compounds present in food. Also, the moderate ingestion of wine is certainly an important factor.

Most studies have been conducted at preclinical levels (*in vitro* and *in vivo*), aiming to elucidate the action mechanisms. Nonetheless, issues, including the absorption and bioconversion, the number of compounds and their subsequent metabolites in blood circulation, their accumulation and distribution on tissues, the chemical shapes capable of acting on specific receptors in the human organism, and so forth, are still not fully understood.

Despite the existing evidence, there is no consensus regarding its acceptance as an alternative, which aids in the prevention of diseases. Hence, more studies at the clinical level, considering a larger number of volunteers of different ethnicities, lifestyles, and health conditions, are certainly required, with the special consideration that these bioactive compounds cannot be used to replace the medicaments, since they do not possess curative properties, rather they are components of a healthy diet that can help to prevent diseases.

Within the potential strategies, some viticulture practices might contribute to improve the synthesis of bioactive compounds during the vine cultivation. Later into the winery, a proper extraction from the grapes, as well as procedures to minimize the loss of such compounds during the fermentation and aging stages, can improve the bioactive profile of produced wines.

Another important issue is the presence of products such as alcohol-free wines in the markets, which have also shown effectiveness due to the high content of bioactive compounds but with the advantage of avoiding the problems associated with excessive ethanol ingestion.

4.1 Emerging technologies

These kinds of technologies have demonstrated their efficacy to improve the extraction of bioactive compounds in pre-fermentation stages although, until now, some disadvantages have been reported during their application. For instance, the HHP, PL, and e-beam irradiation can diminish the content of anthocyanins like cyanidin in treated grapes [78, 81, 82].

In addition, the high extraction of vitisin derivatives at 70°C by using of HHP, as previously reported by Corrales et al. [80], converts the temperature into a critical parameter that limits its applicability in the winery. This fact indicates the need for more studies to optimize the extraction process.

Likewise, during PL applications [81], it is important to ensure a uniform exposition of the berry surface. The authors suggest the use of roller conveyor belts to change the position of the irradiated berry in order to improve the extraction.

Finally, the scaling of these technologies at the industrial level is still a pending issue since most studies have been carried out in small volumes and in static systems at laboratory level. In order to implement such technologies in wineries, more studies concerning large volumes and continuous flow systems, like the one performed by González-Arenzana et al. [113] with PEF, are needed.

4.2 Pyranoanthocyanins and their effects on health

It has been observed that the antioxidant potential of wine may decrease in aged wines as a result of the reduction of anthocyanins, resveratrol, and flavonols and the simultaneous synthesis of condensation products.

In general, the vitisins have shown lower potential to neutralize free radicals like O_2^- with respect to their anthocyanin precursors [7], while the pyruvic adduct of the delphinidin has shown greater ability to neutralize OH^- and O_2^- when compared with other pyranoanthocyanins.

The pyranoanthocyanin synthesis by incorporation of pyruvic acid in positions 4 and 5 of A-ring in the structure of the anthocyanin precursor can decrease the potential to suppress free radicals, which might be related to the loss of -OH from carbon 5, that together with -OH from carbon 7, favors the antioxidant activity of anthocyanins [114]. These condensations can be achieved at the fermentation level, although these mostly happen during the aging of wine. Thus, in accordance with the traditional winemaking process, these would be necessary as a strategy to provide physicochemical and microbiological stability to the wine.

As in anthocyanin precursor state, pyranoanthocyanins have shown antioxidant and anti-inflammatory activities. For example, against pro-oxidant (H_2O_2) and pro-inflammatory (TNF-α) molecules, in addition to neutralizing the secretion of interleukin 8 (IL-8) in cell cultivation of adenocarcinoma from the human colon [17]. Vitisin A has been shown a protective effect against the secretion of monocyte chemoattractant protein-1 (MCP-1) induced by TNF-α factor in human endothelial cell cultures [115], in addition to show great stability in simulated (*in vitro*) gastrointestinal conditions [116], indicating its potential availability and effectiveness in *in vivo* conditions and at clinical level.

5. Conclusions

There is vast evidence regarding the health benefits of wine, especially red wine, that results from higher contents of bioactive compounds, which aid in the prevention of diseases and provide good health benefits when consumed in moderation. Studies carried out at the pre-clinical and clinical stages have been reviewed, mostly at the pre-clinical level. Therefore, the gathered studies contribute to the better understanding of the action mechanisms by which the bioactive compounds may act in the human organism (clinical level) taking advantage of the antioxidant, anti-inflammatory, antitumor, antithrombotic, and antimicrobial activity, among others, to prevent several diseases.

According to the reviewed literature, studies addressing specific procedures to improve the bioactive profile of wine are still scarce. Hence, we described potential technological strategies that may contribute to the increase in, or at least maintenance of, the levels of different bioactive compounds present in wine during the winemaking process. Starting from the production at the vineyard, cultivation strategies can be applied in order to stimulate the greater synthesis of certain compounds. Once into the winery, the pre-fermentative treatments can increase the extraction of bioactive compounds by treating the grapes with HHP, PEF, LP, US, e-beam irradiation, and ozonization. At the fermentative level, yeasts with low adsorption and/or consumption of bioactive compounds, low anthocyanin-β-glucosidase activity, and high production of pyranoanthocyanins and/or precursor molecules of these, among other strategies, can be utilized. Although, in most cases, the content of bioactive compounds can decrease during the aging period, novel strategies like AOL can help to maintain the levels of these compounds in wines. Also, recurrent chemical processes during aging, despite modifying the structures of the grape compounds, have the advantage of allowing the synthesis of pyranoanthocyanins, polymerization of anthocyanins and flavanols, and anthocyanin-tannin condensations, among others, while maintaining the bioactive profile of the wine to a certain degree. All the above are potential strategies to be considered as technological alternatives that are applicable during the winemaking process, which enhance the content of bioactive compounds in the wine, therefore transferring their benefits to the health of the consumer.

Acknowledgements

This work was funded by the Universidad Privada del Norte (UPN), Project UPN-20191001: Valorización de principios bioactivos y desarrollo de nuevas aplicaciones de excedentes y residuos agroindustriales.

Conflict of interest

The authors declare that they have no conflict of interest.

DOI: http://dx.doi.org/10.5772/intechopen.85861

Author details

Ricardo Vejarano*, Angie Gil-Calderón, Valeria Díaz-Silva
and Jackeline León-Vargas
Agroindustrial Engineering Program, Universidad Privada del Norte (UPN),
Trujillo, Peru

*Address all correspondence to: ricardo.vejarano@upn.edu.pe

References

[1] Lingua MS, Fabani MP, Wunderlin DA, Baroni MV. From grape to wine: Changes in phenolic composition and its influence on antioxidant activity. Food Chemistry. 2016;**208**:228-238. DOI: 10.1016/j.foodchem.2016.04.009

[2] Spranger I, Sun B, Mateus AM, Freitas V, Ricardo-da-Silva JM. Chemical characterization and antioxidant activities of oligomeric and polymeric procyanidin fractions from grape seeds. Food Chemistry. 2008;**108**(2):519-532. DOI: 10.1016/j.foodchem.2007.11.004

[3] Pascual-Teresa S, Rivas-Gonzalo JC, Santos-Buelga C. Prodelphinidins and related flavanols in wine. International Journal of Food Science and Technology. 2000;**35**(1):33-40. DOI: 10.1046/j.1365-2621.2000.00338.x

[4] Ferlazzo N, Visalli G, Cirmi S, Lombardo GE, Laganà P, Di Pietro A, et al. Natural iron chelators: Protective role in A549 cells of flavonoids-rich extracts of Citrus juices in Fe^{3+}-induced oxidative stress. Environmental Toxicology and Pharmacology. 2016;**43**:248-256. DOI: 10.1016/j.etap.2016.03.005

[5] Gupta A, Birhman K, Raheja I, Sharma SK, Kar HK. Quercetin: A wonder bioflavonoid with therapeutic potential in disease management. Asian Pacific Journal of Tropical Disease. 2016;**6**(3):248-252. DOI: 10.1016/S2222-1808(15)61024-6

[6] Carocho M, Ferreira ICFR. A review on antioxidants, prooxidants and related controversy: Natural and synthetic compounds, screening and analysis methodologies and future perspectives. Food and Chemical Toxicology. 2013;**51**:15-25. DOI: 10.1016/j.fct.2012.09.021

[7] García-Alonso M, Rimbach G, Sasai M, Nakahara M, Matsugo S, Uchida Y, et al. Electron spin resonance spectroscopy studies on the free radical scavenging activity of wine anthocyanins and pyranoanthocyanins. Molecular Nutrition & Food Research. 2005;**49**(12):1112-1119. DOI: 10.1002/mnfr.200500100

[8] Furlan AL, Jobin ML, Buchoux S, Grélard A, Dufourc EJ, Géan J. Membrane lipids protected from oxidation by red wine tannins: A proton NMR study. Biochimie. 2014;**107**:82-90. DOI: 10.1016/j.biochi.2014.07.008

[9] McAnulty LS, Miller LE, Hosick PA, Utter AC, Quindry JC, McAnulty SR. Effect of resveratrol and quercetin supplementation on redox status and inflammation after exercise. Applied Physiology, Nutrition, and Metabolism. 2013;**38**(7):760-765. DOI: 10.1139/apnm-2012-0455

[10] Cohen HY, Miller C, Bitterman KJ, Wall NR, Hekking B, Kessler B, et al. Calorie restriction promotes mammalian cell survival by inducing the SIRT1 deacetylase. Science. 2004;**305**:390-392. DOI: 10.1126/science.1099196

[11] Rubiolo JA, Mithieux G, Vega FV. Resveratrol protects primary rat hepatocytes against oxidative stress damage: Activation of the Nrf2 transcription factor and augmented activities of antioxidant enzymes. European Journal of Pharmacology. 2008;**591**(1-3):66-72. DOI: 10.1016/j.ejphar.2008.06.067

[12] Rojo AI, Innamorato NG, Martin-Moreno AM, De Ceballos ML, Yamamoto M, Cuadrado A. Nrf2 regulates microglial dynamics and neuroinflammation in experimental Parkinson's disease. Glia. 2010;**58**(5):588-598. DOI: 10.1002/glia.20947

[13] Perez-Vizcaino F, Duarte J, Jimenez R, Santos-Buelga C, Osuna A.

Antihypertensive effects of the flavonoid quercetin. Pharmacological Reports. 2009;**61**(1):67-75. DOI: 10.1016/S1734-1140(09)70008-8

[14] Noguer MA, Cerezo AB, Navarro ED, Garcia-Parrilla M. Intake of alcohol-free red wine modulates antioxidant enzyme activities in a human intervention study. Pharmacological Research. 2012;**65**(6):609-614. DOI: 10.1016/j.phrs.2012.03.003

[15] Chiva-Blanch G, Urpi-Sarda M, Ros E, Arranz S, Valderas-Martínez P, Casas R, et al. Dealcoholized red wine decreases systolic and diastolic blood pressure and increases plasma nitric oxide. Short communication. Circulation Research. 2012;**111**:1065-1068. DOI: 10.1161/CIRCRESAHA.112.275636

[16] Chong E, Chang SL, Hsiao YW, Singhal R, Liu SH, Leha T, et al. Resveratrol, a red wine antioxidant, reduces atrial fibrillation susceptibility in the failing heart by PI3K/AKT/eNOS signaling pathway activation. Heart Rhythm. 2015;**12**(5):1046-1056. DOI: 10.1016/j.hrthm.2015.01.044

[17] Peng Y, Zhang H, Liu R, Mine Y, McCallum J, Kirby C, et al. Antioxidant and anti-inflammatory activities of pyranoanthocyanins and other polyphenols from staghorn sumac (*Rhus hirta* L.) in Caco-2 cell models. Journal of Functional Foods. 2016;**20**:139-147. DOI: 10.1016/j.jff.2015.10.026

[18] Jang M, Cai L, Udeani G, Slowing K, Thomas C, Beecher C, et al. Cancer chemopreventive activity of resveratrol, a natural product derived from grapes. Science. 1997;**275**(5297):218-220. DOI: 10.1126/science.275.5297.218

[19] Gertz M, Nguyen GT, Fischer F, Suenkel B, Schlicker C, Franzel B, et al. A molecular mechanism for direct sirtuin activation by resveratrol. PLoS One. 2012;**7**(11):e49761. DOI: 10.1371/journal.pone.0049761

[20] Yoshizaki T, Schenk S, Imamura T, Babendure JL, Sonoda N, Bae EJ, et al. SIRT1 inhibits inflammatory pathways in macrophages and modulates insulin sensitivity. American Journal of Physiology Endocrinology and Metabolism. 2010;**298**(3):E419-E428. DOI: 10.1152/ajpendo.00417.2009

[21] Zhang F, Liu J, Shi JS. Anti-inflammatory activities of resveratrol in the brain: Role of resveratrol in microglial activation. European Journal of Pharmacology. 2010;**636**(1-3):1-7. DOI: 10.1016/j.ejphar.2010.03.043

[22] Timmers S, Konings E, Bilet L, Houtkooper RH, van deWeijer T, Goossens GH, et al. Calorie restriction-like effects of 30 days of resveratrol supplementation on energy metabolism and metabolic profile in obese humans. Cell Metabolism. 2011;**14**(5):612-622. DOI: 10.1016/j.cmet.2011.10.002

[23] Renaud S, de Logeril M. Wine, alcohol, platelets and the French paradox for coronary heart disease. Lancet. 1992;**339**:1523-1526. DOI: 10.1016/0140-6736(92)91277-F

[24] St. Leger AS, Cochrane AL, Moore F. Factors associated with cardiac mortality in developed countries with particular reference to the consumption of wine. Lancet. 1979;**1**:1017-1020. DOI: 10.1016/S0140-6736(79)92765-X

[25] Iriti M, Varoni EM. Moderate red wine consumption in cardiovascular disease: Ethanol versus polyphenols. In: Preedy V, Watson RR, editors. The Mediterranean Diet. An Evidence-Based Approach. 1st ed. London: Academic Press; 2015. pp. 143-151. DOI: 10.1016/B978-0-12-407849-9.00014-2

[26] Baer DJ, Judd JT, Clevidence BA, Muesing RA, Campbell WS, Brown ED, et al. Moderate alcohol consumption

lowers risk factors for cardiovascular disease in postmenopausal women fed acontrolled diet. The American Journal of Clinical Nutrition. 2002;**75**(3): 593-599. DOI: 10.1093/ajcn/75.3.593

[27] Brien SE, Ronksley PE, Turner BJ, Mukamal KJ, Ghali WA. Effect of alcohol consumption on biological markers associated with risk of coronary heart disease: Systematic review and meta-analysis of interventional studies. British Medical Journal. 2011;**342**:d636. DOI: 10.1136/bmj.d636

[28] Iriti M, Varoni EM. The good health of Bacchus: Melatonin in grapes, the unveiled myth. LWT- Food Science and Technology. 2016;**65**:758-761. DOI: 10.1016/j.lwt.2015.09.010

[29] Cicero A, Borghi C. Evidence of clinically relevant efficacy for dietary supplements and nutraceuticals. Current Hypertension Reports. 2013;**15**(3):260-267. DOI: 10.1007/s11906-013-0333-8

[30] Grossman E, Laudon M, Zisapel N. Effect of melatonin on nocturnal blood pressure: Meta-analysis of randomized controlled trials. Vascular Health and Risk Management. 2011;**7**:577-584. DOI: 10.2147/VHRM.S24603

[31] Favero G, Rodella L, Reiter R, Rezzani R. Melatonin and its atheroprotective effects: A review. Molecular and Cellular Endocrinology. 2013;**382**(2):926-937. DOI: 10.1016/j.mce.2013.11.016

[32] Escurriol V, Cofán M, Serra M, Bulló M, Basora J, Salas-Salvadó J, et al. Serum sterol responses to increasing plant sterol intake from natural foods in the Mediterranean diet. European Journal of Nutrition. 2009;**48**(6):373-382. DOI: 10.1007/s00394-009-0024-z

[33] Marangoni F, Poli A. Phytosterols and cardiovascular health. Pharmacological Research. 2010;**61**(3):193-199. DOI: 10.1016/j.phrs.2010.01.001

[34] Trautwein EA, Demonty I. Phytosterols: Natural compounds with established and emerging health benefits. Oilseeds and Fats, Crops and Lipids. 2007;**14**(5):259-266. DOI: 10.1051/ocl.2007.0145

[35] Dai J, Mumper RJ. Plant phenolics: Extraction, analysis and their antioxidant and anticancer properties. Molecules. 2010;**15**(10):7313-7352. DOI: 10.3390/molecules15107313

[36] Ruitenberg A, van Swieten J, Witteman J, Mehta K, van Duijn C, Hofman A, et al. Alcohol consumption and risk of dementia: The Rotterdam study. Lancet. 2002;**359**(9303):281-286. DOI: 10.1016/S0140-6736(02)07493-7

[37] Renaud S, Beswick A, Fehily A, Sharp D, Elwood P. Alcohol and platelet aggregation: The Caerphilly prospective heart disease study. The American Journal of Clinical Nutrition. 1992;**55**(5):1012-1017. DOI: 10.1093/ajcn/55.5.1012

[38] Wang XB, Zhu L, Huang J, Yin YG, Kong XQ, Rong QF, et al. Resveratrol-induced augmentation of telomerase activity delays senescence of endothelial progenitor cells. Chinese Medical Journal. 2011;**124**(24):4310-4315. DOI: 10.3760/cma.j.issn.0366-6999.2011.24.033

[39] Chondrogianni N, Kapeta S, Chinou I, Vassilatou K, Papassideri I, Gonos ES. Anti-ageing and rejuvenating effects of quercetin. Experimental Gerontology. 2010;**45**(10):763-771. DOI: 10.1016/j.exger.2010.07.001

[40] Szmitko PE, Verma S. Cardiology patient pages. Red wine and your heart. Circulation. 2005;**111**(2):e10-e11. DOI: 10.1161/01.CIR.0000151608.29217.62

[41] Lisabeth L, Bushnell C. Stroke risk in women: The role of menopause and

hormone therapy. Lancet Neurology. 2012;**11**(1):82-91. DOI: 10.1016/S1474-4422(11)70269-1

[42] Jeong SI, Shin JA, Cho S, Kim HW, Lee JY, Kang JL, et al. Resveratrol attenuates peripheral and brain inflammation and reduces ischemic brain injury in aged female mice. Neurobiology of Aging. 2016;**44**:74-84. DOI: 10.1016/j.neurobiolaging.2016.04.007

[43] Li Z, Pang L, Fang F, Zhang G, Zhang J, Xie M, et al. Resveratrol attenuates brain damage in a rat model of focal cerebral ischemia via up-regulation of hippocampal Bcl-2. Brain Research. 2012;**1450**:116-124. DOI: 10.1016/j.brainres.2012.02.019

[44] Huang W, Chen Z, Wang Q, Lin M, Wu S, Yan Q, et al. Piperine potentiates the antidepressant-like effect of trans-resveratrol: Involvement of monoaminergic system. Metabolic Brain Disease. 2013;**28**(4):585-595. DOI: 10.1007/s11011-013-9426-y

[45] Ogle WO, Speisman RB, Ormerod BK. Potential of treating agerelated depression and cognitive decline with nutraceutical approaches: A mini-review. Gerontology. 2013;**59**(1):23-31. DOI: 10.1159/000342208

[46] Pathak L, Agrawal Y, Dhir A. Natural polyphenols in the management of major depression. Expert Opinion on Investigational Drugs. 2013;**22**(7):863-880. DOI: 10.1517/13543784.2013.794783

[47] Costa D, Galvão AM, Di Paolo RE, Freitas AA, Lima JC, Quina FH, et al. Photochemistry of the hemiketal form of anthocyanins and its potential role in plant protection from UV-B radiation. Tetrahedron. 2015;**71**(20):3157-3162. DOI: 10.1016/j.tet.2014.06.092

[48] Decean H, Fischer-Fodor E, Tatomir C, Perde-Schrepler M, Somfelean L, Burz C, et al. *Vitis vinifera* seeds extract for the modulation of cytosolic factors BAX-α and NF-kB involved in UVB-induced oxidative stress and apoptosis of human skin cells. Clujul Medical. 2016;**89**(1):72-81. DOI: 10.15386/cjmed-508

[49] Sudheer AR, Muthukumaran S, Devipriya N, Menon VP. Ellagic acid, a natural polyphenol protects rat peripheral blood lymphocytes against nicotine-induced cellular and DNA damage in vitro: With the comparison of N-acetylcysteine. Toxicology. 2007;**230**(1):11-21. DOI: 10.1016/j.tox.2006.10.010

[50] Larrosa M, Tomás-Barberán FA, Espín JC. The dietary hydrolysable tannin punicalagin releases ellagic acid that induces apoptosis in human colon adenocarcinoma Caco-2 cells by using the mitochondrial pathway. The Journal of Nutritional Biochemistry. 2006;**17**(9):611-625. DOI: 10.1016/j.jnutbio.2005.09.004

[51] Yun JM, Afaq F, Khan N, Mukhtar H. Delphinidin, an anthocyanidin in pigmented fruits and vegetables, induces apoptosis and cell cycle arrest in human colon cancer HCT116 cells. Molecular Carcinogenesis. 2010;**48**(3):260-270. DOI: 10.1002/mc.20477

[52] Feng R, Wang SY, Shi YH, Fan J, Yin XM. Delphinidin induces necrosis in hepatocellular carcinoma cells in the presence of 3-methyladenine, an autophagy inhibitor. Journal of Agricultural and Food Chemistry. 2010;**58**(7):3957-3964. DOI: 10.1021/jf9025458

[53] Hafeez BB, Siddiqui IA, Asim M, Malik A, Afaq F, Adhami VM, et al. A dietary anthocyanidin delphinidin induces apoptosis of human prostate cancer PC3 cells *in vitro* and *in vivo*: Involvement of nuclear factor-kappaB signaling. Cancer Research.

2008;**68**(20):8564-8572. DOI: 10.1158/0008-5472.CAN-08-2232

[54] Mukherjee S, Dudley JI, Das DK. Dose-dependency of resveratrol in providing health benefits. Dose-Response. 2010;**8**(4):478-500. DOI: 10.2203/dose-response.09-015. Mukherjee

[55] Prasad R, Katiyar SK. Grape seed proanthocyanidins inhibit migration potential of pancreatic cancer cells by promoting mesenchymal-to-epithelial transition and targeting NF-κB. Cancer Letters. 2013;**334**(1):118-126. DOI: 10.1016/j.canlet.2012.08.003

[56] Cvorovic J, Tramer F, Granzotto M, Candussio L, Decorti G, Passamonti S. Oxidative stress-based cytotoxicity of delphinidin and cyanidin in colon cancer cells. Archives of Biochemistry and Biophysics. 2010;**501**(1):151-157. DOI: 10.1016/j.abb.2010.05.019

[57] Busquets S, Ametller E, Fuster G, Olivan M, Raab V, Argilés JM, et al. Resveratrol, a natural diphenol, reduces metastatic growth in an experimental cancer model. Cancer Letters. 2007;**245**(1-2):144-148. DOI: 10.1016/j.canlet.2005.12.035

[58] Muñoz-González I, Thurnheer T, Bartolomé B, Moreno-Arribas MV. Red wine and oenological extracts display antimicrobial effects in an oral bacteria biofilm model. Journal of Agricultural and Food Chemistry. 2014;**62**(20):4731-4737. DOI: 10.1021/jf501768p

[59] Queipo-Ortuño MI, Boto-Ordóñez M, Murri M, Gomez-Zumaquero JM, Clemente-Postigo M, Estruch R, et al. Influence of red wine polyphenols and ethanol on the gut microbiota ecology and biochemical biomarkers. The American Journal of Clinical Nutrition. 2012;**95**(6):1323-1334. DOI: 10.3945/ajcn.111.027847

[60] Friedman M. Antibacterial, antiviral, and antifungal properties of wines and winery byproducts in relation to their flavonoid content. Journal of Agricultural and Food Chemistry. 2014;**62**(26):6025-6042. DOI: 10.1021/jf501266s

[61] Just JR, Daeschel MA. Antimicrobial effects of wine on *Escherichia coli* O157:H7 and *Salmonella typhimurium* in a stomach model system. Journal of Food Science. 2003;**68**(1):285-290. DOI: 10.1111/j.1365-2621.2003.tb14154.x

[62] Clouser CL, Chauhan J, Bess MA, Oploo JL, Zhou D, Dimick-Gray S, et al. Anti-HIV-1 activity of resveratrol derivatives and synergistic inhibition of HIV-1 by the combination of resveratrol and decitabine. Bioorganic & Medicinal Chemistry Letters. 2012;**22**(21):6642-6646. DOI: 10.1016/j.bmcl.2012.08.108

[63] Goldberg DM, Ng E, Karumanchiri A, Diamandis EP, Soleas GJ. Resveratrol glucosides are important components of commercial wines. American Journal of Enology and Viticulture. 1996;**47**:415-420

[64] Stockley C, Hoj P. Better wine for better health: Fact or fiction? Australian Journal of Grape and Wine Research. 2005;**11**(2):127-138. DOI: 10.1111/j.1755-0238.2005.tb00284.x

[65] Ruggiero A, Vitalini S, Burlini N, Bernasconi S, Iriti M. Phytosterols in grapes and wine, and effects of agrochemicals on their levels. Food Chemistry. 2013;**141**(4):3473-3479. DOI: 10.1016/j.foodchem.2013.05.153

[66] Song J, Smart R, Wang H, Dambergs B, Sparrow A, Qian MC. Effect of grape bunch sunlight exposure and UV radiation on phenolics and volatile composition of *Vitis vinifera* L. cv. Pinot noir wine. Food Chemistry.

2015;**173**:424-431. DOI: 10.1016/j.foodchem.2014.09.150

[67] Pastore C, Zenoni S, Fasoli M, Pezzotti M, Tornielli G, Filippetti I. Selective defoliation affects plant growth, fruit transcriptional ripening program and flavonoid metabolism in grapevine. BMC Plant Biology. 2013;**13**:30. DOI: 10.1186/1471-2229-13-30

[68] Berli F, D'angelo J, Cavagnaro B, Bottini R, Wuilloud R, Silva MF. Phenolic composition in grape (*Vitis vinifera* L. cv. Malbec) ripened with different solar UV-B radiation levels by capillary zone electrophoresis. Journal of Agricultural and Food Chemistry. 2008;**56**(9):2892-2898. DOI: 10.1021/jf073421+

[69] Bavaresco L, Mattivi F, De Rosso M, Flamini R. Effects of elicitors, viticultural factors, and enological practices on resveratrol and stilbenes in grapevine and wine. Mini Reviews in Medicinal Chemistry. 2012;**12**(13):1366-1381. DOI: 10.2174/13895575112091366

[70] Vitalini S, Gardana C, Zanzotto A, Fico G, Faoro F, Simonetti P, et al. From vineyard to glass: Agrochemicals enhance the melatonin content, total polyphenols and antiradical activity of red wines. Journal of Pineal Research. 2011;**51**(3):278-285. DOI: 10.1111/j.1600-079X.2011.00887.x

[71] Vejarano R, Morata A, Loira I, González MC, Suárez-Lepe JA. Theoretical considerations about usage of metabolic inhibitors as possible alternative to reduce alcohol content of wines from hot areas. European Food Research and Technology. 2013;**237**(3):281-290. DOI: 10.1007/s00217-013-1992-z

[72] Kostadinović S, Wilkens A, Stefova M, Ivanova V, Vojnoski B, Mirhosseini H, et al. Stilbene levels and antioxidant activity of Vranec and Merlot wines from Macedonia: Effect of variety and enological practices. Food Chemistry. 2012;**135**(4):3003-3009. DOI: 10.1016/j.foodchem.2012.06.118

[73] Federico Casassa L, Bolcato EA, Sari SE, Fanzone ML, Jofre VP. Combined effect of prefermentative cold soak and SO_2 additions in Barbera D'Asti and Malbec wines: Anthocyanin composition, chromatic and sensory properties. LWT- Food Science and Technology. 2016;**66**:134-142. DOI: 10.1016/j.lwt.2015.10.026

[74] Morata A, Loira I, Vejarano R, González C, Callejo MJ, Suárez-Lepe JA. Emerging preservation technologies in grapes for winemaking. Trends in Food Science and Technology. 2017;**67**:36-43. DOI: 10.1016/j.tifs.2017.06.014

[75] Santos MC, Nunes C, Cappelle J, Gonçalves FJ, Rodrigues A, Saraiva JA, et al. Effect of high pressure treatments on the physicochemical properties of a sulphur dioxide-free red wine. Food Chemistry. 2013;**141**:2558-2566. DOI: 10.1016/j.foodchem.2013.05.022

[76] Donsi F, Ferrari G, Pataro G. Applications of pulsed electric field treatments for the enhancement of mass transfer from vegetable tissue. Food Engineering Reviews. 2010;**2**:109-130. DOI: 10.1007/s12393-010-9015-3

[77] Castellari M, Matricardi L, Arfelli G, Rovere P, Amati A. Effects of high-pressure processing on polyphenoloxidase enzyme activity of grape musts. Food Chemistry. 1997;**60**:647-649. DOI: 10.1016/S0308-8146(97)00050-2

[78] Morata A, Loira I, Vejarano R, Bañuelos MA, Sanz P, Otero L, et al. Grape processing by high hydrostatic pressure: Effect on microbial populations, phenol extraction and

wine quality. Food and Bioprocess Technology. 2015;**8**(2):277-286. DOI: 10.1007/s11947-014-1405-8

[79] Corrales M, Toepfl S, Butz P, Knorr D, Tauscher B. Extraction of anthocyanins from grape by-products assisted by ultrasonics, high hydrostatic pressure or pulsed electric fields: A comparison. Innovative Food Science & Emerging Technologies. 2008;**9**:85-91. DOI: 10.1016/j.ifset.2007.06.002

[80] Corrales M, Butz P, Tauscher B. Anthocyanin condensation reactions under high hydrostatic pressure. Food Chemistry. 2008;**110**:627-635. DOI: 10.1016/j.foodchem.2008.02.055

[81] Escott C, Vaquero C, del Fresno JM, Bañuelos MA, Loira I, Han SY, et al. Pulsed light effect in red grape quality and fermentation. Food and Bioprocess Technology. 2017;**10**(8):1540-1547. DOI: 10.1007/s11947-017-1921-4

[82] Morata A, Bañuelos MA, Tesfaye W, Loira I, Palomero F, Benito S, et al. Electron beam irradiation of wine grapes: Effect on microbial populations, phenol extraction and wine quality. Food and Bioprocess Technology. 2015;**8**:1845-1853. DOI: 10.1007/s11947-015-1540-x

[83] Yang N, Huang K, Lyu C, Wang J. Pulsed electric field technology in the manufacturing processes of wine, beer, and rice wine: A review. Food Control. 2016;**61**:28-38. DOI: 10.1016/j.foodcont.2015.09.022

[84] López N, Puértolas E, Condón S, Álvarez I, Raso J. Effects of pulsed electric fields on the extraction of phenolic compounds during the fermentation of must of Tempranillo grapes. Innovative Food Science & Emerging Technologies. 2008;**9**:477-482. DOI: 10.1016/j.ifset.2007.11.001

[85] Delsart C, Cholet C, Ghidossi R, Grimi N, Gontier E, Gény L, et al. Effects of pulsed electric fields on Cabernet Sauvignon grape berries and on the characteristics of wines. Food and Bioprocess Technology. 2014;**7**:424-436. DOI: 10.1007/s11947-012-1039-7

[86] Puértolas E, Saldana G, Condon S, Alvarez I, Raso J. Evolution of polyphenolic compounds in red wine from Cabernet Sauvignon grapes processed by pulsed electric fields during aging in bottle. Food Chemistry. 2010;**119**:1063-1070. DOI: 10.1016/j.foodchem.2009.08.018

[87] Puértolas E, López N, Condón S, Raso J, Álvarez I. Pulsed electric fields inactivation of wine spoilage yeast and bacteria. International Journal of Food Microbiology. 2009;**130**:49-55. DOI: 10.1016/j.ijfoodmicro.2008.12.035

[88] Tiwari BK, Patras A, Brunton N, Cullen PJ, O'Donnell CP. Effect of ultrasound processing on anthocyanins and color of red grape juice. Ultrasonics Sonochemistry. 2010;**17**:598-604. DOI: 10.1016/j.ultsonch.2009.10.009

[89] Ghafoor K, Choi YH, Jeon JY, Jo IH. Optimization of ultrasound assisted extraction of phenolic compounds, antioxidants, and anthocyanins from grape (*Vitis vinifera*) seeds. Journal of Agricultural and Food Chemistry. 2009;**57**:4988-4994. DOI: 10.1021/jf9001439

[90] Fava J, Hodara K, Nieto A, Guerrero S, Alzamora SM, Castro MA. Structure (micro, ultra, nano), color and mechanical properties of *Vitis labrusca* L. (grape berry) fruits treated by hydrogen peroxide, UV-C irradiation and ultrasound. Food Research International. 2011;**44**:2938-2948. DOI: 10.1016/j.foodres.2011.06.053

[91] Augustine S, Kudachikar VB, Vanajakshi V, Ravi R. Effect of combined preservation techniques on the stability and microbial quality and retention of anthocyanins in grape

pomace stored at low temperature. Journal of Food Science and Technology. 2013;**50**:332-338. DOI: 10.1007/s13197-011-0325-0

[92] Artés-Hernández F, Aguayo E, Artés F, Tomás-Barberán FA. Enriched ozone atmosphere enhances bioactive phenolics in seedless table grapes after prolonged shelf life. Journal of the Science of Food and Agriculture. 2007;**87**(5):824-831. DOI: 10.1002/jsfa.2780

[93] Sarig P, Zahavi T, Zutkhi Y, Yannai S, Lisker N, Ben-Arie R. Ozone for control of post-harvest decay of table grapes caused by *Rhizopus stolonifer*. Physiological and Molecular Plant Pathology. 1996;**48**(6):403-415. DOI: 10.1006/pmpp.1996.0032

[94] Laureano J, Giacosa S, Río Segade S, Torchio F, Cravero F, Gerbi V, et al. Effects of continuous exposure to ozone gas and electrolyzed water on the skin hardness of table and wine grape varieties. Journal of Texture Studies. 2016;**47**(1):40-48. DOI: 10.1111/jtxs.12158

[95] Paissoni MA, Río Segade S, Giacosa S, Torchio F, Cravero F, Englezos V, et al. Impact of post-harvest ozone treatments on the skin phenolic extractability of red winegrapes cv Barbera and Nebbiolo (*Vitis vinifera* L.). Food Research International. 2017;**98**:68-78. DOI: 10.1016/j.foodres.2016.11.013

[96] Bellincontro A, Catelli C, Cotarella R, Mencarelli F. Postharvest ozone fumigation of Petit Verdot grapes to prevent the use of sulfites and to increase anthocyanin in wine. Australian Journal of Grape and Wine Research. 2017;**23**:200-206. DOI: 10.1111/ajgw.12257

[97] Fernández-Cruz E, Álvarez-Fernández MA, Valero E, Troncoso AM, García-Parrilla MC. Melatonin and derived l-tryptophan metabolites produced during alcoholic fermentation by different wine yeast strains. Food Chemistry. 2017;**217**:431-437. DOI: 10.1016/j.foodchem.2016.08.020

[98] Morata A, Loira I, Suárez-Lepe JA. Influence of yeasts in wine colour. In: Morata A, Loira I, editors. Grape and Wine Biotechnology. London: IntechOpen; 2016. pp. 285-305. DOI: 10.5772/65055

[99] Barcia MT, Pertuzatti PB, Rodrigues D, Gómez-Alonso S, Hermosín-Gutiérrez I, Godoy HT. Occurrence of low molecular weight phenolics in *Vitis vinifera* red grape cultivars and their winemaking by-products from São Paulo (Brazil). Food Research International. 2014;**62**:500-513. DOI: 10.1016/j.foodres.2014.03.051

[100] Wightman JD, Wrolstad RE. D-glucosidase activity in juice-processing enzymes based on anthocyanin analysis. Journal of Food Science. 1996;**61**:544-552. DOI: 10.1111/j.1365-2621.1996.tb13153.x

[101] Morata A, Vejarano R, Ridolfi G, Benito S, Palomero F, Uthurry C, et al. Reduction of 4-ethylphenol production in red wines using HCDC+ yeasts and cinnamyl esterases. Enzyme and Microbial Technology. 2013;**52**(2):99-104. DOI: 10.1016/j.enzmictec.2012.11.001

[102] Bakker J, Timberlake CF. Isolation, identification, and characterization of new color-stable anthocyanins occurring in some red wines. Journal of Agricultural and Food Chemistry. 1997;**45**:35-43. DOI: 10.1021/jf960252c

[103] Benito S, Palomero F, Morata A, Uthurry C, Suárez-Lepe JA. Minimization of ethylphenol precursors in red wines via the formation of pyranoanthocyanins by selected yeasts. International Journal of Food Microbiology. 2009;**132**:145-152. DOI: 10.1016/j.ijfoodmicro.2009.04.015

[104] Morata A, Benito S, Loira I, Palomero F, González MC, Suárez-Lepe JA. Formation of pyranoanthocyanins by *Schizosaccharomyces pombe* during the fermentation of red must. International Journal of Food Microbiology. 2012;**159**:47-53. DOI: 10.1016/j.ijfoodmicro.2012.08.007

[105] Benito S, Morata A, Palomero F, González MC, Suárez-Lepe JA. Formation of vinylphenolic pyranoanthocyanins by *Saccharomyces cerevisiae* and *Pichia guillermondii* in red wines produced following different fermentation strategies. Food Chemistry. 2011;**124**:15-23. DOI: 10.1016/j.foodchem.2010.05.096

[106] Vejarano R. *Saccharomycodes ludwigii*, control and potential uses in winemaking processes. Fermentation. 2018;**4**(3):71. DOI: 10.3390/fermentation4030071

[107] Vejarano R, Gil-Calderón A, Morata A. Effect of metabolic inhibitors on the alcoholic fermentation: Tolerant yeasts. In: Proceedings of the LACCEI International Multi-Conference for Engineering, Education and Technology; 2018; p. 171. DOI: 10.18687/LACCEI2018.1.1.171

[108] Monagas M, Bartolomé B, Gómez-Cordovés C. Evolution of polyphenols in red wines from *Vitis vinifera* L. during aging in the bottle. European Food Research and Technology. 2005;**220**:331-340. DOI: 10.1007/s00217-004-1109-9

[109] Loira I, Vejarano R, Morata A, Ricardo-da-Silva JM, Laureano O, González MC, et al. Effect of *Saccharomyces* strains on the quality of red wines aged on lees. Food Chemistry. 2013;**139**:1044-1051. DOI: 10.1016/j.foodchem.2013.01.020

[110] Sheng R, Gu ZL, Xie ML, Zhou WX, Guo CY. EGCG inhibits cardiomyocyte apoptosis in pressure overload-induced cardiac hypertrophy and protects cardiomyocytes from oxidative stress in rats. Acta Pharmacologica Sinica. 2007;**28**(2):191-201. DOI: 10.1111/j.1745-7254.2007.00495.x

[111] Palomero F, Morata A, Benito S, Calderón F, Suárez-Lepe JA. New genera of yeasts for over-lees aging of red wine. Food Chemistry. 2009;**112**:432-441. DOI: 10.1016/j.foodchem.2008.05.098

[112] Salmon JM. Interactions between yeast, oxygen and polyphenols during alcoholic fermentations: Practical implications. LWT- Food Science and Technology. 2006;**39**:959-965. DOI: 10.1016/j.lwt.2005.11.005

[113] González-Arenzana L, Portua J, López R, López N, Santamaría P, Garde-Cerdán T, et al. Inactivation of wine-associated microbiota by continuous pulsed electric field treatments. Innovative Food Science & Emerging Technologies. 2015;**29**:187-192. DOI: 10.1016/j.ifset.2015.03.009

[114] Rice-Evans CA, Miller NJ. Total antioxidant status in plasma and body fluids. Methods in Enzymology. 1994;**234**:279-293. DOI: 10.1016/0076-6879(94)34095-1

[115] García-Alonso M, Rimbach G, Rivas-Gonzalo JC, de Pascual-Teresa S. Antioxidant and cellular activities of anthocyanins and their corresponding vitisins A–studies in platelets, monocytes, and human endothelial cells. Journal of Agricultural and Food Chemistry. 2004;**52**:3378-3384. DOI: 10.1021/jf035360v

[116] McDougall GJ, Fyffe S, Dobson P, Stewart D. Anthocyanins from red wine–their stability under simulated gastrointestinal digestion. Phytochemistry. 2005;**66**:2540-2548. DOI: 10.1016/j.phytochem.2005.09.003

Chapter 12

Main Operating Conditions That Can Influence the Evolution of Wines during Long-Term Storage

Francesca Venturi, Chiara Sanmartin, Isabella Taglieri, Giuseppe Ferroni, Mike Frank Quartacci, Cristina Sgherri, Xiaoguo Ying, Gianpaolo Andrich and Angela Zinnai

Abstract

Nowadays, among all the possible wine packaging materials, an increasing use of polyethylene terephthalate (PET), multilayer Tetra Brik, and Bag-in-Box containers can be observed. Due to the fact that oxygen is counted among the primary factors which act on wine aging and degradation, a tight control of oxygen is critical during wine making and conservation. Wine protection from external conditions is strictly linked to packaging, which has the basic role to preserve the quality of wine during its evolution and aging. In this chapter the time evolution of different wines will be analyzed according to the storage conditions used. In particular, the following specific cases of study will be discussed: Case of study 1a, 1b, and 1c: influence of storage conditions (storage temperature, packaging material and volume of packaging) on the time evolution of red wine over a storage period of 12 months. Case of study 2: evolution of glass bottled rosé wine as a function of closure (cork stopper with or without aluminum capsule), storage position and brightness regime over a period 12 months.

Keywords: red wine, rosé wine, storage conditions, packaging, bottle position, capsule, antioxidant capacity, kinetic characterization

1. Introduction

1.1 Food packaging, shelf life, and quality decay rate

According to [1], it is possible to highlight four basic functions for traditional food packaging. The most basic function of packaging is *containment*, as food products must be contained before they can be moved from one place to another. Furthermore, for many food products, the *protection* afforded by the package is an essential part of the preservation process. At this regard, packaging protects its contents from the outside environmental effects of water, water vapor, gases, odors, microorganisms, dust, shocks, vibrations, compressive forces, and so on. Packaging allows also primary packages to be assembled into secondary (e.g., cardboard boxes) and tertiary packages (e.g., stretch-wrapped pallets), thus improving the *convenience* throughout the supply chain. In this way, the handling of the material is made more functional because a reduced number of containers and loading operations must be handled or carried out, respectively. Finally, packaging can provide the *communication* necessary for food sailing: as consumers can make purchasing decisions using the numerous clues provided by the graphics and the distinctive shapes of the packaging, there is an old saying that "a package must protect what it sells and sell what it protects."

Overall, packaging is an essential element in food manufacture since it facilitates food management, increases food shelf life, and makes it more acceptable to consumers.

According to [2], "shelf life" can be defined as a finite length of time after production (in some cases after maturation or aging) and packaging during which the food product retains a required level of quality under well-defined storage conditions. In other words, taking for granted the consumer's safety, for any kind of food product, there should be a defined quality level (defined as "acceptability limit") discriminating products that are still acceptable for consumption from those no longer acceptable. Once defined the storage conditions to be used, for each food product, "shelf life" represents the time needed to reach the acceptability limit which is directly influenced by the "quality decay rate" of the stored food.

1.2 Packaging material for wine storage

Nowadays, glass containers are still preferred for wine bottling [3] being them readily recyclable and characterized by a high impermeability to gases and vapors, stability over time, and transparency [4]. On the other hand, because of some objective limitations for the extensive use of glass containers in food industry (i.e., heavy weight, fragility to internal pressure, impact and thermal shock, etc.) [5], there is a worldwide growing demand for alternative solutions to glass also for wine bottling [6]. This with the aim to propose inexpensive packaging resources, practical to use and often marketed as "eco-friendly," particularly in relation to their contributions to waste prevention [3, 7, 8].

For the above reasons, starting from the past two decades, among all the possible packaging materials, an increased utilization of polymeric materials also for wine packaging, including polyethylene terephthalate (PET) bottles, multilayer Tetra Brik®, and Bag-in-Box (BiB)-type containers, has been observed [1, 9]. Some of the main advantages and disadvantages of typical materials used in wine packaging are reported in **Table 1**.

1.3 Main storage conditions affecting the quality decay rate of wines

According to [15], wine aging can be defined as the time that goes from the end of winemaking (during which wine is subjected to different operations depending on both the vine and usual winery methodology) to its final consumption. In bottles, the proper aging of wine is linked to the presence of reduced conditions that lead to color changes and to the establishment of desired sensory (olfactory and tasteful) characteristics. During evolution and aging, the contact of wine with oxygen should be limited to the minimum. The time needed to develop such transformation differs among wines and is a function of both starting chemical composition and storage conditions.

Among all the operating conditions that can be selected during long-term wine storage, the main ones involved in the quality decay rate of wines are described below.

Material	Brief description	Advantages	Disadvantages
Glass [10, 11]	Soda-lime glass, composed of about 75% silicon dioxide (SiO_2), calcium oxide (CaO), sodium oxide (Na_2O), and several minor additives	Impermeable to gases and vapors Odorless and chemically inert Useful for heat sterilization Good insulation Produced in different shapes Variations in glass color can protect light-sensitive contents Transparent Reusable and recyclable	Brittleness Fragility to internal pressure, impact, and thermal shock Needs a separate closure Limitation in thin glass Heavyweight Transportation costs
PET [12]	Polyethylene terephthalate (PET) is combined with terephthalic acid and ethylene glycol	Fluid and moldable Produced in different shapes Flexible Variations in PET color can protect light-sensitive contents Transparent Inexpensive Lightweight Wide range of physical and optical properties Easy to print Integrated into production processes where the package is formed, filled, and sealed in the same production line Easy handling by consumers Needs a separate closure	Variable permeability to light Limited reuse Poor barrier to gases and vapors Not suitable to protect wine for long periods of time Migration of chemicals from PET to food
Tetra Brik® [13]	Tetra Brik® packaging is made up of three raw materials: duplex paper (about 75%), aluminum (about 5%), and low-density polyethylene (about 20%)	Good barrier properties to light Integrated into production processes where the package is formed, filled, and sealed in the same production line Lightweight Recyclable Efficient, low-cost protection Easy handling by consumers	Impacts the organoleptic quality Poor barrier to gases and vapors Not suitable to protect wine for long periods of time When used as primary packaging, it is coated or laminated to improve functional and protective properties Migration of chemicals from internal coating to the content Hard to recycle
Bag-in-Box® (BiB) [14]	The product is sealed in a bag comprising one or more plies of high barrier flexible films, mechanically supported by an external paperboard carton. A valve fitment is attached to the bag through which the product is filled and dispensed	Good barrier properties to light Integrated into production processes where the package is formed, filled, and sealed in the same production line Lightweight Improved distribution efficiency Enhanced end-use convenience Increased cost-effectiveness Easy handling by consumers	Impacts the organoleptic quality Poor barrier to light, gases and vapors Easily sorbs aroma compounds, particularly if hydrophobic Incomplete air tightness of the valve Not suitable to protect wine for long periods of time

Table 1.
Advantages and disadvantages of typical materials used in wine packaging.

1.3.1 pO_2 in storage atmosphere

During wine storage, spontaneous clearing, color stabilization, and reactions that lead to the formation of more complex compounds are observed [16]. In red and rosé wines, reactions of copigmentation and polymerization of anthocyanins (Ant) take place as the storage time in bottle increases [17]. These reactions cause the formation of more stable compounds responsible for the change from the bluish-red hues of young wines to the orange-red ones characteristic of aged wines [18].

As oxygen is one of the main factors affecting wine evolution as well as its deterioration [3, 19–22], changes occurring after fermentation are partly driven by chemical oxidations deriving from winemaking and storage [23].

During storage in glass bottle, the only barrier against the external atmosphere is represented by the closure system, and the evolution of phenolic compounds in the development of wine color and mouthfeel mainly depends on the transfer of oxygen through the bottle stopper [24]. In this condition, oxygen diffusion into the bottled wine is strongly dependent on the effective sealing of the closure [25, 26]. Indeed, oxygen permeability may greatly change from cap to cap, and this heterogeneity is one of the main factors affecting variation among bottles [23].

Furthermore, as recently reported by [27, 28], the combination of aluminum capsule with cork stopper as well as the storage position used during bottle aging can deeply influence the oxygen intake through the closure system and then the quality decay rate of the stored wine.

1.3.2 Storage temperature

Arrhenius equation describes the relationship between the kinetic constant of a reaction and temperature [29]:

$$k = A \bullet e^{-\frac{E_a}{R \times T}} \quad (1)$$

where k, kinetic constant; A, pre-exponential factor, constant for temperature variations not too high, the value of which depends on the frequency of collisions and the steric factor; E_a, activation energy, also constant for temperature variations not too high; R, gas constant 8.3144 J/(mol K); T, absolute temperature (K).

Based on this equation, it can be assumed that, as the temperature rises, there is an increase in the rate of occurring reactions.

In this context, the reaction mechanisms involved in wine aging as well as their activation energy are very sensitive to temperature, and increasing storage temperature involves an acceleration of the aging process of wine thus influencing its shelf life. In particular, high temperature is a particularly unfavorable condition during storage as the rate of quinone formation enhances with the increase in temperature, although the kinetics of this reaction is temperature independent [30–34].

Besides affecting the kinetics of degradative reactions and particularly the oxidative ones [35–37], temperature also influences the amount of oxygen dissolved in wine. At temperatures of 5–35°C, the amount of O_2 needed for the saturated wine ranged from 10.5 to 5.6 mg/L, the lowest concentration being dissolved at the highest temperature [38]. Furthermore, temperature influences the oxygen permeability of thermoplastic polymers [1, 34, 39, 40].

Other parameters affected by temperature are some physical features of wines, such as viscosity and density: Košmerl and Abramovič [41] characterized 40 samples of bottled Slovenian wines by standard chemical analyses, in order to analyze the effect of temperature (from 20 to 50°C) on their density and viscosity. They concluded that wine behaved as Newtonian fluids so that their density and viscosity were dependent on temperature and decreased nonlinearly with increasing temperature. In particular, they observed a very strong effect of temperature on the viscosity of wines in samples with a high reducing sugar concentration. Yanniotis et al. [42] measured the viscosity of commercial red, white, and sweet wines as well as of model aqueous ethanol and glycerol solutions; they observed that the viscosity decreased with the increase in temperature, and this trend could be fully described by the Arrhenius equation. It was also observed that alcohol and dry extract were the two main factors influencing the viscosity of wines [42].

1.3.3 Brightness level

Exposure of bottled wine to light tends to occur in retail outlets or in domestic situations where artificial (including fluorescent) lighting generates short wavelength (low visible and ultraviolet) radiations. As widely reported in the literature and, in particular, by Dias and coworkers [43], both off-odor production and pigmentation enhancement occur following light exposure.

Most of transparent glass bottles do not guarantee an adequate protection from long-wave radiations, thus exposing wine (mainly white and rosé) to the negative effects of photooxidation. Such reaction is often supported and potentiated by high temperatures [43] which are often detected on the shelves of some supermarkets.

1.4 Main parameters useful to describe the quality decay of wine over storage time

1.4.1 Chemical evolution of stored wine: Kinetics of SO_2 and anthocyanin degradation

As SO_2 plays an important protective role against oxidation in wine, the chemical degradation of this compound during storage may represent a good index of the oxidative processes occurring in the product as a function of the packaging used [39, 44].

Generally, in wine, SO_2 can exist in many interconvertible forms represented by a variety of "free" (FSO_2) and "bound" (BSO_2) forms. The actual protective concentration of SO_2 during wine evolution and aging depends on many factors (i.e., pH, level and type of binding compounds, oxygen concentration, and so on). Thus, the total SO_2 concentration ($TSO_2 = FSO_2 + BSO_2$) can be considered an index of the oxidative damage caused by storage conditions. Indeed, FSO_2 is an intermediate product which concentration is influenced by various chemical reactions different from the oxidative ones.

As reported in [26], the time evolution of TSO_2 concentration could be described by a first-order kinetic equation:

$$-d[TSO_2]_{t=t}/dt = kTSO_2 \bullet [TSO_2]_{t=t} \tag{2}$$

where k_{TSO2} is the kinetic constant related to TSO_2 degradation and $[TSO_2]_{t=t}$ is the concentration of total SO_2 at the generic reaction time t = t.

After integration, the following equation can be obtained:

$$[TSO_2]_{t=t} = [TSO_2]_{t=0} \bullet e^{-kTSO2 \bullet t} \tag{3}$$

where the two functional parameters k and $[TSO_2]_{t=0}$ may be considered a valid measure of the effect induced by oxidation during wine storage as a function of the packaging and storage temperature used.

Color is one of the most important organoleptic characteristics of red wines and affects the quality evaluation of the product [45]. Anthocyanins (Ant) are the most important molecules responsible of the young red wines' color. The color change from red-purple to brick-red hues is strongly related to the concentration of oxygen present in the stored wine [46].

The same experimental approach reported above to describe TSO_2 time evolution can be also followed to describe the time evolution of total anthocyanin concentration (TAnt) that may represent a second index of oxidative degradation of the product as a function of packaging.

1.4.2 Chemical evolution of stored wine: Antioxidant capacity

As polyphenols are widely known to play a protective action on the organism against cardiovascular and degenerative diseases [47], the moderate consumption of wine, especially red and rosé ones, has been associated with the reduction of mortality caused by many chronic diseases, a phenomenon that is commonly known as the "French paradox" [48]. In this context, the health properties of wines have been mainly interpreted on the basis of the antioxidant properties of the flavonoid fraction, which are related to both free radical scavenging and transition metal chelating mechanisms [49].

1.4.3 Sensorial evolution of stored wine

In the field of sensory science, sensory analysis was initially adopted as a tool for quality control [50]. Since then, it has evolved in one of the most diffused and sophisticated toolkits, allowing to achieve an exhaustive description of the characteristics of the products [51]. According to Stone et al. [52], "Sensory evaluation is a scientific discipline used to evoke, measure, analyze and interpret reactions referable to those characteristics of products as they are perceived by the senses of sight, smell, taste, touch, and hearing" [50].

In this context, it is possible to introduce the "sensory shelf life" concept of a product [53]. This can be defined as the storage time at which overall quality, or the intensity of a specific sensory attribute, reaches a predetermined value or "failure criterion," assuming that once the product has reached this point, it is no longer saleable [54].

As a function of specific characters, sensory analysis should also be performed in parallel with microbiological and/or chemical-physical shelf life analysis to monitor the sensory profile of the product for potential deleterious sensory attribute changes [53]. Thus, sensory variables used during sensory shelf life testing could include the monitoring of specific sensory attributes related to visual, aroma, and taste attributes which can be used as indices of sensory quality.

As reported by Jackson [55], most sensory changes that negatively affect wine shelf life are those associated with oxidation and hydrolysis of esters. Such changes are involved in reduction, polymerization, structural rearrangement, and volatility modifications; their relative importance depends on wine style, production techniques, varietal origin, storage conditions, and consumer expectation [55].

2. Experimental evidences

With the aim to better understand the time evolution of wines during bottle aging as a function of storage conditions, among the literature available on the topic, we selected and discussed two real case reports recently developed by our group (**Figure 1**; **Table 2**).

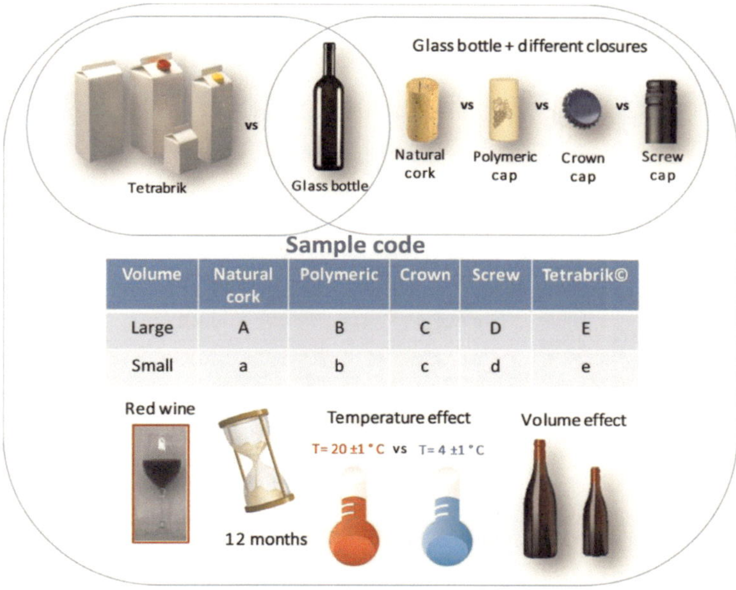

Figure 1.
Cases of study 1a, 1b, and 1c: graphical abstract—Experimental setup.

2.1 Case of study 1a, 1b, and 1c: Influence of storage conditions (temperature, packaging material, and volume of packaging) on the time evolution of a red wine over a storage period of 12 months

The red wine (**Table 2**) was packed in different packaging materials at the same time in a commercial winery bottling line using a fully automated bottling/filling station, as described in **Figure 1**.

Parameter	Mean value ± c.i.*
Alcohol (%v/v)	11.46 ± 0.06
pH	3.62 ± 0.01
Titratable acidity (g/L as tartaric acid)	4.82 ± 0.70
Net volatile acidity (g/L as acetic acid)	0.550 ± 0.003
Total SO_2 (TSO_2) (g/L)	0.106 ± 0.001
Total phenols (g/L as gallic acid)	2.140 ± 0.064
Total anthocyanins (g/L as malvin)	0.470 ± 0.006

**c.i., confidence interval = $P < 0.05$.*

Table 2.
Initial chemical composition of the red wine.

2.1.1 Case of study 1a: Influence of storage temperature

As reported in **Table 3**, after 12 months of storage, it can be observed that the aging of red wine was significantly delayed at the lowest temperature, regardless of the packaging solution adopted. The only exception was represented by the wine stored in glass bottles closed by natural corks [56].

Sample	k_{TSO2} (months^{-1})	$[TSO_2]_{t=0}$ (mg L^{-1})	r^2
A (T = 20 ± 1°C)	0.056^{a*}	106.8	0.95
A (T = 4 ± 1°C)	0.052^{a}	106.8	0.96
B (T = 20 ± 1°C)	0.060^{a}	105.7	0.97
B (T = 4 ± 1°C)	0.054^{b}	105.7	0.82
C (T = 20 ± 1°C)	0.053^{a}	105.3	0.96
C (T = 4 ± 1°C)	0.045^{b}	105.3	0.81
D (T = 20 ± 1°C)	0.061^{a}	106.2	0.93
D (T = 4 ± 1 °C)	0.052^{b}	106.2	0.82
E (T = 20 ± 1 °C)	0.070^{a}	105.5	0.96
E (T = 4 ± 1 °C)	0.043^{b}	105.5	0.70

**Within the same sample, values with different letters are significantly different ($P < 0.05$).*

Table 3.
*TSO_2 degradation constant (k_{TSO2}) and initial total SO_2 concentration $[TSO_2]_{t=0}$ as a function of storage temperature (time = 12 months). Each sample was identified by code letter ranging from A/a to E/e as described in **Figure 1**.*

2.1.2 Case of study 1b: Influence of volume (two volumes for each packaging) on the chemical evolution of stored wine

As shown in **Table 4**, after 12 months of storage, it can be observed that the TSO_2 degradation rate significantly increased when the volume of the container decreased, regardless of the packaging solution used. In this case, the only exception was represented by the wine stored in glass bottles closed with screw caps.

Sample	k_{TSO2} (months^{-1})	$[TSO_2]_{t=0}$ (mg L^{-1})	r^2
A	0.056^{b*}	106.8	0.95
a	0.073^{a}	106.8	0.97
B	0.060^{b}	105.7	0.97
b	0.068^{a}	105.7	0.95
C	0.053^{b}	105.3	0.96
c	0.069^{a}	105.3	0.93
D	0.061^{a}	106.2	0.93
d	0.059^{a}	106.2	0.98
E	0.070^{b}	105.5	0.96
e	0.082^{a}	105.5	0.98

**Within the same sample, values with different letters are significantly different ($P < 0.05$). Samples represented with upper case letters refer to samples stored in packages with larger volume.*

Table 4.
*TSO_2 degradation constant (k_{TSO2}) and initial total SO_2 concentration $[TSO_2]_{t=0}$ as a function of package volume (T = 20 ± 1°C, storage time = 12 months). Each sample was identified by code letter ranging from A/a to E/e as described in **Figure 1**.*

2.1.3 Case of study 1c: Influence of the packaging material (glass bottles provided with different closures, bag-in-box containers and Tetra Brik®) on the chemical and sensorial evolution of stored wine

As evidenced in **Tables 3** and **4**, the effects of packaging on both SO_2 degradation (**Table 5**) and sensorial characteristics (**Table 6**) were investigated during time

Sample	k_{TSO2} (months^{-1})	$[TSO_2]_{t=0}$ (mg L^{-1})	r^2
a	0.073^{b*}	106.8^{a*}	0.97
b	0.068^{c}	105.7^{a}	0.95
c	0.069^{c}	105.3^{a}	0.93
d	0.059^{d}	106.2^{a}	0.98
e	0.082^{a}	105.5^{a}	0.98

**In each column, the values labeled with different superscript letters show statistically significant differences ($P < 0.05$).*

Table 5.
*Kinetic parameters describing the time evolution of TSO_2 concentration as a function of the packaging used during storage (small volume packages, T = 20°C, storage time = 12 months). Each sample was identified by code letter ranging from A/a to E/e as described in **Figure 1**.*

in wines stored at room temperature (T = 20 ± 1°C) and in small containers. Among all the parameters evaluated, the concomitance of these two conditions together led to a faster degradation.

As reported in **Table 5**, the oxidative degradation occurring in the red wine stored in containers at room temperature (T = 20 ± 1°C) for 12 months was strongly dependent on the packaging, being the TSO_2 degradation rate statistically significant.

In particular, in the wine stored in Tetra Brik®, the reduction of TSO_2 concentration occurred at a faster rate compared to the wine in glass bottles, independently of the closure. This result may be explained with the fact that glass protected wine from oxidative reactions better than the multilayer material. As regards the closures, the lowest TSO_2 degradation rate was observed with screw caps.

Table 6 shows the main sensorial parameters evaluated in the red wines contained in various packages during storage in order to follow the development during time of the organoleptic characteristics. Apart from the closure, after 12 months the wine stored in glass bottles presented high values for the positive sensorial attributes "frankness," "harmony of odor," and "overall pleasantness." On the contrary, the wine stored in Tetra Brik® showed a worsening of the organoleptic characteristics, with high values for "degree of oxidation" and "aftertaste."

2.1.4 Conclusions related to case of study 1a, 1b, and 1c

The results show how the characteristics of packaging affect wine bouquet and flavor as a function of the storage conditions, suggesting that their rational

Sample	Degree of oxidation	Frankness	Harmony of odor	Aftertaste	Overall pleasantness
Wine at starting time	0.7^{b*}	6.0^{a}	4.7^{ab}	2.2^{b}	3.8^{a}
a	4.8^{a}	3.7^{ab}	4.2^{ab}	3.3^{ab}	4.8^{a}
b	4.3^{ab}	4.5^{ab}	5.3^{a}	3.7^{ab}	4.5^{a}
c	3.7^{ab}	4.8^{ab}	5.3^{a}	2.3^{b}	4.5^{a}
d	3.8^{ab}	4.5^{ab}	4.8^{ab}	4.2^{ab}	4.8^{a}
e	4.8^{a}	1.8^{b}	2.0^{b}	6.3^{a}	1.0^{b}

**In each column, the values labeled with different superscript letters show differences statistically significant ($P < 0.05$).*

Table 6.
*Sensorial evolution of red wine as a function of the packaging used during storage (small volume packages, T = 20 ± 1°C, storage time = 12 months). Each sample was identified by code letter ranging from A/a to E/e as described in **Figure 1**.*

optimization, based on experimental data, could improve the shelf life of wine and enhance the consumer's enjoyment during tasting.

Among all the experimental conditions, the rate of wine aging was higher when the volume of the containers decreased and storage temperature increased. Furthermore, after 12 months of storage, glass bottles generally better preserved wine from oxidation than multilayer materials, regardless of the closure characteristics.

To highlight the fact that the rate of TSO_2 degradation may represent a chemical index of the aging degree of the red wine during storage, the TSO_2 degradation kinetic constant (**Table 5**) was correlated for all packaging conditions with the sensory attributes (see **Table 6**). The correlation coefficients are reported in **Table 7**.

Parameter	k_{TSO2}
Frankness	**−0.84**
Harmony of odor	**−0.80**
Aftertaste	0.53
Degree of oxidation	**0.75**
Overall pleasantness	**−0.80**

Note: The correlation coefficients that indicate a strong correlation (≥0.6) are reported in boldface.

Table 7.
Correlation matrix relating the kinetic constant describing TSO_2 degradation to wine attributes (storage time = 12 months; T = 20°C; small volume packages).

According to Paula and Conti-Silva [57], a correlation coefficient of about 0.70 indicates a fairly strong correlation. Thus, data reported for this case of study evidenced that the TSO_2 degradation rate (k_{TSO2}) is strongly inversely correlated to positive sensorial attributes such as "frankness" and "harmony of odor" as well as the hedonic parameter "overall pleasantness," whereas the negative attribute "degree of oxidation" is directly correlated with k_{TSO2}.

Based on the above observations, an integrated approach deriving from the merging of both chemical and sensorial data can be used to identify the best packaging and storage conditions necessary to extend the shelf life of red wines. In this context, k_{TSO2} represents a useful index to describe the chemical evolution of red wines in combination with the main sensorial attributes generally associated with oxidative evolution.

The preliminary results obtained after 12 months of storage indicate that wine evolution during storage could be greatly influenced by the packaging characteristics (i.e., materials and volumes). Furthermore, also temperature imposed during the storage period seems to play a key role in the evolution of wine, since it can directly influence the oxygen permeability of the system "wine + package."

2.2 Case of study 2: Evolution of glass bottled rosé wine as a function of closure (cork stopper with or without aluminum capsule), storage position, and brightness regime over a period of 12 months

The samples reported in **Figure 2** are identified by code letters composed of a capital letter, which represents the closure type (C = with capsule) and the storage position (H = horizontal; V = vertical) and of a small letter, which indicates the light conditions. In particular, the letter "d" indicates that wines were stored in the dark, while "l" means that wines were stored under a cool fluorescent lamp (645 lux), considered as the common lighting of most supermarkets (**Table 8**) [28].

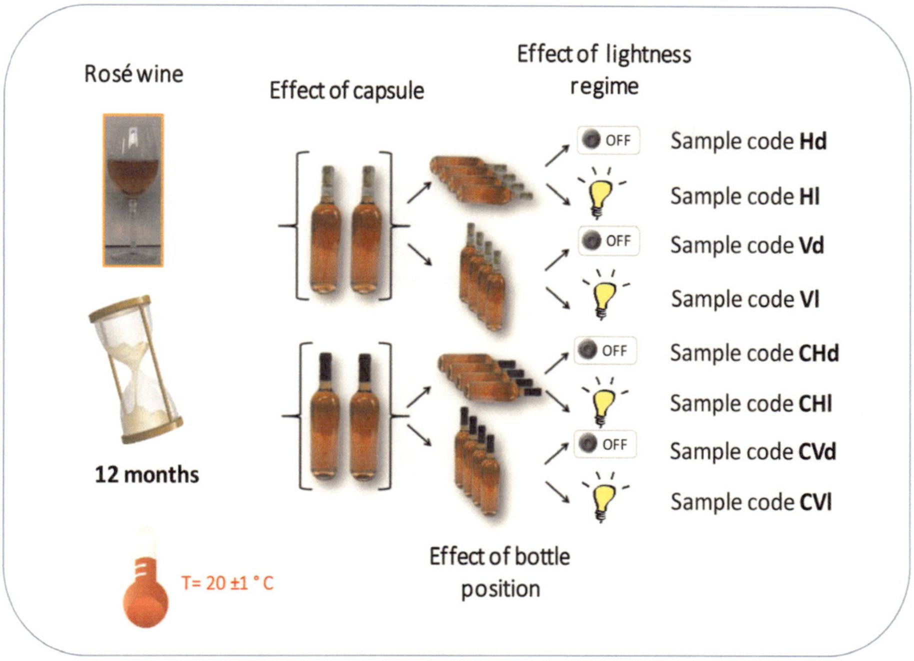

Figure 2.
Case of study 2: graphical abstract—Experimental setup.

Parameter	Mean value ± C.I. ($p < 0.05$)
Alcohol (%v/v)	11.33 ± 0.06
pH	3.32 ± 0.01
Titratable acidity (g/L as tartaric acid)	4.92 ± 0.01
Net volatile acidity (g/L as acetic acid)	0.33 ± 0.01
Total SO_2 (g/L)	0.133 ± 0.009
Total phenols (g/L as gallic acid)	0.332 ± 0.004
Not flavonoid phenols (g/L as gallic acid)	0.219 ± 0.009
Total anthocyanins (g/L as malvin)	0.087 ± 0.002

Table 8.
Chemical composition of the rosé wine utilized for the experimental runs (t = 0).

2.2.1 Influence of storage conditions on antioxidant capacity of stored wine

To highlight the influence of storage conditions on the time evolution of the rosé wine, the antioxidant capacity of all the stored samples was determined after 6 and 12 months from bottling by the ABTS assay according to Sgherri et al. [58]. As shown in **Figure 3**, following the first observation period (6 months after bottling), only small changes in the antioxidant capacity of wines were observed, whereas after 12 months of storage, conditions significantly affected this parameter.

In particular, the antioxidant capacity of wine was better preserved when the bottles were closed with capsules and stored in the dark in a horizontal position. Furthermore, the storage in the dark delayed the decrease of the antioxidant capacity of wine regardless of the other parameters. The influence exerted by the light exposure reached its maximum when the bottles were closed with cork stoppers and stored in a vertical position.

2.2.2 Influence of storage conditions on kinetics of TSO_2 and TAnt degradation

To better evidence the possible effects of the closure system (with or without capsule) and of the storage position (horizontal versus vertical) on the chemical deterioration of wine, the values of the kinetic constants k_{TSO2} and k_{TAnt} (**Table 9**) were carried out for bottles stored in brightness conditions. This is because changes in the antioxidant capacity of wine were faster when it was stored under a cool fluorescent lamp (see **Figure 3**).

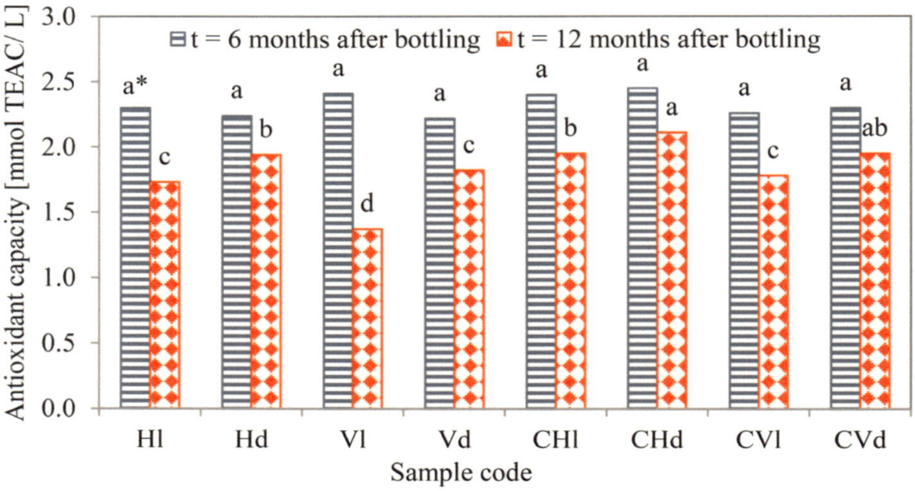

Figure 3.
*Evolution of antioxidant capacity during storage. *In each couple of data, the values labeled with different superscript letters show statistically significant differences (P < 0.05).*

As reported in **Table 9**, after 12 months from bottling the differences induced by both the closure system and the storage position on the degradation rate of TSO_2 as well as TAnt were statistically significant, evidencing that these storage conditions were among those that affect the oxidation rate of the rosé wine. In particular, wine degradation rate was the highest when the rosé wine was stored in glass bottles closed with natural corks without the application of a capsule, regardless of the position (vertical or horizontal) used during storage. Furthermore, independently

Sample*	k_{TSO2} ± c.i. (months^{-1}) × 10^2	$[TSO_2]_{t=0}$ ± c.i. (mg/L)	r^2	k_{Tant} (months^{-1}) × 10^2	$[TAnt]_{t=0}$ ± c.i. (mg/L)	r^2
Al	2.54 ± 0.06	133.8 ± 0.4	0.98	2.99 ± 0.07	87.3 ± 0.01	0.81
Bl	2.66 ± 0.06	135.7 ± 0.4	0.94	3.31 ± 0.07	87.0 ± 0.01	0.91
Cl	2.03 ± 0.06	132.5 ± 0.4	0.65	2.39 ± 0.07	87.5 ± 0.01	0.85
Dl	2.44 ± 0.06	130.0 ± 0.4	0.88	2.62 ± 0.07	86.5 ± 0.01	0.85

**Al = glass + natural cork without capsule, horizontal storage position, fluorescent lamp. Bl = glass + natural cork without capsule, vertical storage position, fluorescent lamp. Cl = glass + natural cork + capsule, horizontal storage position, fluorescent lamp. Dl = glass + natural cork + capsule, vertical storage position, fluorescent lamp.*

Table 9.
Kinetic parameters describing the time evolution of TSO_2 and TAnt concentration as a function of the storage conditions.

from the closure system, the time evolution of the rosé wine during storage was delayed when bottles were stored in the horizontal position.

2.2.3 Conclusions related to case of study 2

To confirm the convenience in using the rates of TSO_2 and total anthocyanin degradation as parameters effectively describing the oxidative evolution during storage of a rosé wine, the kinetic constants k_{TSO2} and k_{Ant} (**Table 9**) as well as their combination ($k_{TSO2} + k_{Ant}$) were correlated with the wine antioxidant capacity. This was performed over time and for all storage conditions. The Pearson's correlation coefficients are reported in **Table 10**.

Kinetic constant (months^{-1})	TEAC (L^{-1})
k_{TSO2}	**−0.86**
k_{Ant}	**−0.94**
$k_{TSO2} + k_{Ant}$	**−0.93**

Note: The correlation coefficients that indicate a strong correlation (≥0.6) are reported in boldface.

Table 10.
Correlation matrix relating the kinetic constant describing TSO_2 (k_{TSO2}), total anthocyanins (k_{Ant}), degradation, and a combination of them ($k_{TSO2} + k_{Ant}$) to wine antioxidant capacity (storage time = 12 months).

The results (**Table 10**) highlight that all the degradation kinetic constants were strictly inversely correlated with the antioxidant capacity of wine.

Notwithstanding k_{TAnt} did not result a good index for monitoring the chemical evolution of a red wine stored in the same conditions used in this research study [27], the correlation between k_{TAnt} and the antioxidant capacity showed by the rosé wine was higher than that determined when k_{TSO2} was considered.

Furthermore, k_{TSO2} confirmed to be a suitable index for the description of the oxidative evolution of different wines, regardless of the wine style (i.e., white, rosé, full-bodied red) and the operating conditions (i.e., packaging, storage or tasting conditions), according to what is reported in [59–61].

It can be concluded that also antioxidant capacity could be considered a useful index to describe the chemical evolution of the rosé wine under investigation, when correlated with the total anthocyanin decay rate constant (k_{TAnt}) and, at a lower extent, with the TSO_2 decay rate constant (k_{TSO2}).

3. Conclusions

Based on the analysis of recent papers available in international literature as well as on the experimental results discussed above, the main issues related to wine storage could be outlined in some main topics useful to better clarify the role played by both packaging and storage conditions on the evolution of the most diffused kinds of wines.

Firstly, packaging characteristics (i.e., material and volume) deeply influence wine evolution: glass bottles generally preserved wine better than multilayer materials; larger volumes slow down the wine deterioration rate over time regardless the kind of packaging selected.

Regardless the material utilized for packaging, the storage temperature plays a key role in the evolution of wine, since it can directly influence the oxygen permeability of the system "wine + package": lower temperature allows to improve the shelf life.

When the wine is stored in glass bottle, its quality decay rate appears significantly influenced by the kind of stopper, closure system, and storage position. In particular, when traditional cork stopper is utilized, the longer shelf life can be allowed by the combination of stopper with the extra-closure provided by an aluminum capsule. Moreover, the storage in glass bottles maintained in the dark and or in horizontal position further slows down the wine degradation, regardless the closure applied to the glass bottle.

Depending on the wine chemical composition, TSO_2 and TAnt decay rate constants (k_{TSO2}, k_{Ant}) together with antioxidant capacity can be considered the main chemical indexes to describe the wine evolution.

In conclusion, a new "integrated approach" deriving from the merging of chemical, kinetic, and sensorial data can be applied in order to identify the best storage conditions to preserve the quality of wines, improve their shelf life, and enhance the consumer's enjoyment during tasting.

Conflict of interest

Authors have no conflict of interest to declare.

Funding

This research was funded by the University of Pisa, project SALI (*Strumenti di sostenibilità delle produzioni alimentari*, Sustainability tools for food production) code PRA_2017_34.

Author details

Francesca Venturi[1,2]*, Chiara Sanmartin[1,2], Isabella Taglieri[1], Giuseppe Ferroni[1], Mike Frank Quartacci[1,2], Cristina Sgherri[1], Xiaoguo Ying[3], Gianpaolo Andrich[1,2] and Angela Zinnai[1,2]

1 Department of Agriculture, Food and Environment, University of Pisa, Pisa, Italy

2 Interdepartmental Research Center "Nutraceuticals and Food for Health", University of Pisa, Pisa, Italy

3 Zhejiang Provincial Key Laboratory of Health Risk Factors for Seafood, College of Food and Pharmacy, Zhejiang Ocean University, China

*Address all correspondence to: francesca.venturi@unipi.it

References

[1] Robertson GL. Food Packaging: Principles and Practice. 3rd ed. Boca Raton, FLA: Taylor and Francis/CRC Press; 2012

[2] Nicoli MC. An introduction to food shelf life: Definitions, basic concepts and regulatory aspects. In: Shelf Life Assessment of Food. 1st ed. London: CRC Press/Taylor and Francis Group; 2012. pp. 1-16

[3] Ghidossi R, Poupot C, Thibon C, Mietton-Peuchot M. The influence of packaging on wine conservation. Food Control. 2012;**23**:302-311. DOI: 10.1016/j.foodcont.2011.06.003

[4] Mentana A, Pati S, La Notte E, Del Nobile MA. Chemical changes in Apulia table wines as affected by plastic packages. LWT-Food Science and Technology. 2009;**42**:1360-1366. DOI: 10.1016/j.lwt.2009.03.022

[5] Wani AA, Singh P, Langowski H-C. Food technologies: Packaging. In: Encyclopedia of Food Safety. Vol. 3: Food, Materials, Technologies and Risks. Massachussetts, USA: Academic Press; 2014. pp. 211-218

[6] Charters S, Pettigrew S. The dimensions of wine quality. Food Quality and Preference. 2007;**18**: 997-1007. DOI: 10.1016/j.foodqual.2007.04.003

[7] Aarnio T, Hämäläinen A. Challenges in packaging waste management in the fast food industry. Resources, Conservation and Recycling. 2008;**52**: 612-621. DOI: 10.1016/j.resconrec.2007.08.002

[8] Cleary J. Life cycle assessments of wine and spirits packaging at the product and the municipal scale: A Toronto, Canada case study. Journal of Cleaner Production. 2013;**44**: 143-151. DOI: 10.1016/j.jclepro.2013. 01.009

[9] Revi M, Badeka A, Kontakos S, Kontominas MG. Effect of packaging material on enological parameters and volatile compounds of dry white wine. Food Chemistry. 2014;**152**: 331-339. DOI: 10.1016/j.foodchem.2013.11.136

[10] Copley GJ. The Composition and Manufacture of Glass and its Domestic and Industrial Applications. London: Taylor & Francis; 2001

[11] Marsh K, Bugusu B. Food packaging —Roles, materials, and environmental issues: Scientific status summary. Journal of Food Science. 2007;**72**:R39-R55. DOI: 10.1111/j.1750-3841.2007.00301.x

[12] Sanmartin C, Venturi F, Taglieri I, Nari A, Andrich G, Zinnai A. Study of the migration of chemicals from conventional food contact materials into food and environment: Hurdles and limits for a whole hazard identification and risk assessment. Agrochimica. 2018; **62**:179-200. DOI: 10.12871/00021857 201827

[13] Korkmaz A, Yanik J, Brebu M, Vasile C. Pyrolysis of the tetra pak. Waste Management. 2009;**29**: 2836-2841. DOI: 10.1016/j.wasman.2009.07.008

[14] Petersen K, Væggemose Nielsen P, Bertelsen G, Lawther M, Olsen MB, Nilsson NH, et al. Potential of biobased materials for food packaging. Trends in Food Science and Technology. 1999;**10**: 52-68. DOI: 10.1016/S0924-2244 (99)00019-9

[15] Cassino C, Tsolakis C, Bonello F, Gianotti V, Osella D. Wine evolution during bottle aging, studied by ^{1}H NMR spectroscopy and multivariate statistical analysis. Foodservice Research International. 2019;**116**:566-577. DOI: 10.1016/j.foodres.2018.08.075

[16] Marquez A, Serratosa MP, Merida J. Influence of bottle storage time colour, phenolic composition and sensory properties of sweet red wines. Food Chemistry. 2014;**1**:507-514. DOI: 10.1016/j.foodchem.2013.09.103

[17] Eiro MJ, Heinonen M. Anthocyanin colour behaviour and stability during storage: Effect to intermolecular copigmentation. Journal of Agricultural and Food Chemistry. 2002;**50**: 7461-7466. DOI: 10.1021/jf0258306

[18] Atanasova V, Fulcrand H, Cheyner V, Moutonet M. Effect of oxygenation on polyphenols changes occurring in the course of wine-making. Analytica Chimica Acta. 2002;**458**:15-27. DOI: 10.1016/S0003-2670(01)01617-8

[19] Vidal JC, Moutounet M. Monitoring of oxygen in the gas and liquid phases of bottles of wine at bottling and during storage. Journal International des Sciences de la Vigne et du Vin. 2006;**40**: 35-45

[20] Kwiatkowski MJ, Skouroumounis GK, Lattey KA, Waters EJ. The impact of closures, including screw cap with three different headspace volumes, on the composition, colour and sensory properties of a cabernet sauvignon wine during two years' storage. Australian Journal of Grape and Wine Research. 2007;**13**:81-94. DOI: 10.1111/j.1755-0238.2007.tb00238.x

[21] Dombre C, Rigou P, Wirth J, Chalier P. Aromatic evolution of wine packed in virgin and recycled PET bottles. Food Chemistry. 2015;**176**:376-387. DOI: 10.1016/j.foodchem.2014.12.074

[22] Crouvisier-Urion K, Bellat JP, Gougeon RD, Karbowiak T. Gas transfer through wine closures: A critical review. Trends in Food Science and Technology. 2018;**78**:255-269. DOI: 10.1016/j.tifs.2018.05.021

[23] Wirth J, Caillé S, Souquet JM, Samson A, Dieval JB, Vidal S, et al. Impact of post-bottling oxygen exposure on the sensory characteristics and phenolic composition of Grénache rosé wines. Food Chemistry. 2012;**132**: 1861-1871. DOI: 10.1016/j.foodchem.2011.12.019

[24] Gao Y, Tian Y, Liu D, Li Z, Zhang XX, Li JM, et al. Evolution of phenolic compounds and sensory in bottled red wines and their co-development. Food Chemistry. 2015;**172**:565-574. DOI: 10.1016/j.foodchem.2014.09.115

[25] Skouroumounis GK, Kwiatkowski MJ, Francis IL, Oakey H, Capone DL, Duncan B, et al. The impact of closure type and storage conditions on the composition, colour, and flavor properties of a Riesling and a wooded chardonnay wine during five years' storage. Australian Journal of Grape and Wine Research. 2005;**11**:369-377. DOI: 10.1111/j.1755-0238.2005.tb00036.x

[26] Venturi F, Sanmartin C, Taglieri I, Xiaoguo Y, Andrich G, Zinnai A. The influence of packaging on the sensorial evolution of white wine as a function of the operating conditions adopted during storage. Agrochimica. 2016;**60**:150-160. DOI: 10.12871/0021857201627

[27] Venturi F, Sanmartin C, Taglieri I, Xiaoguo Y, Quartacci MF, Sgherri C, et al. A kinetic approach to describe the time evolution of red wine as a function of packaging conditions adopted: Influence of closure and storage position. Food Packaging and Shelf Life. 2017;**13**:44-48. DOI: 10.1016/j.fpsl.2017.07.001

[28] Sanmartin C, Taglieri I, Ying X, Quartacci MF, Sgherri C, Andrich G, et al. Main operating conditions that can influence the evolution of a rosé wine during long-term storage: A kinetic approach. Agrochimica. 2018;**62**: 253-262

[29] Labuza TP. Enthalpy/entropy compensation in food reactions. Food Technology. 1980;**34**:67-77

[30] Margalit Y. Concepts in Wine Chemistry. 3rd ed. Sawn Francisco, CA: The Wine Appreciation Guild; 2012. p. 544

[31] Ribéreau-Gayon P, Glories Y, Maujean A, Dubourdieu D. Handbook of Enology, Vol. 2: The Chemistry of Wine Stabilization and Treatments. 2nd ed. Ltd England: Wiley; 2000. p. 441. DOI: 10.1002/0470010398.index

[32] Vivas de Gaulejac N, Nonier MF, Absalon C, Bourgeois G. Study and quantification of monomeric flavan-3-ol and dimeric procyanidin quinonic forms by HPLC/ESI-MS. application to red wine oxidation. Journal of the Science of Food and Agriculture. 2001;**81**:1172-1179. DOI: 10.1002/jsfa.926

[33] Hopfer H, Buffon PA, Ebeler SE, Heymann H. The combined effects of storage temperature and packaging on the sensory, chemical, and physical properties of a cabernet sauvignon wine. Journal of Agricultural and Food Chemistry. 2013;**61**:3320-3334. DOI: 10.1021/jf3051736

[34] Hopfer H, Ebeler SE, Heymann H. The combined effects of storage temperature and packaging type on the sensory and chemical properties of chardonnay. Journal of Agricultural and Food Chemistry. 2012;**60**:10743-10754. DOI: 10.1021/jf302910f

[35] Zinnai A, Venturi F, Sanmartin C, Quartacci MF, Andrich G. Chemical and laccase catalysed oxidation of gallic acid: Determination of kinetic parameters. Research Journal of BioTechnology. 2013;**8**:62-65

[36] Venturi F, Andrich G, Quartacci MF, Sanmartin C, Andrich L, Zinnai A. A kinetic method to identify the optimum temperature for β-glucanase activity. South African Journal of Enology and Viticulture. 2013;**34**: 281-286

[37] Sanmartin C, Venturi F, Sgherri C, Nari A, Macaluso M, Flamini G, et al. The effects of packaging and storage temperature on the shelf-life of extra virgin olive oil. Heliyon. 2018;**4**: e00888. DOI: 10.1016/j.heliyon.2018.e00888

[38] Du Toit WJ, Marais J, Pretorius IS, Du Toit M. Oxygen in must and wine: A review. South African Journal of Enology and Viticulture. 2006;**27**:76-94. DOI: 10.21548/27-1-1610

[39] Fu Y, Lim LT, McNicholas PD. Changes on enological parameters of white wine packaged in bag in box during secondary shelf life. Journal of Food Science. 2009;**74**:608-618. DOI: 10.1111/j.1750-3841.2009.01316.x

[40] Makhotkina O, Pineau B, Kilmartin PA. Effect of storage temperature on the chemical composition and sensory profile of sauvignon Blanc wines. Australian Journal of Grape and Wine Research. 2012;**18**:91-99. DOI: 10.1111/j.1755-0238.2011.00175.x

[41] Košmerl T, Abramovič H, Klofutar C. The rheological properties of Slovenian wines. Journal of Food Engineering. 2000;**46**:165-171. DOI: 10.1016/S0260-8774(00)00079-0

[42] Yanniotis S, Kotseridis G, Orfanidou A, Petraki A. Effect of ethanol, dry extract and glycerol on the viscosity of wine. Journal of Food Engineering. 2007;**81**:399-403. DOI: 10.1016/j.jfoodeng.2006.11.014

[43] Dias DA, Smith TA, Ghiggino KP, Scollary GR. The role of light, temperature and wine bottle colour on pigment enhancement in white wine. Food Chemistry. 2012;**135**:2934-2941. DOI: 10.1016/j.foodchem.2012.07.068

[44] Venturi F, Sanmartin C, Taglieri I, Xiaoguo Y, Andrich G, Zinnai A. A kinetic approach to describe the time evolution of red wine as a function of packaging and storage conditions. Acta

Alimentaria. 2017;**46**:336-345. DOI: 10.1556/066.2017.0001

[45] Bimpilas A, Panagopoulou M, Tsimogiannis D, Oreopoulou V. Anthocyanin copigmentation and color of wine: The effect of naturally obtained hydroxyl cinnamic acids as cofactors. Food Chemistry. 2016;**197**:39-46. DOI: 10.1016/j.foodchem.2015.10.095

[46] Avizcuri JM, Sàenz-Navajas MP, Echàvarri JF, Ferreira V, Fernàndez-Zurbano P. Evaluationof the impact of initial red wine, composition on changes in color and anthocyanin content during bottle storage. Food Chemistry. 2016; **216**:123-134. DOI: 10.1016/j.foodchem.2016.06.050

[47] Venturi F, Sanmartin C, Taglieri I, Nari A, Andrich G, Terzuoli E, et al. Development of phenol-enriched olive oil with phenolic compounds extracted from wastewater produced by physical refining. Nutrients. 2017;**9**:916. DOI: 10.3390/nu9080916

[48] Đorđević NO, Pejin B, Novaković MM, Dalibor M, de Stanković J, Snežan MJ, et al. Some chemical characteristics and antioxidant capacity of novel merlot wine clones developed in Montenegro. Scientia Horticulturae. 2017;**225**: 505-511. DOI: 10.1016/j.scienta.2017.07.045

[49] Tuberoso CIG, Serreli G, Congiu F, Montoro P, Fenu MA. Characterization, phenolic profile, nitrogen compounds and antioxidant activity of Carignano wines. Journal of Food Composition and Analysis. 2017;**58**:60-68. DOI: 10.1016/j.jfca.2017.01.017

[50] Bailetti LI, Santini C, Supino S. Sensory sciences and competitiveness in the wine business. In: Santini C, Cavicchi A, editors. Case Studies in the Wine Industry. 1st ed. Cambridge: Woodhead Publishing Series; 2019. pp. 75-85. DOI: 10.1016/B978-0-08-100944-4.00006-9

[51] Varela P, Ares G. Sensory profiling, the blurred line between sensory and consumer science. A review of novel methods for product characterization. Food Research International. 2012;**48**: 893-908. DOI: 10.1016/j.foodres.2012.06.037

[52] Stone H, Bleibaum R, Thomas HA. Sensory Evaluation Practices. 4th ed. Massachussetts, USA: Academic Press; 2012. p. 446

[53] O'Sullivan MG. Shelf life and sensory quality of foods and beverages. In: A Handbook for Sensory and Consumer-Driven New Product Development. 1st ed. Cambridge: Woodhead publishing; 2017. pp. 103-123. DOI: 10.1016/B978-0-08-100352-7.00006-3

[54] Lawless HT, Heymann H. Sensory Evaluation of Food. 2nd ed. Berlin: Springer-Verlag; 2010. p. 596. DOI: 10.1007/978-1-4419-6488-5

[55] Jackson R. Shelf life of wine. In: The Stability and Shelf Life of Food. 2nd ed. Cambridge: Woodhead Publishing; 2016. pp. 311-346. DOI: 10.1016/B978-0-08-100435-7.00011-3

[56] Venturi F, Sanmartin C, Taglieri I, Xiaoguo Y, Deng S, Andrich G, et al. The influence of packaging on the time evolution of red wine. Agro Food Industry Hi Tech. 2017;**28**:46-49

[57] Paula AM, Conti-Silva AC. Texture profile and correlation between sensory and instrumental analysis of extruded snacks. Journal of Food Engineering. 2014;**121**:9-14. DOI: 10.1016/j.jfoodeng.2013.08.007

[58] Sgherri C, Micaelli F, Andreoni N, Baldanzi M, Ranieri A. Retention of phenolic compounds and antioxidant properties in potato bread obtained from a dough enriched with a powder from the purple cv. Vitelotte.

Agrochimica. 2016;**60**:312-328. DOI: 10.12871/00021857201646

[59] Venturi F, Andrich G, Sanmartin C, Taglieri I, Scalabrelli G, Ferroni G, et al. Glass and wine: A good example of the deep relationship between drinkware and beverage. Journal of Wine Research. 2016;**27**:153-171. DOI: 10.1080/09571264.2016.1160879

[60] Venturi F, Zinnai A, Fantoni G, Gabelloni D, Razionale AV. Glass and wine: The indissoluble marriage. In: Proceedings of the 19th International Conference on Engineering Design (ICED 2013); 19-22 August 2013; Seoul, South Korea. Vol. 7 DS75-07. 2013. pp. 497-506

[61] Venturi F, Andrich G, Sanmartin C, Scalabrelli G, Ferroni G, Zinnai A. The expression of a full-bodied red wine as a function of the characteristics of the glass utilized for the tasting. CyTA Journal of Food. 2014;**12**:291-297. DOI: 10.1080/19476337.2013.848473

Chapter 13

Air Depleted and Solvent Impregnated Cork Powder as a New Natural and Sustainable Wine Fining Agent

Luís Filipe da Mota Ribeiro,
Maria Fernanda Gil Cosme Martins
and Fernando Hermínio Ferreira Milheiro Nunes

Abstract

We recently proposed a simple methodology to improve cork powder waste adsorption properties through vacuum degassing and solvent impregnation, to use this abundant and cheap material as a new wine fining agent. Its applicability was first shown for red wine 4-ethylphenol (4-EP) and 4-ethylguaiacol (4-EG) reduction. Nowadays, the presence of 4-EP and 4-EG is a serious problem in the wine industry, known as "Brett character", by the negative aroma imparted by these volatile phenols (VPs) to red wine. There are only some curative treatments to remove these compounds without impacting negatively on wine quality. Optimised cork powder was used successfully as a new treatment for the reduction of these negative VPs (41–75% for 4-EP and 40–69% for 4-EG) increasing at the same time wine sensory performance. Wine treated with cork powder reduced 6.9% phenolic acids and catechin and 2.3% monomeric anthocyanins without any significant change in colour intensity. In this chapter, the cork complex structure is discussed, besides the impact of its use in wine containing VPs on physicochemical composition and quality. This new application of this natural, abundant and cheap material has the potential of being a new wine fining agent with low environmental impact.

Keywords: cork composition, adsorption properties, red wine, volatile phenols, aroma, phenolic compounds, sensory attributes

1. Introduction

Cork, the outside part of the oak (*Quercus suber* L.), is a natural, renewable, sustainable raw material, which is periodically harvested from the tree, usually every 9–12 years, depending on the cultivation region [1]. *Quercus suber* L. is a tree that grows slowly in same regions of the western Mediterranean (Portugal, Spain, Southern France, part of Italy and North Africa) and China [2–4]. Portugal is the main cork producer, transforming about 75% of all the cork [3, 4]. Industrial transformation of cork generates up to 25 wt.% of cork dusts as by-product [5, 6].

Cork wastes and cork powders have been used as bioadsorbents for removing pesticides and other pollutants from wastewaters with promising results [7].

Biosorption is an emergent technology expected to show strong growth soon because it offers high cost effectiveness, although further improvements in its performance are needed [1]. Environmental protection legislation is becoming progressively important and effective solutions will be at premium [8].

The cork material is compact, devoid of intercellular spaces and with a regular honeycomb organisation (**Figure 1**). This material is composed by dead parenchymatous cells with voids, prismatic, air-filled interiors, hexagonal on average and are arranged base-to-base in an alignment oriented in the tree's radial direction [9].

The cells are small and have sizes under those of synthetic foams. The area of the prism base is 4–6 $\times 10^{-6}$ cm with a mean prism edge of 13–15 μm; prim height is usually in the range of 30–40 μm. The mean cell volume is approximately 2×10^{-8} cm^3 and the number of cells per unit is 4–7 $\times 10^7$ cm^{-3}. The cell walls are thin with a thickness of 1–1.5 μm. The solid mass volume fraction of the cork is only about 10%.

Cork powder maintains the cork cellular structure intact [10], and its adsorption properties can be improved by removing the air and simultaneous impregnation with ethanol rendering the cell wall components more accessible to the adsorbates [10]. This simple treatment was shown to increase cork powder adsorption capacity of 4-EP and 4-EG by at least 4 times in a real wine matrix, with the cork powder

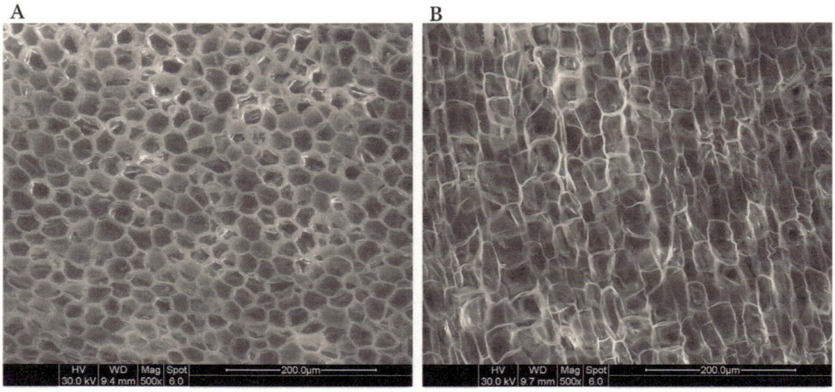

Figure 1.
Structure of cork as observed by SEM in the two main sections: (A) tangential section, perpendicular to the tree's radial direction; (B) transverse section, perpendicular to the tree's axial direction.

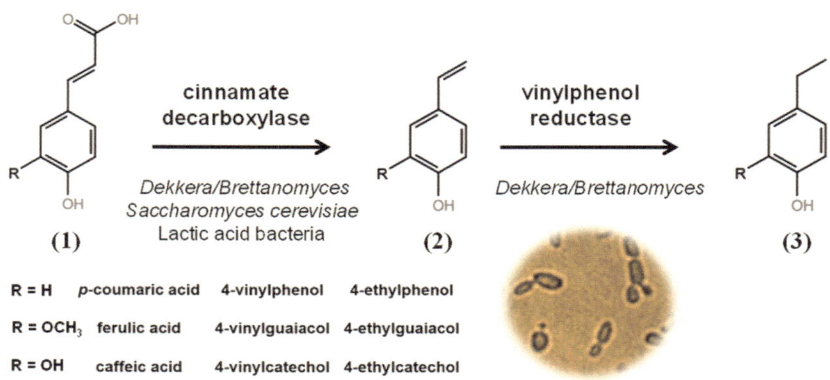

Figure 2.
Formation of volatile phenols from hydroxycinnamate precursors or their degradation products (vinylphenols) in wines by Dekkera/Brettanomyces.

adsorption capacity increasing with the increase in concentration of these wine contaminants [10].

In red winemaking, especially those aged in wood barrels, the contamination and growth of *Dekkera/Brettanomyces* yeasts result in the formation of 4-EP and 4-EG by decarboxylation of *p*-coumaric and ferulic acids present in wine and subsequent reduction of the correspondent vinylphenols (**Figure 2**) [11, 12].

These VPs are responsible for negative aromatic notes like horsy sweat, smoky, barnyard and medicinal [11, 13]. This important sensory defect has been reported in several wine styles around the world, especially, premium wines [14, 15], considered negative by professionals, consumers and wine industry [16, 17], and thus, VPs are a generalised problem in red winemaking.

For these reasons, several treatments to avoid or to reduce compounds have been tested. Preventive action includes, for example, the maintenance of adequate levels of sulphur dioxide throughout the winemaking process, reduction/elimination of oxygen levels in wine, use of dimethyl dicarbonate (DMDC) before bottling and the addition of fungal chitosan, which are some of the measures that have found some degree of success [18, 19]. Several remediation treatments have also been developed to eliminate the already formed VPs from wine or to decrease the headspace content by decreasing their partition coefficients to the gas phase without changing the total wine VP content. Of these methods, those tested in wines presenting good removal efficiency at practical application doses are activated carbons [20, 21], potassium caseinate [22], egg albumin [22] and esterified cellulose [23]. Nevertheless, although they are efficient in reducing the total amount of VPs in wines, the use of potassium caseinate and egg albumin presented the risk of the potential allergenicity of these fining agents and therefore it is mandatory to label the wine bottle if the residual concentration is higher than 0.25 mg/L (EU Regulation 579/2012). For the decrease of headspace abundance of VPs chitosans has been shown to be effective [24].

The success of cork powder in adsorption of VPs from such a complex matrix as wine without affecting the wine quality significantly in terms of phenolic composition is certainly due to the structure and chemical composition of its main components namely suberin, lignin and cell wall polysaccharides.

2. Cork chemical composition

The chemical composition of cork has been widely examined [25–33] and presented some variability that depends on factors such as geographic origin, soil

Principal components (%)	Range	Average
Suberin	40–53	45.8
Lignin	21–29	24.4
Polysaccharides	10–16	12.5
Extractives	6–19	12.6
Tannins	6–7	6.5
Ash	0.85–2.1	1.4

Adapted from [1, 25, 34–36].

Table 1.
Chemical composition of cork.

and climate conditions, genetic origin, tree dimensions, age and growth conditions (**Table 1**). Cork from *Quercus suber L.* has specific properties such as low permeability and great elasticity; this is the result, at least partially, from its specific chemical composition (and more especially from that of suberin) [26, 29, 31–33]. The cork cell wall structure consists in a thin internal primary cork cell wall rich in lignin and a thick secondary wall rich in suberin, alternating with a wax lamella and a thin tertiary wall of polysaccharides.

2.1 Suberin

Suberin, a natural aliphatic-aromatic crosslinked polyester, is the major component of cork, accounting for 30–50% of its weight. It is a very important structural component of the cell wall and its removal destroys cell integrity. Suberin polymeric structure is mainly composed by two types of monomers, glycerol and long-chain fatty acids and alcohols, which are linked by ester bonds, **Figure 3** [9].

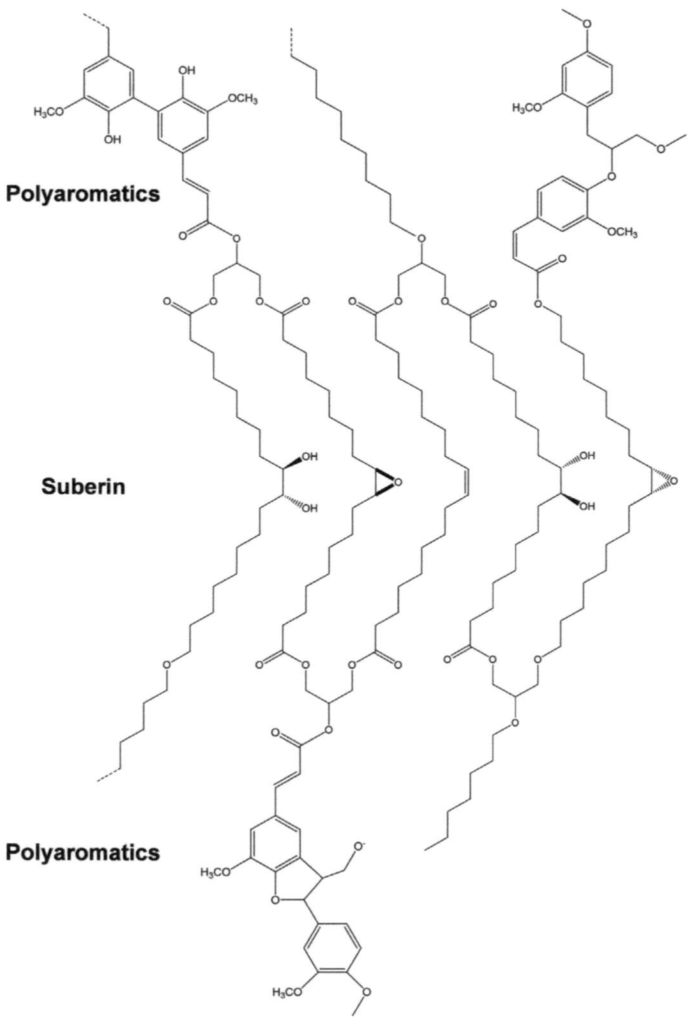

Figure 3.
Schematic representation of suberin structure (adapted from Graça [37]).

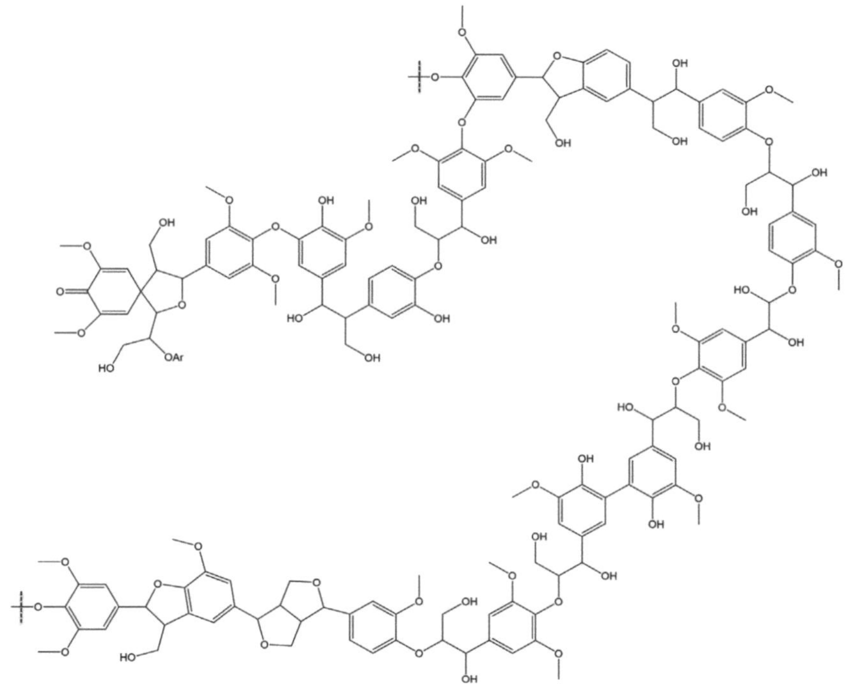

Figure 4.
Schematic representation of lignin structure (adapted from Achyuthan et al. [39]).

2.2 Lignin

Lignin is the second most important component in cork cell walls accounting for 15–30% of its weight [9]. It is a crosslinked polymer of aromatic nature. Due to the importance of lignin, many studies were done in wood pulping and more recently, for biomass deconstruction [38]. Lignin is a polymer made up by three monomer types of phenyl propane (*p*-coumaryl, coniferyl and sinapyl alcohols) linked through a free-radical reaction started via enzymatic phenoxy radical formation (**Figure 4**). The inter-unit linkages in the polymer can be of several kinds: β-*O*-4′, α-*O*-4′, β-β′, β-5′, 5-5′, 4-*O*-5′ or β-1′. The specific relation of the monomers and intermonomeric linkages depend on the material [9]. In cork, lignin also contributes to the mechanical support and rigidity of the cell walls. If lignin is selectively removed from cell walls, a total collapse of the cells is observed.

2.3 Polysaccharides

In cork, the cell wall polysaccharides, cellulose and hemicelluloses, represent approximately 20% of its weight. Cellulose is in the primary and tertiary cell walls of cork, accounting for nearly 10% [40]. There is less information concerning the molecular weight, crystallinity and chain orientation of cork cellulose. Cellulose is water insoluble due to an extensive intermolecular hydrogen bonding between adjacent polymers, and interaction with water often only occurs in the amorphous regions. The hemicelluloses are another water insoluble group of polysaccharides present in cork cell walls. The main known hemicellulose polysaccharides comprise three different groups of polysaccharides (**Figure 5**), the 4-*O*-methylglucuronoxylan, arabino-4-*O*-methylglucuronoxylan and 4-*O*-methylglucurono-arabinogalactoglucoxylan [41–44]. Xylans in the cell walls are amorphous and the

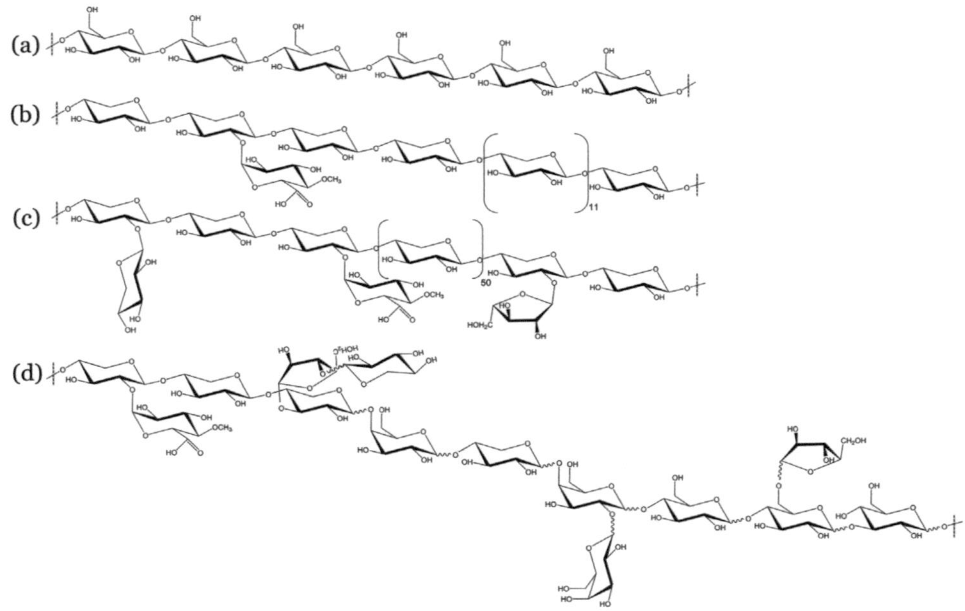

Figure 5.
Schematic structures of main cork cell wall polysaccharides: (a) cellulose, (b) 4-O-methylglucuronoxylan, (c) arabino-4-O-methylglucuronoxylan and (d) 4-O-methylglucurono-arabinogalactoglucoxylan.

irregular occurrence of branching of the main chain does not permit strong intermolecular association by hydrogen bonding; nevertheless, they are extracted using strong alkaline solutions (4–10% w/v NaOH). Pectins also exist in low quantities in cork, approximately 1.5%, placed in the middle lamella [45].

2.4 Extractable components

Cork contains 8–20% of low molecular weight compounds including fatty acids, terpenes, long-chain aliphatic compounds and saccharides, collectively known as extractives [34, 46]. Cork contains also about 6% of tannins [36]. The most important of these components are waxes and tannins [31]. Waxes are extracted by low polarity solvents, such as benzene, chloroform, ethyl acetate [47], hexane [36] and ether [26]. The waxes are responsible for the cork impermeability. The waxes extracted were found to consist of two fractions: neutral and acidic. The neutral fraction is mostly composed of fatty alcohols (C_{18}—C_{26}) with some unsaturated groups and triterpenes.

The acid fraction is essentially composed of fatty acids (C_{14}—C_{24}) with unsaturated ω-hydroxyacids, 18-hydroxy-9,12-octadienoic and 18-hydroxy-9-octadecenoic acids. More or less 50% of the waxes are triterpenes from friedelin and lupine families including friedelin, 3-α-hydroxyfriedelan-2-one, botulin, betulinic acid, β-sitosterol and sitost-4-en-3-one [48]. Cork extractable phenolic compounds include ellagic acid and some quantities of gallic acid, protocatechuic acid/aldehyde, aesculetin, vanillic acid, caffeic acid, vanillin, scopoletin, ferulic acid, coniferyl aldehyde and sinapaldehyde [49, 50]. The extraction of tannins can be done by polar solvents such as water [51] and ethanol [52]. Cork tannins include roburins A and E, grandini, vescalagin and castalagin. The yields of these two components change in function of the nature of the cork (virgin or reproduction) where significant variation is found in the bibliography [1].

3. Optimised cork powder (CKP) as a wine fining agent to remove negative volatile phenols in contaminated red wine

The air removal of the cork powder cell structure and simultaneous impregnation with ethanol with or without previous removal of cork extractives increased significantly the 4-ethylphenol and 4-ethylguaiacol adsorption performance (**Table 2**).

Although a significant removal of wine VPs was observed, the overall quality of the treated wine cannot be accessed only by the decrease in these negative aroma compounds, as the impact on the other wine positive aroma components is important to define the final overall sensory olfactory quality [15, 20, 21, 22, 24]. The red wine colour characteristics are important for consumer acceptance of the treated wine, because there is straight relation between the colour and the wine's phenolic composition, namely anthocyanins, whose concentration can be changed by the fining procedure.

In order to have a deeper insight on the impact of optimised cork powder in the wine chemical composition besides the removal efficiency of the VPs, the change in the headspace aroma abundance of wine, phenolic composition and chromatic characteristics were studied and the overall impact on the wine sensory characteristics was evaluated by an expert panel.

3.1 Impact of optimised cork powder on the wine aroma headspace abundance

Air removal and ethanol impregnation of cork samples with and without extractive removal decreased the total headspace aroma abundance (CKNI 32% and CKFI 37%) significantly. The decrease in the particle size of the CKF did not differ significantly on the removal of headspace aroma compounds, although there was an average decrease of 3.7% in relation to CKF (**Table 3**). The duplication in application dose of CKFI75 resulted in a significant decrease of the total abundance of headspace aroma by more 29% (**Table 3**). There was a significant correlation ($r = 0.731$, $n = 14$, $p < 0.003$) between the headspace aroma abundance and the octanol-water partition coefficient (log P) of the aroma compounds, strongly

Factors		**Wine spiked levels**			
		Medium		**High**	
A	**B**	**4-EP**	**4-EG**	**4-EP**	**4-EG**
No impregnation	CKN	85.3 ± 2.7^a	9.2 ± 0.2^a	109.6 ± 5.1^a	10.5 ± 0.6^a
	CKF	168.8 ± 4.2^b	19.2 ± 2.7^b	738 ± 36.9^b	71.5 ± 5.4^b
Vacuum impregnation	CKNI	270.9 ± 11.8^c	43.4 ± 2.1^c	888.0 ± 16.3^c	133.8 ± 2.0^c
	CKFI	306.0 ± 2.3^d	60.5 ± 1.6^d	1036.5 ± 18.1^d	149.1 ± 3.3^d
A		0.0000001	0.000011	0.0000001	0.0000001
B		0.0000001	0.0000001	0.0000001	0.0000001
A × B		0.0029	0.083033	0.0000001	0.000018

[a]*Values are presented as mean ± standard deviation; medium spiking levels: 750 μg/L 4-EP and 150 μg/L 4-EG; high 1500 μg/L 4-EP and 300 μg/L 4-EG. Means within a column followed by the same letter are not significantly different ANOVA and Tukey post-hoc test ($p < 0.05$).*

Table 2.
Amount of 4-EP and 4-EG (μg/L) removed from wines at two spiked levels[a] of natural cork powder (CKN) and dichloromethane and ethanol extractive free cork powder (CKF) before and after air removal and impregnation with ethanol (CKNI and CKNFI) [10].

Compounds	ID	RI calculated	RI	Odour descriptor	ODT (mg/L)	T0	TF	CKNI	CKFI	CKFI75250	CKFI75500
Ethyl acetate	—	725	715	Fruity	7.5	23.97 ± 1.67^a	23.29 ± 2.97^a	14.32 ± 0.88^b	14.13 ± 0.42^b	14.70 ± 1.84^b	8.53 ± 1.49^c
3-Methylbutan-1-ol-acetate	std	1121	1126	Banana	0.03	65.09 ± 15.64^a	67.59 ± 19.86^a	44.36 ± 3.22^a	39.81 ± 4.81^{ab}	35.30 ± 3.50^{ab}	16.93 ± 3.95^b
2-Methyl-1-butanol	std	1229	1223	Malty	0.48	112.02 ± 8.60^a	109.52 ± 8.6^a	77.47 ± 2.20^c	71.24 ± 3.46^c	76.13 ± 5.49^c	50.10 ± 8.03^d
Ethyl hexanoate	std	1229	1262	Green apple	0.014	27.62 ± 16.17^a	27.87 ± 15.68^a	25.65 ± 0.67^{ab}	22.45 ± 1.30^{ab}	20.69 ± 4.13^b	12.92 ± 2.43^b
Ethyl octanoate	std	1441	1429	Fruity	0.005	164.67 ± 19.71^a	167.17 ± 10.85^a	104.94 ± 6.78^b	90.56 ± 12.53^b	75.94 ± 27.94^b	43.14 ± 1.49^c
Ethyl decanoate	std	1648	1646	Fruity	1.5	78.55 ± 3.63^a	78.95 ± 5.89^a	67.12 ± 6.21^{ab}	53.94 ± 7.51^b	36.93 ± 23.29^c	19.61 ± 2.20^c
Diethyl succinate	std	1683	1698	Light fruity	7.5	66.90 ± 1.14^a	64.40 ± 3.94^a	41.95 ± 1.74^{cd}	41.22 ± 1.81^d	39.72 ± 3.78^d	23.56 ± 5.20^e
Phenylethyl acetate	std	1809	1833	Flowery	0.25	31.16 ± 4.66^a	30.16 ± 3.51^a	22.82 ± 0.89^b	20.54 ± 1.13^b	14.80 ± 6.94^c	8.49 ± 2.96^c
2-Phenylethanol	std	2000	1911	Roses	14.0	365.23 ± 19.85^a	362.73 ± 31.57^a	247.32 ± 7.78^b	240.73 ± 15.31^a	216.92 ± 39.38^b	136.16 ± 60.27^c
4-Ethylguaiacol	std	2012	1989	Smoke	0.15	n.d.	16.62 ± 0.89^a	9.21 ± 0.41^c	9.05 ± 0.68^c	8.00 ± 2.25^c	3.70 ± 1.07^d
Reduction SPME (%)								**44.6**	**45.6**	**50.6**	**77.8**
Octanoic acid	std	2036	2030	Rancid	0.5	17.87 ± 0.57^a	17.62 ± 0.73^a	6.00 ± 3.79^{cd}	8.98 ± 0.97^{bc}	7.19 ± 1.32^{cd}	2.30 ± 1.20^d
4-Ethylphenol	std	2084	2142	Phenolic	0.4	n.d.	18.25 ± 1.23^a	10.37 ± 0.31^c	9.83 ± 0.89^c	7.94 ± 2.19^c	3.37 ± 1.35^d
Reduction SPME (%)								**43.1**	**46.1**	**56.5**	**81.5**
Decanoic	std	2129	2196	Rancid	1.0	4.14 ± 0.45^a	4.19 ± 0.43^a	1.49 ± 0.17^{ab}	2.53 ± 2.76^{ab}	1.24 ± 0.62^{ab}	n.d
Dodecanoic	std	2136	2156	Waxy	6.1	5.48 ± 0.36^a	5.23 ± 0.59^a	0.93 ± 0.15c	0.90 ± 0.41^c	0.90 ± 0.13^c	n.d
Total area-VPs						**962.7 ± 28.0^a**	**958.7 ± 46.7^a**	**652.9 ± 10.1^{cd}**	**607.0 ± 32.1^d**	**571.1 ± 68.3^d**	**296.7 ± 71.1^e**
Reduction (%)						—	—	**31.9**	**36.7**	**40.4**	**69.1**

Results are expressed in absolute area ($\times 10^5$). Data are presented as $\overline{X} \pm s$; ID, identification; std, standard; and RI, retention index [60–62], MW, molecular weight; LOD, limit of olfactory detection). Odour descriptor [63–65]. Values within a line followed by the same letter are not significantly different ANOVA and Tukey post-hoc test (p 0.05). n.d., not detected; uncontaminated (T0) spiked red wine (TF) and wines treated with corks.

$p < 0.05$.

Table 3.
Headspace aroma profile of red wines before (VP-free To and VP-spiked with 750 μg/L of 4-EP and 150 μg/L of 4-EG, TF) and after treatment with natural cork and dichloromethane and ethanol extractive free cork after air removal and ethanol impregnation (CKNI and CKFI) and cork powders with a particle size below 75 μm at two application doses (250 and 500 g/hL) (CKF75250 and CKFI75500).

suggesting that the interaction of the volatile compounds including the VPs with the cork powder is of hydrophobic nature as observed for the interaction of other molecules with cork [7, 53, 54]. When compared to activated carbons applied at 100 g/hL, CKFI75250 (250 g/hL) showed a lower impact on the headspace aroma abundance (40 vs. 75%) and even CKFI75500 (500 g/hL) resulted in a lower reduction of 69%. Therefore, cork powder decreased the wine headspace aroma compounds lesser than the activated carbons [21].

3.2 Impact of optimised cork powder on wine chromatic characteristics and phenolic compounds

Application of optimised cork powder results in a decrease of the colour intensity, although being only significantly different from the control for the CKFI and CKFI75500. For the L* and a*, the same was observed (**Table 4**). These variations for the colour intensity are not due to a decrease in the concentration of monomeric anthocyanins that generally did not change by the use of all cork powders (**Table 5**). For the individual phenolic acids overall, their levels did not change significantly, or their decrease was significant but small, and these decreases occurred mainly for the CKFI75 at the two application doses (decreased for *trans*-caftaric acid—5.6%; coutaric acid—5.9%; caffeic acid—20%; ferulic acid—12% and coumaric acid ethyl ester—19%) (**Table 5**). For catechin, there was no change in its levels for all cork powders applied. These results show that optimised cork powders, either with or without extractive removal, have a low impact on wine phenolic composition; nevertheless, the ethanol impregnated extractive free corks had a significant impact on wine colour intensity, suggesting that these corks influence wine polymeric pigments as no significant changes on monomeric anthocyanins were observed. The impact for cork powders on wine phenolic composition and colour intensity of wines was lower than that generally observed for activated carbons used at 100 g/hL [20].

3.3 Impact of optimised cork powder on wine sensory attributes

To validate the impact of natural and extractive free ethanol impregnated cork powder samples on the headspace VP decrease and its effect on the sensory perception and quality of wines, CKNI, CKFI and CKFI75—treated wines at the two application doses (250 and 500 g/hL, CKFI75250 and CKFI75500, respectively) were subjected to sensory analysis by an expert panel. As expected, the presence of these VPs affect the aroma profile of spiked wine (TF) significantly and negatively (**Table 6**), by the increase of the phenolic attribute, decreasing the wine fruity and floral attributes significantly [20, 24, 55]. The panel consensus on each wine attribute was accessed through the percentage of variance explained by the first PCA [56] applied to the panel scores for each attribute. The variance explained by PC1 ranged from 45 to 87%, yielding the *C*-indexes presented in **Table 6**. Similar values have been reported for trained panels assessing different attributes and different products [20, 24, 62]. Colour intensity, floral, fruity, phenolic, acidity, balance and persistence wine attributes resulted in a consensus among judges (**Table 6**). For the colour hue, limpidity, oxidised (visual), vegetable, oxidised (aroma) and body, the judges attributed identical scores. There is no consensus on the other sensory wine attributes that could be due to the low difference of the attributes among samples or changes in motivation, sensitivity and psychological answer behaviour [57].

In accordance with the instrumental colour intensity, sensory colour intensity of the wines treated with ethanol impregnated extractive free cork powders was significantly lower than T0 and TF, with the increase in the application dose

Samples	L*	a*	b*	C*	h°	ΔE	Colour intensity (A.U.)	Hue
T	10.84 ± 0.49^{a}	40.55 ± 0.68^{a}	35.62 ± 0.31^{ab}	53.97 ± 0.72^{ab}	0.72 ± 0.00^{a}	—	10.04 ± 1.04^{a}	0.66 ± 0.00^{a}
CKNI	13.17 ± 0.18^{b}	43.32 ± 0.23^{b}	35.47 ± 0.00^{ab}	55.99 ± 0.18^{b}	0.69 ± 0.00^{c}	3.54 ± 0.29^{ab}	9.09 ± 0.07^{ab}	0.72 ± 0.00^{b}
CKFI	13.87 ± 0.08^{c}	44.37 ± 0.01^{b}	36.22 ± 0.29^{b}	57.28 ± 0.19^{b}	0.68 ± 0.00^{c}	4.88 ± 0.01^{b}	8.97 ± 0.17^{b}	0.71 ± 0.01^{b}
CKFI75250	10.85 ± 0.72^{a}	40.72 ± 1.14^{a}	34.08 ± 0.99^{ab}	53.10 ± 1.51^{ab}	0.70 ± 0.00^{b}	1.99 ± 0.48^{a}	9.40 ± 0.07^{ab}	0.74 ± 0.00^{c}
CKFI75500	14.84 ± 0.20^{c}	45.32 ± 0.34^{b}	36.17 ± 0.27^{b}	57.98 ± 0.44^{b}	0.67 ± 0.00^{c}	6.35 ± 0.41^{b}	8.34 ± 0.01^{b}	0.76 ± 0.01^{c}

Data are presented as $\overline{X} \pm s$; data within a column followed by the same letter are not significantly different (Tukey, p 0.05). L, lightness; a*, redness; b*, yellowness; ΔE*, colour difference. The values corresponding to ΔE* were obtained taking as a reference the untreated wine (T) and wines treated with corks (CKNI, CKFI, CKFI75250 and CKFI75500). A.U., absorbance units.*
$p < 0.05$.

Table 4.
Chromatic characteristics of red wines before (TF) and after treatment with natural cork and dichloromethane and ethanol extractive free cork after air removal and ethanol impregnation (CKNI and CKFI) and cork powders with a particle size below 75 μm at two application doses (250 and 500 g/hL).

Samples	Del-3-Glc	Cya-3-Glc	Pet-3-Glc	Peo-3-Glc	Mal-3-Glc	Del-3-AcGlc	Cya-3-AcGlc	Mal-3-AcGlc	Del-3-CoGlc	Cya-3-CoGlc	Mal-3-CoGlc
T	1.18 ± 0.00	8.24 ± 0.17^{a}	13.58 ± 0.30^{a}	4.77 ± 0.07^{a}	155.84 ± 0.69^{a}	1.58 ± 0.02^{a}	0.98 ± 0.00^{a}	26.04 ± 0.46^{a}	1.37 ± 0.02^{a}	1.49 ± 0.15^{a}	13.70 ± 0.78^{a}
CKNI	1.12 ± 0.02	7.66 ± 0.12^{b}	13.10 ± 0.06^{a}	4.55 ± 0.04^{b}	150.37 ± 1.21^{b}	1.24 ± 0.01^{d}	0.79 ± 0.03^{b}	25.66 ± 0.30^{a}	1.24 ± 0.02^{a}	1.43 ± 0.04^{a}	12.09 ± 0.53^{ab}
CKFI	1.14 ± 0.03	8.00 ± 0.05^{ab}	13.59 ± 0.03	4.78 ± 0.03^{a}	156.65 ± 1.05^{a}	1.50 ± 0.00^{b}	0.87 ± 0.01^{ab}	26.69 ± 0.35^{a}	1.31 ± 0.02^{b}	1.23 ± 0.04^{a}	12.61 ± 0.14^{ab}
CKFI75250	1.15 ± 0.02	7.71 ± 0.19^{ab}	13.66 ± 1.15	4.53 ± 0.01^{b}	154.04 ± 0.36^{ab}	1.42 ± 0.04^{c}	0.83 ± 0.04^{ab}	25.89 ± 0.91^{a}	1.29 ± 0.05^{a}	1.08 ± 0.28^{a}	11.89 ± 0.48^{b}
CKFI75500	1.12 ± 0.01	7.93 ± 0.06^{ab}	13.42 ± 0.23	4.67 ± 0.08^{ab}	154.08 ± 2.30^{ab}	1.34 ± 0.02^{c}	0.84 ± 0.05^{ab}	26.68 ± 0.56^{a}	1.30 ± 0.04^{a}	1.18 ± 0.01^{a}	12.04 ± 0.32^{ab}

	Gallic acid	Catechin	*t*-Caftaric acid	Coutaric acid	Caffeic acid	*p*-Coumaric acid	Ferulic acid	Caffeic acid ethyl ester	Coumaric acid ethyl ester
T	24.11 ± 1.55^{a}	9.28 ± 1.53^{a}	30.02 ± 0.25^{a}	11.23 ± 0.18^{ab}	0.84 ± 0.00^{a}	0.72 ± 0.01^{a}	0.57 ± 0.00^{a}	0.39 ± 0.00^{a}	3.89 ± 0.04^{a}
CKNI	21.92 ± 1.79^{a}	8.55 ± 1.16^{a}	29.61 ± 0.28^{a}	11.02 ± 0.20^{ab}	0.80 ± 0.08^{a}	0.57 ± 0.03^{b}	0.54 ± 0.00^{b}	0.33 ± 0.04^{ab}	3.37 ± 0.18^{b}
CKFI	22.50 ± 0.30^{a}	8.48 ± 1.10^{a}	29.23 ± 0.02^{ab}	10.81 ± 0.01^{ab}	0.69 ± 0.01^{b}	0.55 ± 0.00^{b}	0.52 ± 0.01^{c}	0.33 ± 0.00^{ab}	3.28 ± 0.03^{b}
CKFI75250	22.77 ± 0.07^{a}	8.27 ± 0.26^{a}	28.57 ± 0.44^{b}	10.56 ± 0.29^{b}	0.67 ± 0.03^{b}	0.68 ± 0.02^{a}	0.50 ± 0.01^{c}	0.31 ± 0.02^{ab}	3.15 ± 0.15^{b}
CKFI75500	25.02 ± 0.09^{a}	7.74 ± 0.61^{a}	28.09 ± 0.56^{b}	10.57 ± 0.18^{b}	0.66 ± 0.00^{b}	0.67 ± 0.01^{a}	0.50 ± 0.01^{c}	0.31 ± 0.01^{b}	3.12 ± 0.02^{b}

Values are presented as $\overline{X} \pm s$; means within a column followed by the same letter are not significantly different (Tukey, p 0.05). Del-3-Glc, delphinidin-3-glucoside, Cya-3-Glc, cyanidin-3-glucoside, Pet-3-Glc, petunidin-3-glucoside, Peo-3-Glc, peonidin-3-glucoside, Mal-3-Glc, malvidin-3-glucoside, Del-3-AcGlc, delphinidin-3-acetylglucoside, Cya-3-AcGlc, cyanidin-3-acetylglucoside, Pet-3-AcGlc, petunidin-3-acetylglucoside, Peo-3-AcGlc, peonidin-3-acetylglucoside, Mal-3-AcGlc, malvidin-3-acetylglucoside, Del-3-CoGlc, delphidin-3-coumaroylglucoside, Cya-3-CoGlc, cyanidin-3-coumaroylglucoside, Pet-3-CoGlc, petunidin-3-coumaroylglucoside, Peo-3-CoGlc, peonidin-3-coumaroylglucoside and Mal-3-CoGlc, malvidin-3-coumaroylglucoside. Data within a column followed by the same letter are not significantly different ANOVA and Tukey post-hoc test ($p < 0.05$).
$p < 0.05$.

Table 5.
Monomeric anthocyanin and phenolic acid composition of spiked red wines (TF) and after treatment with natural cork and dichloromethane and ethanol extractive free cork after air removal and ethanol impregnation (CKNI and CKFI) and cork powders with a particle size below 75 μm at two application doses (250 and 500 g/hL) (CKFI75250 and CKFI75500).

	T0	TF	CKNI	CKFI	CKFI75250	CKFI75500	*p*	C-index[1]
Intensity	3.70 ± 0.48^{a}	3.70 ± 0.48^{a}	3.60 ± 0.52^{ab}	3.40 ± 0.52^{b}	3.40 ± 0.52^{b}	2.70 ± 0.48^{c}	$p < 0.000001$	3.3
Hue	3.40 ± 0.52^{a}	3.40 ± 0.52^{a}	3.40 ± 0.52^{a}	3.30 ± 0.48^{a}	3.40 ± 0.52^{a}	3.00 ± 0.67^{b}	$p < 0.000001$	—
Limpidity	3.40 ± 0.84	3.40 ± 0.84	3.40 ± 0.84	3.40 ± 0.84	3.40 ± 0.84	3.40 ± 0.84	$p = 1.0$	—
Oxidised (visual)	1.80 ± 0.42	1.80 ± 0.42	1.80 ± 0.42	1.80 ± 0.42	1.80 ± 0.42	2.00 ± 0.00	$p < 0.811$	—
Fruity	3.60 ± 0.52^{a}	1.70 ± 0.48^{b}	2.20 ± 1.14^{c}	2.20 ± 0.79^{c}	2.60 ± 1.07^{d}	2.20 ± 1.23^{c}	$p < 0.000001$	1.6
Floral	2.50 ± 0.53^{a}	1.30 ± 0.65^{b}	1.40 ± 0.52^{b}	1.60 ± 0.84^{c}	1.80 ± 1.03^{d}	1.40 ± 0.52^{b}	$p < 0.000001$	2.5
Vegetable	1.40 ± 0.52^{a}	1.80 ± 0.42^{b}	1.80 ± 0.42^{b}	1.70 ± 0.48^{b}	1.70 ± 0.48^{b}	1.80 ± 0.42^{b}	$p < 0.00016$	—
Phenolic	1.10 ± 0.32^{a}	3.80 ± 0.63^{b}	2.70 ± 0.48^{c}	2.70 ± 0.48^{c}	2.80 ± 0.42^{c}	2.80 ± 0.42^{c}	$p < 0.000001$	6.3
Oxidised (aroma)	2.00 ± 0.67^{a}	2.60 ± 0.52^{b}	2.40 ± 0.84^{ab}	2.80 ± 0.42^{b}	2.40 ± 0.84^{ab}	2.80 ± 0.42^{b}	$p < 0.066$	—
Bitterness	1.70 ± 0.67^{a}	2.30 ± 0.48^{b}	2.20 ± 0.42^{b}	2.20 ± 0.42^{b}	2.20 ± 0.42^{b}	2.00 ± 0.67^{a}	$p < 0.0307$	0.9
Acidity	2.20 ± 0.79	2.70 ± 0.48	2.60 ± 0.52	2.60 ± 0.52	2.60 ± 0.52	2.30 ± 0.48	$p < 0.0605$	1.1
Astringency	2.00 ± 0.67^{a}	2.60 ± 0.52^{b}	2.80 ± 0.42^{c}	2.40 ± 0.52^{d}	2.40 ± 0.52^{d}	2.50 ± 0.85^{bd}	$p < 0.000001$	0.8
Body	2.70 ± 0.48	2.50 ± 0.53	2.40 ± 0.52	2.50 ± 0.53	2.40 ± 0.84	2.50 ± 0.53	$p < 0.333$	—
Balance	3.00 ± 0.67^{a}	2.20 ± 0.42^{b}	2.0 ± 0.67^{b}	2.40 ± 0.52^{b}	2.40 ± 0.52^{b}	2.20 ± 0.42^{b}	$p < 0.0037$	1.1
Persistence	3.00 ± 0.67^{a}	2.00 ± 0.67^{b}	2.20 ± 0.42^{c}	2.40 ± 0.52^{d}	2.40 ± 0.52^{d}	2.40 ± 0.52^{d}	$p < 0.000001$	3.1

[1]*Consonance analysis results—no variance observed for most panellists. Data are presented as the* $\overline{X} \pm s$ *(n = 12); data within a line followed by the same letter are not significantly different (Duncan p < 0.05).*
$p < 0.05$.

Table 6.
Mean scores of each attribute after sensory analysis of volatile phenol-free (T0) and volatile phenol-spiked (TF) red wine after treatment with natural cork and dichloromethane and ethanol extractive free cork after air removal and ethanol impregnation (CKNI and CKFI) and cork powders with a particle size below 75 μm at two application doses (250 and 500 g/hL) (CKFI75250 and CKFI75500).

(CKFI75500) presenting a significantly lower score than CKFI75250 and CKFI. This decrease in colour intensity in the CKFI75500 is also followed by a decrease in the sensory hue, being in accordance with the significant change in h° and L* for this sample. Neither natural nor extractive free cork powders changed significantly the limpidity and oxidised visual sensory attributes.

For VP-spiked wine, the application of all cork powders in two application doses (250 and 500 g/hL) of CKFI75 decreased the negative phenolic attribute significantly compared to the spiked wine (TF); however, the scores obtained were also significantly higher than those observed for the initial unspiked wine (T0). For the fruity aroma attribute, the application of all cork powder allowed recovering significantly the fruity aroma attribute in relation to the VP-spiked wine (TF); nevertheless, the scores were also significantly lower than that observed for the original unspiked wine (T0). The fruity aroma attribute was significantly higher for the CKFI75250 than for all other cork powder samples even higher than CKFI75500. This could be due to the higher decrease in headspace aroma abundance responsible for the fruity notes for this application dose as discussed previously.

For the floral attribute, only CKFI and CKFI75250 allowed increasing significantly this sensory attribute in relation to the TF, and again the scores obtained for the cork-treated wines were significantly lower than that obtained for T0. As observed for the fruity attribute, also for the floral attribute the increase in application dose of CKFI75 decreased the floral attribute of the wine (**Table 6**). The TF wine presented an increased vegetable attribute that did not decreased with the application of cork powder samples, nevertheless, the scores observed was very low (**Table 6**). No significant differences were observed for the oxidised aroma attribute in all samples (T0, TF and cork powder treated wines).

The application of cork powder did not change the acidity and body of the wine samples significantly; however, significant differences were obtained for bitterness, astringency, balance and persistence (**Table 6**). The spiking wine resulted in a significant increase in the bitterness attribute in relation to the T0. Except for CKFI75500, the other cork powders did not decrease bitterness to the levels observed for T0. For astringency, spiking of wine with VPs increase this sensory attribute, and no cork-powder sample decreased the astringency to the initial levels (T0), nevertheless CKFI and CKFI75250 were able to decrease significantly the astringency in relation to TF. For CKNI, a significant increase of astringency in relation to TF was observed, and this can be explained probably by a migration of phenolic compounds from this cork-powder [58, 59]. For balance, TF significantly decreased this sensory attribute, and the application of all cork powders did not lead to scores significantly different from the TF. For persistence, the application of cork powders to TF significantly increased the persistence of wine; however, the scores obtained were significantly lower than the persistence of T0 (**Table 6**).

3.4 Impact of wine chemical composition on sensory profile of red wine treated with extractive free and ethanol impregnated cork powder and application doses

The sensory scores provided by the expert panel for aroma (**Figure 6**), taste and tactile/textural descriptors (**Figure 7**) and the chemical composition of wines, concerning the abundance of headspace aroma compounds and phenolic compounds, respectively, were subjected to multiple factor analysis. From the variable map, it can be concluded that for the first and second factors, both groups of variables contribute almost equally (53 and 46%, and 36 and 64% for the sensory and chemical data for the first and second factors, respectively) (**Figure 6b**).

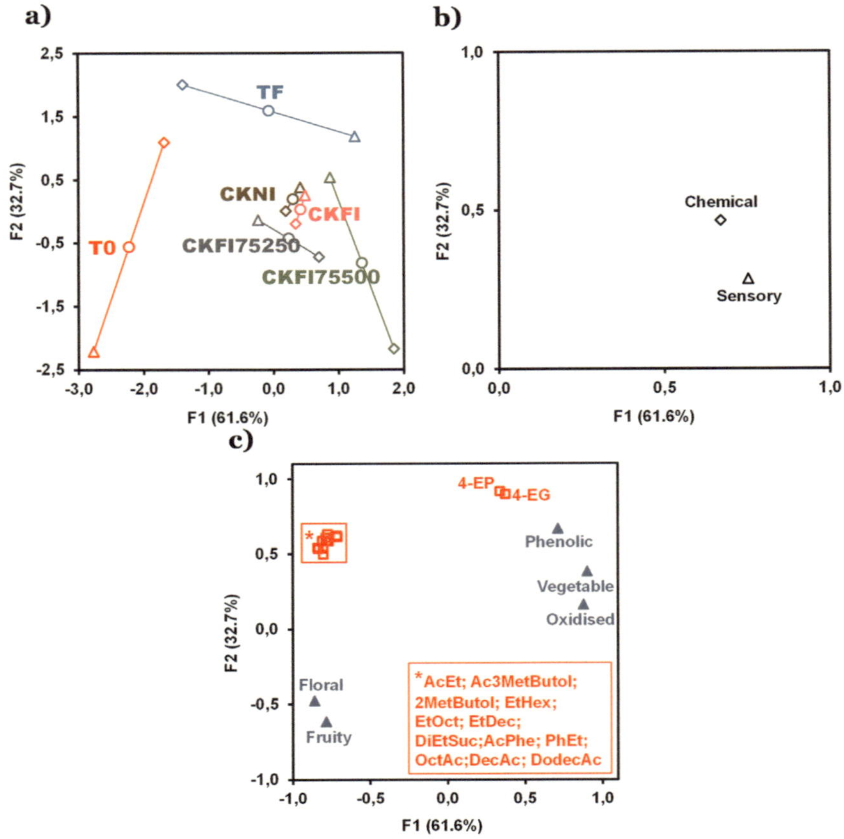

Figure 6.
Multiple factorial analysis of aroma sensory and chemical data: (a) representation of wine samples and clouds; (b) representation of groups (tables) of variables and (c) distribution of variables. VP-free (T0) and VP-spiked (TF) red wines and after treatment with natural cork and dichloromethane and ethanol extractive free cork after air removal and ethanol impregnation (CKNI and CKFI) and cork powders with a particle size below 75 µm at two application doses (250 and 500 g/hL, CKFI75250 and CKFI75500, respectively). Centroid (○); sensory data (Δ); chemical data (◊). AcEt, ethylacetate; Ac3 MetBut, 3-methylbutan-1-ol acetate; 3-MetButol, 3-methylbutan-1-ol; EtOct, ethyl octanoate; EtDec, ethyl decanoate; DiEtSuc, diethyl succinate; PhEt, 2-phenylethanol; 4-EG, 4-ethylguaiacol; 4-EP, 4-ethylphenol; OctAc, octanoic acid; DecAc, decanoic acid; DodAc, dodecanoic acid.

The phenolic negative attribute and the 4-EP and 4-EG headspace abundance were positively correlated with F1, showing that the reduction of the headspace abundance of 4-EP and 4-EG caused by CKNI, CKFI, CKFI75250 and CKFI75500 was important for the decrease of this wine defect. The fruity and floral positive attributes were negatively correlated with F1, showing that the decrease of the headspace abundance of these VPs was important for their perception. However, the abundance of the other headspace aroma compounds was also important for their perception, as they also present negative F1 score. These results are in accordance with previous works that verified that the absence of wine aroma defects, including VPs, was more important for the final wine aroma profile, where that negative off-odorants exert a strong aroma suppression impact on fruity aroma [20, 21, 24, 61, 66].

The phenolic composition of wines although changed significantly, especially after application of the CKFI75 at the two levels, the decrease was not high; nevertheless, significant differences were observed for bitterness, astringency, balance and persistence by sensory analysis, parameters usually linked to the phenolic

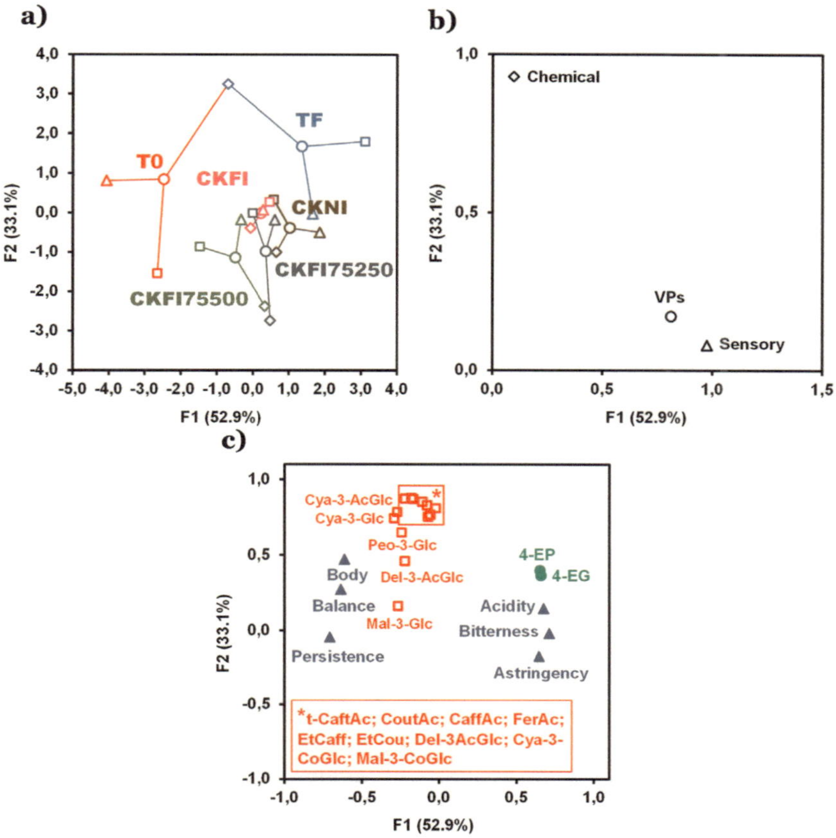

Figure 7.
Multiple factorial analysis of taste and tactile/textural sensory data, phenolic compound chemical data and volatile phenol headspace abundance: (a) representation of wine samples and clouds; (b) representation of groups (tables) of variables and (c) distribution of variables. VP-free (T0) and VP-spiked (TF) red wines and after treatment with natural cork, dichloromethane and ethanol extractive free after air removal and ethanol impregnation (CKNI and CKFI) and cork powders with a particle size below 75 µm at two application doses (250 and 500 g/hL, CKFI75250 and CKFI75500, respectively). Centroid (○); sensory data (Δ); chemical data (◊); VP headspace abundance (•). TotPhe, total phenols; FlavPhe, flavonoid phenols; t-CaftAc, trans-*caftaric acid; CoutAc, coutaric acid; Del-3-Glc, delphinidin-3-O-glucoside, Cya-3-Glc, cyanidin-3-O-glucoside, Peo-3-AcGlc, peonidin-3-O-(6-O-acetyl)-glucoside; VPs, volatile phenols; 4-EP, 4-ethylphenol; 4-EG, 4-ethylguaiacol.*

composition of wines [67]. By the phenolic composition of treated wines, the headspace abundance of 4-EP and 4-EG was also used for MFA, because is actually known that the aroma can interact with the perceived bitterness and astringency of foods, where wines are included [24, 67]. The first factor was important to describe the sensory and VP headspace abundance variables (**Figure 7b**). In the case of the chemical variables, only the second factor was important for its description. The correlation maps of observations and variables (**Figure 7c**) show that the persistence, body and balance attributes were correlated with F1 in the negative direction. However, acidity, bitterness and astringency attributes were correlated with F1 in the positive direction, and there was also a positive correlation between VP headspace abundance with this factor. The correlation of bitterness and astringency, unpleasant wine sensory attributes, with the headspace abundance of VPs, responsible for the negative phenolic aroma, can be explained by the relationship between several aroma compounds with the bitterness and astringency of foods, shown also for wine [24, 68]. The significant decrease observed in some phenolic compounds after application of ethanol impregnated cork-powders does not seem to be responsible for the change in the taste/tactile descriptors observed after wine treatment.

The results obtained from MFA supported the results from sensory analysis of the wines obtained after treatment with the different ethanol impregnated cork powders at the applied doses, highlighting the efficiency of extractive free cork-powders, especially cork powder with a lower particle size at 250 g/hL application dose (CKFI75250), for decreasing the levels of 4-EP and 4-EG in wines and for recovery of fruity and floral aroma attributes. A decrease in phenolic, bitterness and astringency attributes was also observed. The results obtained for visual (colour), aroma, taste and tactile/textural descriptors determined by the expert panel, validated by the wine chemical composition after treatment with ethanol impregnated cork powders show that the wine treated with CKFI75250 resulted in a significant increase in the sensory quality compared to TF, although not identical to T0 wine. This is explained by the efficient removal of VPs and no negative impact on the wine phenolic composition and a lower impact on the headspace aroma compounds when compared to CKFI75500.

4. Conclusions

Optimised cork powder can be a new, cheap, sustainable and efficient fining agent for removal of VPs from wines presenting the unpleasant "Brett character". Its efficiency is shown by the capacity to adsorb significant amounts of 4-EP and 4-EG from a real red wine matrix, presenting a lower impact on the headspace positive aroma compounds when compared to other oenological solutions, already tested. The low impact on the phenolic composition of wines, especially on the monomeric anthocyanins, makes its impact on wine colour limited. Contaminated wines treated with optimised cork powder (extractive free and solvent impregnation) show a significant decrease of the negative phenolic attribute and a significant increase in the positive sensory fruity and floral attributes. This natural product can, in the near future, represent a new oenological fining solution with low environmental impact, contributing to a more sustainable wine industry.

Acknowledgements

We acknowledge Aveleda SA for supplying wine, SAI Enology for providing CKP samples and Lisete Fernandes for the SEM images We appreciate the financial support provided to CQ-VR (PEst-OE/QUI/UI0616/2014) by FCT and COMPETE.

Conflict of interest

No conflict of interest.

Author details

Luís Filipe da Mota Ribeiro[1,2]*, Maria Fernanda Gil Cosme Martins[2] and Fernando Hermínio Ferreira Milheiro Nunes[1]

1 Chemistry Research Centre—Vila Real, (CQ-VR), Food and Wine Chemistry Laboratory, Department of Chemistry, University of Trás-os-Montes e Alto Douro, Vila Real, Portugal

2 Chemistry Research Centre—Vila Real, (CQ-VR), Food and Wine Chemistry Laboratory, Department of Biology and Environment, University of Trás-os-Montes e Alto Douro, Vila Real, Portugal

*Address all correspondence to: fmota@utad.pt

References

[1] Silva SP, Sabino MA, Fernandes EM, Correlo VM, Boese LF, Reis RL. Cork: Properties, capabilities and applications. International Materials Reviews. 2005; **50**:345-365

[2] Barberis A, Dettori S, Filiggheddu MR. Management problems in Mediterranean cork oak forests: Post-fire recovery. Journal of Arid Environments. 2003;**54**:565-569

[3] Costa A, Pereira H, Oliveira A. Variability of radial growth in cork oak adult trees under cork production. Forest Ecology and Management. 2003; **175**:239-246

[4] Fialho C, Lopes F, Pereira H. The effect of cork removal on the radial growth and phenology of young cork oak trees. Forest Ecology and Management. 2001;**141**:251-258

[5] Gil L. Cork powder waste: An overview. Biomass and Bioenergy. 1997; **13**:59-61

[6] Godinho MH, Martins AF, Belgacem MN, Gil L, Cordeiro NMA. Properties and processing of cork powder filled cellulose derivatives composites. In: Macromolecular Symposia. Vol. 169. 2000. pp. 223-228

[7] Pintor AMA, Ferreira CIA, Pereira JC, Silva PC, Vilar SP, Botelho VJP, et al. Use of cork powder and granules for the adsorption of pollutants: A review. Water Research. 2012;**1**:1-5

[8] Volesky B. Detoxification of metal-bearing effluents: Biosorption for the next century. Hydrometallurgy. 2001; **59**:203-216

[9] Pereira H. Rationale of cork properties. BioResources. 2015;**10**:1-23

[10] Filipe-Ribeiro L, Cosme F, Nunes FM. A simple method to improve cork powder waste adsorption properties: Valorisation as a new sustainable wine fining agent. ACS Sustainable Chemistry & Engineering. 2019;7: 1105-1112. DOI: 10.1021/acssuschemeng.8b04775

[11] Chatonnet P, Dubourdieu D, Boidron J, Pons M. The origin of ethylphenols in wines. Journal of the Science of Food and Agriculture. 1992;**60**: 165-178. DOI: 10.1002/jsfa.2740600205

[12] Chatonnet P, Viala C, Dubourdieu D. Influence of polyphenolic components of red wines on the microbial synthesis of volatile phenols. American Journal of Enology and Viticulture. 1997;**48**:443-448

[13] Ribéreau-Gayon P, Glories Y, Maujean A, Dubourdieu D. Handbook of Enology. Vol. 2: The Chemistry of Wine Stabilization and Treatments. 2nd ed. Bordeaux: John Wiley & Sons; 2006

[14] Campolongo S, Siegumfeldt H, Aabo T, Cocolin L, Arneborg N. The effects of extracellular pH and hydroxycinnamic acids influence the intracelular pH of Brettanomyces bruxellensis DSM 7001. LWT-Food Science and Technology. 2014;**59**: 1088-1092. DOI: 10.1016/j.lwt.2014.06.006

[15] Milheiro J, Filipe-Ribeiro L, Vilela A, Cosme F, Nunes FM. 4-Ethylphenol, 4-ethylguaiacol and 4-ethylcatechol in red wines: Microbial formation, prevention, remediation and overview of analytical approaches. Critical Reviews in Food Science and Nutrition. 2017. DOI: 10.1080/10408398.2017.1408563

[16] Schumaker MR, Chandra M, Malfeito-Ferreira M, Ross CF. Influence of brettanomyces ethylphenols on red wine aroma evaluated by consumers in the United States and Portugal. Food

Research International. 2017;**100**: 161-167

[17] Tempère S, Schaaper MH, Cuzange E, de Lescar R, de Revel G, Sicard G. The olfactory masking effect of ethylphenols: Characterization and elucidation of its origin. Food Quality and Preference. 2016;**50**:135-144

[18] Portugal C, Sáenz Y, Rojo-Bezares B, Zarazaga M, Torres C, Cacho J, et al. Brettanomyces susceptibility to antimicrobial agents used in winemaking: In vitro and practical approaches. European Food Research and Technology. 2014;**238**:641-652

[19] Suárez R, Suárez-Lepe JA, Morata A, Calderón F. The production of ethylphenols in wine by yeasts of the genera Brettanomyces and Dekkera: A review. Food Chemistry. 2007;**102**:10-21

[20] Filipe-Ribeiro L, Milheiro J, Matos CC, Cosme F, Nunes FM. Reduction of 4-ethylphenol and 4-ethylguaiacol in red wine by activated carbons with different physicochemical characteristics: Impact on wine quality. Food Chemistry. 2017;**229**:242-251

[21] Filipe-Ribeiro L, Milheiro J, Matos CC, Cosme F, Nunes FM. Data on changes in red wine phenolic compounds, headspace aroma compounds and sensory profile after treatment of red wines with activated carbons with different physicochemical characteristics. Data in Brief. 2017;**12**: 188-202

[22] Milheiro J, Filipe-Ribeiro L, Cosme F, Nunes MF. A simple, cheap and reliable method for control of 4-ethylphenol and 4-ethylguaiacol in red wines. Screening of fining agents for reducing volatilee phenols levels in red wines. Journal of Chromatography B. 2017;**1041-1042**:183-190

[23] Larcher R, Puecher C, Rohregger S, Malacarne M, Nicolini G. 4-Ethylphenol and 4-ethylguaiacol depletion in wine using esterified cellulose. Food Chemistry. 2012;**132**:2126-2130

[24] Filipe-Ribeiro L, Cosme F, Nunes FM. Reducing the negative sensory impact of volatile phenols in red wine with different chitosans: Effect of structure on efficiency. Food Chemistry. 2018;**242**:591-600

[25] Gil L. Cortiça: Produção, Tecnologia e Aplicação. Lisbon: INETI; 1998

[26] Pereira H. Chemical composition and variability of cork from *Quercus suber* L. Wood Science and Technology. 1988;**22**:211-218. DOI: 10.1007/BF00386015

[27] Bento MFS, Pereira H, Cunha MA, Moutinho AMC, van den Berg KJ, Boon JJ, et al. Fragmentation of suberin and composition of aliphatic monomers released by methanolysis of cork from *Quercus suber* L. analysed by GC-MS SEC and MALDI-MS. Holzforschung. 2001;**55**:487-493. DOI: 10.1515/HF.2001.080

[28] Caldas MM, Ferreira JL, Borges M. Boletim do Instituto dos Produtos Florestais, Cortiça. Vol. 5781986. pp. 339-342

[29] Cordeiro N, Neto CP, Gandini A, Belgacem MN. Characterization of the cork surface by inverse gas chromatography. Journal of Colloid and Interface Science. 1995;**174**:246-249

[30] Pereira H. Boletim Do Instituto dos Produtos Florestais, Cortiça. Vol. 5501984. pp. 237-240

[31] Pereira H. Constituição Química da Cortiça-Estado Actual dos Conhecimentos, Cortiça. Vol. 4831979. pp. 259-264

[32] Pereira H. Anais Instituto Superior Agronomia. Vol. 401981. pp. 17-25

[33] Bernards MA, Lewis NG. The macromolecular aromatic domain in

suberized tissue: A changing paradigm. Phytochemistry. 1998;**47**:915-933. DOI: 10.1016/S0031-9422(98)80052-6

[34] Pereira H. Variability of the chemical composition of cork. BioResources. 2013;**8**:2246-2256

[35] Jové P, Olivella MA, Cano L. Study of the variability in chemical composition of bark layers of *Quercus suber* L. from different production areas. BioResources. 2011;**6**:1806-1815

[36] Gil L, Moiteiro C. Ullmann: Encyclopedia of Industrial Chemistry. Vol. 9. Weinheim: Wiley-VCH; 2003. pp. 503-522

[37] Graça J. Suberin: The biopolyester at the frontier of plants. Frontiers in Chemistry. 2015;**3**:62. DOI: 10.3389/fchem.2015.00062

[38] Achyuthan KE, Achyuthan AM, Adams PD, Dirk SM, Harper JC, Simmons BA, et al. Supramolecular self-assembled chaos: Polyphenolic lignin's barrier to cost effective lignocellulosic basic biofuels. Molecules. 2010;**15**:8641-8688. DOI: 10.3390/molecules 15118641

[39] Zakzeski J, Bruijnincx PC, Jongerius AL, Weckhuysen BM. The catalytic valorization of lignin for the production of renewable chemicals. Chemical Reviews. 2010;**110**:3552-3599. DOI: 10.1021/cr900354u

[40] Pereira H. Boletim do Instituto dos Produtos Florestais, Cortiça. Vol. 6001988. pp. 15-18

[41] Asensio A. *Quercus suber* polysaccharides. 1. Structural studies of the hemicellulose-A from the cork of *Quercus suber*. Carbohydrate Research. 1987;**161**:167-170

[42] Asensio A. *Quercus suber* polysaccharides. 2. Structural studies of a hemicellulose-B fraction from the cork of *Quercus suber*. Carbohydrate Research. 1987;**16**:134-138

[43] Asensio A. Polysaccharides from the Cork of *Quercus suber*, II. Hemicellulose. Journal of Natural Products. 1988;**51**: 488-491

[44] Asensio A. Structural studies of a hemicellulose B fraction (B-2) from the cork of *Quercus suber*. Canadian Journal of Chemistry. 1988;**66**:449-453

[45] Rocha SM, Goodfellow BJ, Delgadillo I, Neto CP, Gil AM. Enzymatic isolation and structural characteristics of polymeric suberic of cork from *Quercus suber* L. International Journal of Biological Macromolecules. 2001;**28**:107-119

[46] Gandini A, Pascoal C, Silvestre AJD. Suberin: A promising renewable resource for novel macromolecular materials. Progress in Polymer Science. 2006;**31**:878-892

[47] Llorach R, Espín JC, Tomás-Barberán FA, Ferreres F. Valorization of cauliflower (*Brassica oleracea* L. var. botrytis) by-products as a source of antioxidant phenolics. Journal of Agricultural and Food Chemistry. 2003; **51**:2181-2187

[48] Castola V, Bighelli A, Rezzi S, Melloni G, Gladiali S, Desjobert J-M, et al. Composition and chemical variability of the triterpene fraction of dichloromethane extracts of cork (*Quercus suber* L.). Industrial Crops and Products. 2002;**15**:15-22

[49] Conde E, Cadahía E, Garcia-Vallejo MC, Simón BF. Low molecular weight polyphenols in cork of *Quercus suber*. Journal of Agricultural and Food Chemistry. 1997; **45**:2695-2700

[50] Mazzoleni V, Caldentey P, Silva A. Phenolic compounds in cork used for

production of wine stoppers as affected by storage and boiling of cork slabs. American Journal of Enology and Viticulture. 1988;**49**:6-10

[51] Moure A, Cruz JM, Franco D, Dominguez JM, Sineiro F, Domíngues H, et al. Natural antioxidants from residual sources. Food Chemistry. 2001; **72**:145-171. DOI: 10.1016/S0308-8146 (00)00223-5

[52] Pereira H. Química da Cortiça III. Extracção da Cortiça Com Solventes Orgânicos e Água; Boletim do Instituto dos Produtos Florestais Cortiça. Vol. 5091981. pp. 57-59

[53] Domingues V, Alves A, Cabral M, Delerue-Matos C. Sorption behaviour of bifenthrin on cork. Journal of Chromatography. A. 2005;**1069**:127-132

[54] Olivella MÀ, Bazzicalupi C, Bianchi A, Fiol N, Villaescusa I. New insights into the interactions between cork chemical components and pesticides. The contribution of ππ interactions, hydrogen bonding and hydrophobic effect. Chemosphere. 2015;**119**:863-870

[55] Ferreira V, San-Juan F, Escudero A, Culleré L, Fernández-Zurbano P, Sáenz-Navajas MP, et al. Modeling quality of premium Spanish red wines from gas chromatography-olfactometry data. Journal of Agricultural and Food Chemistry. 2009;**57**:7490-7498

[56] Dijksterhuis G. Assessing panel consonance. Food Quality and Preference. 1995;**6**:7-14

[57] Lundahl DS, McDaniel MR. Influence of panel inconsistency on the outcome of sensory evaluations from descriptive panels. Journal of Sensory Studies. 1991;**6**:145-157

[58] Santos SAO, Pinto PCRO, Silvestre AJD, Neto CP. Chemical composition and antioxidant activity of phenolic extracts of cork from *Quercus suber* L. Industrial Crops and Products. 2010;**31**: 521-526

[59] Santos SAO, Villaverdea JJ, Sousa AF, Coelho JFJ, Neto CP. Silvestre AD phenolic composition and antioxidant activity of industrial cork by-products. Industrial Crops and Products. 2013;**47**: 262-269

[60] Vás GL, Gál J, Harangi A, Dobó K, Vékey A. Determination of volatile compounds of Blaufrankisch wines extracted by solid-phase microextration. Journal of Chromatographic Science. 1998;**36**:505-510

[61] Bailley S, Jerkovic V, Marchand-Brynaert J, Collin S. Aroma extraction diluition analysis of sauternes wines. Key of role polyfunctional thiols. Journal of Agricultural and Food Chemistry. 2006;**54**:7227-7234. DOI: 10.1021/jf060814k

[62] Czerny M, Brueckner R, Kirchoff E, Schmitt R, Buettner A. The influence of molecular structure on odor qualities and odor detection thresholds of volatile alkylated phenols. Chemical Senses. 2011;**36**:539-553

[63] Perestrelo R, Fernandes A, Albuquerque FF, Marques JC, Câmara JS. Analytical characterization of the aroma of Tinta Negra mole red wine: Identification of the main odorants compounds. Analytica Chimica Acta. 2006;**563**:154-164

[64] Dragone GS, Mussato I, Oliveira JM, Teixeira JA. Characterization of volatile compounds in an alcoholic beverage produced by whey fermentation. Food Chemistry. 2009;**112**:929-935

[65] Jiang B, Zhang Z. Volatile compounds of young wines from cabernet sauvignon, cabernet gernischet and chardonnay varieties grown in the loess plateau region of China. Molecules. 2010;**15**:9184-9196

[66] Petrozziello M, Asproudi A, Guaita M, Borsa D, Motta S, Panero L, et al. Influence of the matrix composition on the volatility and sensory perception of 4-ethylphenol and 4-ethylguaiacol in model wine solutions. Food Chemistry. 2014;**149**:197-202

[67] Ferrer-Gallego R, Hernández-Hierro JM, Rivas-Gonzalo JC, Escribano-Bailón MT. Sensory evaluation of bitterness and astringency sub-qualities of wine phenolic compounds: Synergistic effect and modulation by aromas. Food Research International. 2014;**62**: 1100-1107

[68] Sáenz-Navajas M, Campo E, Fernández-Zurbano P, Valentin D, Ferreira V. An assessment of the effects of wine volatiles on the perception of taste and astringency in wine. Food Chemistry. 2010;**121**:1139-1149

Chapter 14

Alcohol Reduction by Physical Methods

Matthias Schmitt and Monika Christmann

Abstract

Alcohol reduction of wine has gained significance worldwide. There are several technologies available to reduce the alcohol content in a targeted way. This chapter explains the principles of alcohol reduction by physical methods. Different membrane processes such as osmotic distillation and the two-step dealcoholization process of reverse osmosis combined with osmotic distillation are compared with distillation processes such as vacuum rectification and spinning cone column. An alternative approach the membrane coupling of ultra- and nanofiltration is described as well. All those technologies appear more or less suitable to reduce the alcohol content in a targeted way. Nevertheless improper handling can cause severe quality losses for the wine. Therefore, enologists should have a thorough understanding of the technologies to avoid negative impact on wine quality through the treatment.

Keywords: alcohol reduction, dealcoholized wine, low- and reduced-alcohol wine, vacuum rectification, osmotic distillation, spinning cone column

1. Introduction

Studies from several countries show rising alcohol contents for wine over the last decades. There are many factors contributing to that phenomenon. Better viticultural practices and improved plant material lead to elevated sugar levels in grapes. In higher alcohol yields of selected yeast strains, modern vinification techniques furthermore lead to an increase in alcohol. The other driving factor for that development is the climate change which cannot be turned back that easily as the other factors. With rising alcohol contents, some wines appear outbalanced and alcoholic which can lead to the consumers' rejection. Additionally, enologists run into fermentation problem caused by elevated sugar contents of grape must and excessive alcohol contents at the end of fermentation. Especially the production of sparkling wine requires moderate alcohol contents to avoid problems with second fermentation.

As a result alcohol management has taken a new direction, from mainly maximizing alcohol contents to minimizing alcohol levels, as well. There are several physical methods available for reducing the alcohol content to a targeted level. They are either based on membrane processes such as osmotic distillation and reverse osmosis coupled with another treatment or on distillation under vacuum. The physical methods for alcohol reduction allow a targeted optimization of the alcohol content according to marked demand or to adapt to taxation and import tariffs based on actual alcohol content of the wine.

2. Comparison of different methods for alcohol reduction

There are several strategies available to produce wine with less alcohol. The most interventions take place before the wine status either in the vineyard, prior, or during fermentation (**Table 1**).

The strategies based on grapevine breeding and selection of clones as well as all strategies in viticulture are preventative and require a certain plan in advance. If, contrary to the assumption, the weather conditions for grape ripeness are very unfavorable, the desired maturity delay or reduced sugar storage in the berry is counterproductive.

In the field of microbiology, two different approaches are possible to produce less alcohol from the initial sugar present.

One possible way is to reduce the sugar content before fermentation by using the enzyme glucose oxidase. The glucose present in the must is converted to gluconic acid in the presence of molecular oxygen by the enzyme glucose oxidase (GOX). The challenge with this process is to reduce the oxidation of other constituents of the must and to reduce excessive acidity in wine [50, 53].

Another microbiological strategy is the use of special yeasts with lower alcohol yield. These yeasts usually show a higher content of fermentation by-products. Due to these other metabolites, the quality as well as the typicality of the wines produced may suffer. The use of genetically modified yeasts is probably seen as very critical by most consumers [60].

Also the metabolism of yeast can be rearranged by taking advantage of the so-called Pasteur effect. For this purpose, a yeast culture is kept in a solution with always less than 5 g/l of sugar. However, the control and addition of must has to be very precise in this process. Automatic measurement and control technology should help to facilitate this process for the user.

2.1 Sugar reduction through membrane coupling

Sugar reduction through membrane coupling can be seen as a unique technological approach for reducing elevated alcohol levels. Before problems arise due to excessive sugar levels in must, fermentation problems are prevented by a selective intervention. The sugar reduction of must is performed in two steps. Subsequent

Grapevine breeding	Viticulture	Microbiology	Enology
New varieties with reduced sugar content	Early harvest at lower sugar levels	Yeasts with reduced alcohol yield	Membrane processes before fermentationSugar reduction by membrane coupling
Clones with reduced sugar accumulation	Adaptation by different training systems	Alternative metabolization of sugar (e.g., enzymatically by glucose oxidase)	Distillation treatments(a) Vacuum distillation(b) Vacuum rectification(c) Spinning cone column
	Canopy management like defoliation or shading		Membrane processes after fermentation(a) Osmotic distillation(b) Nanofiltration or reverse osmosis coupled with second treatment

Table 1.
Overview of strategies to achieve wines with lower alcohol content [57].

treatment with ultrafiltration and nanofiltration removes sugar from the must. Consequently, the fermentation produces a wine with lower alcohol. This technology may help to prevent stuck and sluggish fermentations due to high sugar contents and consequently elevated alcohol levels. These high alcohol levels also have a negative influence on malolactic fermentations [5, 49].

First membranes for ultrafiltration were commercialized in 1926 by membrane filter GmbH [1]. The surface of the membranes is porous, and the pore sizes in ultrafiltration are 10–1000 Å. The retained particles are usually 0.1–10 μm and larger. Common applications of ultrafiltration in food production are dairy processing in milk processing plants and clear filtration in fruit juice production. The use of ultrafiltration for protein removal is conceivable in winemaking [22, 23, 29, 62].

Nanofiltration was developed in the late 1980s. It has been described as a technique between ultrafiltration and reverse osmosis. Nanofiltration usually retains molecules such as sugars and organic acids. The pore size of the membranes is 1–10 nm, and the molecular weight cutoff (MWCO) is at 100–500 Da. The usual working pressure is up to 40 bar. Nanofiltration has many possible applications in winemaking. It is used to remove volatile acidity or to reduce the amount of malic acid. Nanofiltration is also used to concentrate must and wine. If nanofiltration is coupled with another process, the alcohol content of wine as well as the sugar content in the must can be reduced [11, 16, 17, 19, 26, 40, 44, 57].

In this case the permeate of an ultrafiltration is separated in the first step. This fraction contains besides water, acid, and sugar only a few anthocyanins and tannins. During the second step, this fraction is concentrated by nanofiltration. The permeate of the nanofiltration contains then mainly water, some acids, and barely sugar. This aqueous solution is finally blended back to the retentate of the ultrafiltration. The sugar content of the must is thereby reduced after the treatment. The byproduct of that process is the retentate of nanofiltration. This fraction is viscous and high in sugars. The ratio of fructose and glucose is maintained because nanofiltration withholds equal amounts of fructose and glucose. Tartaric acid and potassium are retained only to a small extent, whereby the acidity and pH value are not or hardly changed. Anthocyanins and polyphenols are concentrated in the retentate of nanofiltration due to their molecular size. Consequently, they would be missing in the treated must. Therefore, it is important for red wine to perform the procedure before maceration. A “saignée” has to be done before fermentation. This fraction has to be clarified and treated by the two-step process to avoid color and tannin losses. This pre-clarified fraction is then reduced in sugar content and finally added to the original red wine mash [25, 26, 57].

2.2 Osmotic distillation

The English-language literature contains various synonyms for osmotic distillation, such as membrane distillation, transmembrane distillation, capillary distillation, or pervaporation. Other sources also speak of isothermal membrane distillation [28, 36].

In the process of osmotic distillation, two liquids are separated by a microporous, non-wettable membrane. Both fluids are directed along this membrane, with none of the fluids permeating the membrane pores. Only the volatile components present in the respective liquids can pass the membrane by evaporating and permeating through the pores of the membrane. This gas phases then go into solution of the other side of the membrane. Due to the hydrophobic nature of the membrane, water cannot penetrate the pores of the membrane. Thus, ions, colloids, and macromolecules that do not evaporate and diffuse through the membrane are completely retained.

Osmotic distillation is an isothermal membrane process at atmospheric pressure. The driving force for the molecule passage is the vapor pressure difference of a substance between the two sides of the membrane. The volatile components permeate from the membrane side with higher vapor pressure to the side with lower vapor pressure until equilibrium sets [12, 13].

In osmotic distillation for the partial reduction of alcohol in wine, water is used as strip medium. Apart from possible losses of volatile aroma components, the ethanol flux is of considerable interest. The flux is the amount of permeate that pass through the membrane per unit time. In osmotic distillation, it can be described as follows:

$$Je = Kov\,\Delta Pb \tag{1}$$

In this equation, Je (kg/m^2 h) is the ethanol flux, ΔPb is the vapor pressure difference in terms of ethanol (mmHg) on both sides of the membrane, and Kov is the mass transfer coefficient (m/s). The ethanol flux is influenced by a number of factors. Higher feed and strip media speeds will increase the alcohol transfer through the membrane. Furthermore, the temperature has an influence. As temperatures rise, the flux of volatile components increases. For the efficiency of osmotic distillation, it is important that both sides of the membrane are sufficiently hydrophobic. The pores should not get wet, and no water should penetrate the membrane by capillary action [36, 64].

The gas and vapor passage through the membrane pores takes place by diffusion. The permeation of the volatile molecules through the air space of the membrane pores can be described, depending on the pore radius, by Knudsen and Fick's diffusion. Various references suggest that simultaneous water transfer takes place between both sides of the membrane. The higher the process temperature, the higher the water transfer is. If the so-called stripping water is degassed before treatment in order to avoid an undesirable gas input into the wine, the water transfer is also increased. If the wine temperature is higher than the water temperature, the water transfer increases. In their work, Varavuth et al. [64] proved a water transfer to up to 3 l/m^2 per hour. If the membrane is damaged in its hydrophobic property by improper cleaning and storage, it can be assumed that the transfer of water increases. The water vapor permeating the membrane is relatively more composed of light oxygen atoms. The oxygen isotope ratio (O16/18) is a globally recognized indicator of water addition to wine, according to OIV Resolution OENO 353/2009 [1, 28, 36, 64].

Even if relatively small amounts of water are released into the wine, the osmotic distillation for alcohol reduction could simulate significantly higher levels of water in the wine. The technique of osmotic distillation is widely used in various industries. It can be used both for the degassing of liquids and for the alcohol reduction of beer and wine. The targeted addition of gases or degassing of wine is also summarized as gas management. In contrast to alcohol reduction, a vacuum or a gas is applied to the side opposite the wine. As a result, gas can be specifically added to or removed from the wine. Alcohol reduction of wine by osmotic distillation has been studied by a number of other authors [6, 14, 20, 27, 36–39, 42, 45–48, 51, 56–59, 64, 66].

2.3 Reverse osmosis/nanofiltration and other process

Reverse osmosis is a process for the concentration of liquids, which have a low content of solid components. The passage through the membrane takes place by diffusion through a semipermeable membrane. Consequently the passage takes place against a concentration gradient. During the treatment by reverse osmosis, pressure must be applied that exceeds the osmotic pressure of the solution to be concentrated. The separation of various substances is due to retention in terms of molecular size and by the solution-diffusion mechanism. Originally, reverse osmosis was developed for water treatment or desalination, but nowadays a number of other applications in the food and beverage industry are possible [43, 67].

Common applications of reverse osmosis in the food and beverage industry are the use in dairies or the concentration of juice. Various processes based on reverse osmosis are known in the wine industry.

For must concentration, the reverse osmosis is carried out without further process step. Other enological applications based on reverse osmosis require a second process. Depending on the purpose of the application, various other procedures are used for this subsequent step. When reverse osmosis is used to lower the alcohol content of wine, a permeate is separated in the first step. In addition to alcohol, this aqueous solution contains only a few volatile aroma components. Then, in a second step, this fraction is reduced in its alcohol content by another technology. This is done either by further membrane process, e.g., the osmotic distillation or by a common distillation at atmospheric pressure. Another approach could be replacing the permeate of the reverse osmosis by water, but that so-called diafiltration would mean the addition of water. In many countries the dilution of water is not allowed [10, 15, 18, 65].

Another approach for alcohol reduction is described by Bui et al. [7]. In experiments, they couple two reverse osmosis treatments by differentiating membrane cutoff. In the first step, a permeate with an alcohol content of about 6 vol.% separated. In a second step, this permeate is reduced by a second reverse osmosis treatment to an alcohol content of only 2 vol.%. This fraction is blended back to the initial retentate of the first treatment step to give a reduced-alcohol wine. To date, this approach has not been successful in practice, or there is no plant manufacturer pursuing this approach.

Nanofiltration is a process similar to reverse osmosis. The separation limit of the membranes is usually between 100 and 500 Da. The pore size is between 1 and 10 μm depending on the membrane, and the usual working pressure is 10–30 bar, in some applications also at 40 bar [44].

Compared to reverse osmosis, nanofiltration operates at lower pressure, producing more permeate per m^2 of membrane area. This is due to the membrane structure and the pore size of the membranes. However, other wine components permeate in a higher extent through the nanofiltration membrane. Due to that higher losses of aroma components could occur during the alcohol reduction of the permeate.

Reverse osmosis can also be used in winemaking to reduce volatile acidity [63]. Here, a permeate is separated in the first process step. In addition to ethanol and water, this also contains proportionally more volatile acid. This solution is then passed through an ion exchanger in the second process step. The volatile acid content is thus reduced, and sensory errors can be remedied to a certain extent [68, 69].

Other problems in wines can also be reduced by using reverse osmosis. Fudge et al. [24] describe a method in which off-flavors caused by smoke from larger forest fires can be remedied.

This treatment requires the separation of a permeate first. Then this fraction passes in a second process step: a column with adsorber resins. This significantly reduces volatile phenols such as guaiacol and 4-methylguaiacol. A similar approach was used by Ugarte et al. [65] to remove off-flavors caused by volatile phenols formed by *Brettanomyces* yeasts. Generally speaking, reverse osmosis offers a barrier so that the desired wine constituents are not that widely lost in further treatment steps. Consequently, reverse osmosis in winemaking can be described as a universal membrane process [8].

2.4 Vacuum rectification

Distillation is a thermal separation process in which liquids are vaporized and the vapor then condensed. Generally, distillation is a process that separates substances according to their relative volatility. The relative volatility is a measure of the separability of a distillation with respect to two components to be separated. The relative volatility of two components (α) is calculated from the quotient of the K values of the respective substances [32, 34]:

$$\propto i,j = \frac{\text{K - Value Substance i}}{\text{K - Value Substance j}} \tag{2}$$

The volatility of a substance, in turn, depends on the K value. The K value of a substance describes the tendency of a substance to volatilize [32]:

$$K_i = \frac{(\text{mole fraction substance i in vapor phase})}{(\text{mole fraction substance i in liquid phase})} \tag{3}$$

The higher the *K* value, the higher the amount of the respective substance in the vapor phase. The *K* value depends on the temperature, pressure, and composition of the liquid [32].

Higher temperatures greatly increase the vapor pressure, so the K value of the substance increases as well. If the vapor pressure of the liquid mixture is equal to the ambient pressure in the distillation unit, the liquid begins to boil. The vapor pressure of the liquid mixture is composed according to Dalton's law from the vapor pressures of the individual components, also called partial pressures together. Depending on the nature of the composition of the liquid mixture, the boiling point shifts [34].

The alcohol content of the rising vapors during distillation increases when the boiling liquid contains more alcohol. In addition, the boiling point is lower with increasing alcohol content of the liquid. On the other hand, it can be seen that the gain factor decreases as the alcohol content of the solution increases. The gain factor describes the amount in which the alcohol content increases from the starting liquid until the distillate. The vacuum distillation achieves lower boiling points by applying a vacuum in the column. By lowering the pressure inside the plant, the volatility of the components is increased, and thus the boiling point of the ethanol is reduced. Consequently, the energy required to boil from the ethanol decreases. As a result, the thermal load on the ingredients of the treated liquid is minimized. Alcohol reduction of wine takes place at around 26–35°C [14].

To increase the alcohol content in the distillate, the rising vapors in the distillation column are amplified. This is done by allowing the ascending vapors to flow through the so-called caps of the column against an incoming liquid. The vapor is enriched with volatile components such as ethanol, while the incoming liquid is enriched with high-boiling components from the steam. Depending on the field of application, the columns have different numbers of amplifier caps. This countercurrent distillation or rectification mentioned method is cheaper and less expensive apparatus, as multiple repetitions of single-stage distillation [30].

In general, the alcohol content in the distillate can be up to a content of 97.2 vol.% increase. Then a so-called azeotrope occurs. With an aqueous alcohol solution of 97.2 vol.%, the boiling point at atmospheric pressure is 78.15°C and thus below the usual boiling point of ethanol. Since the rising vapors from this mixture have the same composition as the starting liquid, the gain factor is 1.0, and so no further concentration is realized [34].

In industrial vacuum rectification plants, no further reduction in temperature can be detected during evaporation when the pressure is lowered below 1 mbar. The pressure losses caused by the flow in the pipelines between distillation column and condenser are in charge of that. In order to reduce the loss of aroma during distillation, the condensate is passed to the so-called aroma leaching in countercurrent to the nonalcoholic wine following the rectification. Some of the flavors from the distillate are returned to the nonalcoholic wine [4, 33].

2.5 Spinning cone column

A special form of vacuum rectification is the spinning cone column. This unit is used in the food and beverage industry in various areas for aroma separation and aroma recovery [8].

Unlike conventional columns for vacuum rectification, no static installations are used. Within the cylinder of the spinning cone, there are pairs of a fixed and a movable cone installed. The wine running down the column from the top forms a thin film due to the rotation of the cones. On the underside of the movable cones, there are fins, which swirl the rising vapors and thus allow an increased exchange between the wine and the so-called strip phase.

The special design of the spinning cone column helps to overcome the disadvantage of conventional columns for vacuum rectification. The mass transfer in the column is reduced by the application of the vacuum that instead of turbulent flow, only a laminar flow of the boiling gases prevails. This general disadvantage of distillation under vacuum is qualified by the fact that rotating inserts are mounted in the column. The liquid running down is transformed by its rotation into a thin liquid film. On average, this liquid film is less than 1 mm thick. This results in a very efficient contact between vapors and liquid, whereby the necessary residence time is reduced in the column. In addition, the construction of the spinning cone column, unlike columns for vacuum rectification, can also work with viscous or slurries with a high solid matter content [9, 35].

2.6 Further treatments

A number of further enological methods are conceivable to reduce the alcohol content of wine such as:

- Dialysis
- Pervaporation
- Adsorption of ethanol by organic resins
- Dilution

Except from dilution, all of these are of a technical nature. However, none of these methods have been really successful so far. The reasons for this can be seen either from an economic point of view or in legal aspects. The dilution with water is probably the oldest form of wine fraud and was formerly often used for volume increase. Nowadays the targeted addition of water to reduce the sugar content in must and so to reduce the alcohol content in wine is not legal in most wine-producing countries.

Nevertheless, water addition is legal under certain requirements in some countries. Article 17,010 of the California Administrative Code has the following wording: "...no water in excess of the minimum amount necessary to facilitate normal fermentation, may be used in the production or cellar treatment of any grape wine...."

This provides the enologist a simple and cost-effective way to avoid the stress of the yeast due to high sugar levels and also increase alcohol levels toward the end of fermentation. In addition, unwanted aroma components in the wine are reduced by the dilution.

Another method to achieve wines with less alcohol is the blend with low alcoholic wine. However, the wine law regulations on waste and labeling rights must be observed.

3. Critical evaluation of different technologies for alcohol reduction

The authors did several trials during the last years. The following subchapter will summarize and compare economic and user-oriented considerations [54–60].

3.1 Sugar reduction by membrane coupling

The reduction of the sugar content at must stage by membrane coupling has significant advantages in terms of later fermentation. Excessive sugar levels can be reduced directly before fermentation problems occur due to osmotic stress in the beginning of fermentation or toxic stress due to elevated alcohol at the end of fermentation. Furthermore possible stress for malolactic bacteria is reduced as well. Several trials showed that the treated lots started fermentation faster and continued the fermentation earlier and to a more complete extent.

The quicker and complete fermentation can be seen positive from an economic point of view, as the fermentation tank capacity can be used more efficiently. Moreover stuck and sluggish fermentations are clearly negative in terms of quality and economic consideration [5].

The batch treatment of ultra- and nanofiltration goes along with a certain labor need during harvest, which is in fact the most labor-intensive time during wine production. Possible automation and scale-up of such plants might help to overcome that disadvantage. This treatment could be interesting to be offered by mobile service providers. In that case no additional labor is needed, no investment is necessary, and the regular cleaning and storage of the membrane is needed.

Improper cleaning and storage over several months could cause off-flavors. Even with careful cleaning, membranes can develop an off-flavor from organic matter in the fouling layer. The application of membrane coupling appears more difficult than white wine. The ultrafiltration as the first step of the treatment requires a certain clarification level; otherwise the membranes get clogged. If red mash should be treated, a careful clarification is necessary. In that case a "saignée" is made. That subset is clarified and can be treated. During that time the remaining mash remains with a high content of solids and due to that oxidation and microbiological spoilage can cause later problems. After the membrane treatment, the liquid subset is blended back.

Compared to other treatments for alcohol reduction, the sugar reduction goes along with relatively high volume losses. The reduction of 17 g/l, which corresponds to approximately 1 vol.% less alcohol, means a volume loss of 7% from the initial volume. A further useful application of the nanofiltration retentate could be the sweetening of other wines. Even with a sugar content of 500 g/l, care must be taken to ensure sterile storage. Unlike treatments to remove alcohol, this technology is not in conflict with regulations for distillation.

3.2 Osmotic distillation

Osmotic distillation is a technically simple approach to partial alcohol reduction. Membrane contactors are used in the wine industry in numerous processes such as aeration and degassing of wine. These membranes are more and more widely used by many manufacturers for the preparation of wine and semi-sparkling wine. Such systems are usually based on a membrane contactor with a membrane area of 20 m^2. Depending on the equipment and the degree of automation, the costs for such systems are quite low. Simple systems with manual control valves start from approximately 7000 €. The durability of the membranes highly depends on the care of the membranes and is thus an important factor determining the economic efficiency of the plants. With proper cleaning and storage, the membrane contactors, which are the main cost of the equipment, can be used up to for 5 years before being exchanged by a new membrane. So the method of partial alcohol reduction can be used inexpensively in many businesses. The treatment by the osmotic distillation for alcohol reduction is relatively easy to perform if significant parameters are considered. The amount of previously degassed strip water must be limited to avoid harming the wine quality too much during the treatment. It is advisable to circulate the strip water in a closed and inert system. In many trials it could be shown that an alcohol reduction by 8 g/l should go a long with 14% of the wine volume as strip water. This proved to be the ideal compromise between a quick and aroma-saving treatment.

In order to prevent membrane fouling, the wine to be treated should first be subjected to a wine filtration of min. Separation limit of 5 μm.

The work required to clean and preserve the membranes can be compared to that of conventional cross-flow filtration. Nevertheless the hydrophobic property of the membranes does not allow backflush or use of cleaning enhancers and surfactants.

The alcohol reduction by osmotic distillation is continuous and needs little or no supervision during treatment. If the alcohol reduction is to be carried out close to the maximum permissible limit, it is advisable to reduce a portion of the alcohol content strongly and then adjust the alcohol content precisely by blending with the initial wine.

The performance of the alcohol reduction is not constant as the driving force; the vapor pressure difference between both sides of the membrane gets lower during the treatment. So the alcohol permeation rate decreases during the treatment. The strip water accumulates in the alcohol content. In the experiments, it had alcohol contents between 4 and 7 vol.% [27].

Due to alcohol reduction, the density of the treated wine increases. During the treatment of larger containers, the change in density can cause certain layer formation in the tank. Before assessing the final degree of alcohol reduction, the tank has to be homogenized carefully. Without this mixing, it can lead to errors in the measurement of the alcohol content, and thus a wine may be treated in too high extent. Since the systems for osmotic distillation are relatively small and mobile, it is conceivable to perform such a treatment with a mobile plant. For this purpose, the wine does not need to be brought to a plant as is the case for common systems based on distillation-based processes. The treatment can be carried out within the winery.

If alcohol is separated from the wine, a number of custom regulations might be affected even if the separated alcohol fraction is not very high in alcohol (4–7 vol.%), and so it is not economically interesting to separate the ethanol further in another distillation process. The recycling of the strip water as brandy is neither economically interesting nor from quality aspects to be recommended.

3.3 Reverse osmosis/nanofiltration and other process

Reverse osmosis or nanofiltration alone does not lower the alcohol content of wine. The permeate from that treatment has to be reduced in alcohol content by another step. This alcohol reduced fraction is finally blended with the concentrate from the first step.

The plant from the company Oenodia (Pertuis, France) is a mobile system that combines reverse osmosis, respectively, nanofiltration with osmotic distillation and is used as mobile service in wineries.

In the first step, permeate is reduced in alcohol by osmotic distillation.

The strip water for the osmotic distillation is not pumped in a closed circuit; there is a continuous flow of heated water through the membrane contactor. These process parameters are chosen so that as much alcohol as possible can be separated with this system per time. The alcohol transfer through the membrane is increased by elevated temperatures, and the vapor pressure difference of the respective substances is significantly higher with continuous supply of new strip water than with a closed strip water cycle and limited water amounts [14, 27].

The first step of treatment by reverse osmosis or nanofiltration reduces fouling at the membrane contactor for osmotic distillation as the permeate is free of solids and low in colloid content. In comparison to the expensive and complex membrane contactors, the membranes for the first step can be cleaned more easily. In addition, their prices are much lower.

The oxygen uptake was measured during several treatments and was between 0.6 and 0.8 mg/l on average for the two-stage process. In comparison to that, the single-step osmotic distillation for alcohol reduction showed on average an oxygen uptake of 1.4 mg/l. So the alcohol reduction by membrane systems can be compared with a common gentle wine filtration. In both cases the strip water was degassed. Without degassing the oxygen uptake could have been 4 mg/l and more [61].

The resulting strip water from the second step had similar alcohol content as in the direct osmotic distillation of wine. The alcohol content was in a range of 5–7 vol.%. Compared to treatments based on distillation, the membrane treatments are compact build and mobile. They just require electricity and water of certain softness. Furthermore small lots can be treated, allowing pretrials to assess the final sensory character of the wine.

3.4 Vacuum rectification

Vacuum rectification is a continuous process, and the systems which are used in the beverage industry have a capacity of 300 l/h upward.

Corresponding plants already exist in Germany for more than 100 years.

The number of companies offering dealcoholization based on vacuum rectification has grown significantly during the last years. Common systems are designed for flow rates of 1000–5000 l/h of wine. The respective rectification columns are on site, and the legal settlement terms in distillation are in charge of the service-offering company.

The usual minimum quantity to be treated is 1000 l. At the end of the treatment, the alcohol content of the wine is below 0.5 vol.%.

For example, if 1000 l wine with 14 vol.% are treated, 135 l of pure ethanol are separated. According to the operation of the column, the spirit fraction has an alcohol content up to 80 vol.% Values above that are not to be recommended as the hazard of explosion increases by such high ethanol contents. Assuming an alcohol content of 80 vol.%, 168 l of spirit are separated. Approximately 830 l of alcohol-free wine are remaining that can be used for diluting the alcohol content of the initial wine to any value desired. The alcohol-free wine fraction is very susceptible to microbiological spoilage as the content of free SO2 is reduced by 75% of the initial content and the microbiological effect of ethanol is missing as well. Within hours alcohol-free wine can develop a flor yeast layer. To avoid microbiological contamination and resulting off-flavors in wine, a blending within the next day to a common alcohol level should be done. The losses of SO2 should be replaced again as well.

3.5 Spinning cone column

The spinning cone column is generally used for the separation of volatile components from different liquid–solid mixtures.

The universal applicability of this plant explains its widespread use in various areas of the flavor, food, and beverage industries. In the wine industry, it is used for desulfurization, dealcoholization, and partial alcohol reduction.

Similar to vacuum rectification systems, the spinning cone column is due to its size and infrastructure requirements not suitable for mobile use. Already the pilot plant for trials has a height of 4 m and a weight of 5 t. The need for steam is approximately 85 kg/h with required working pressures of 6–8 bar for a problem-free operation. These parameters are very difficult to realize with common steam generators and pipes applied in the beverage industry. For optimal operation a cooling system of 60 kW is recommended. Systems of that size are to be found just in bigger wineries or cooperatives. Corresponding aggregates for cooling and steam can be rented as mobile systems, but this will generate further costs.

The treatment takes place in two passages at different process temperatures. The performance of the SCC is therefore significantly reduced compared to a conventional rectification column. The spirit fraction resulting from the spinning cone column treatment has just an alcohol content of about 50 vol.% For the commercialization in bulk, a further distillation step, to increase the alcohol content, is recommended. This would be easy to realize with another distillation stage directly at the plant. This could also reduce the loss of volume by returning the nonvolatile residue to the wine. The two passages through the spinning cone column allow a recovery of a very volatile fraction that is separated and blended back to the alcohol fraction after the second passage. Due to that practice, the most volatile components are recovered and are not lost in the ethanol fraction. The declaration of that pre-run as aroma is irritating and led to many misinterpretations of the process. The pre-run of the process is not selectively positive. It is coined by descriptors such as pungend, sulfur coined, and solvent.

From a business perspective, the use of the spinning cone column in the wine industry is conceivable above all as a contracted service. Permanently installed it is used for dealcoholization, partial alcohol reduction, and desulfurization.

3.6 Water addition

The dilution of must with water is the simplest and cheapest solution to reduce the sugar content and thus the subsequent alcohol content. The addition of water dilutes all wine components. This concerns the positive and negative sensory aspects. The water used is not really a cost factor. On the other hand, the volume increase by adding water can have a significant impact in terms of sales. In order to avoid possible negative influences on the subsequent wine quality, the amount of water should be minimized and neutral in terms of taste, free of microorganisms and microbiologically active substances.

Similar to sugar reduction by membrane coupling, the key benefits to be seen are improved fermentation kinetics with less residual sugar in the end. In some wine-producing countries, this practice is legal.

4. Sensory impact of partial alcohol reduction

A detailed assessment of the processes for partial alcohol reduction of wine should include a sensory evaluation as well.

First of all it is important to understand to what degree of alcohol reduction makes the wines different from the initial wine. Furthermore the changes in terms of sensory characteristics should be pointed out. A comparative study of the different physical methods helps to assess what technology is more gentle in terms of wine quality. Several sources report that a wine with an alcohol reduction by 2 vol.% is not differed from the initial wine [2, 3, 14, 39, 41, 52, 56–58].

The extensive investigations of the authors substantiate these results. A total of 39 discriminative tests with a trained panel did not show a significant difference between untreated wines and corresponding samples with 2 vol.% less alcohol. Here the grape variety, the initial alcohol content, and wine style were irrelevant and not influencing the results. These discriminative sensory tests did not show significant differences with several white and red varieties. Even trials with sparkling wines showed that 2 vol.% alcohol difference is not perceived as a significant difference in discriminative tests [59].

The treatment goes along with several collateral damages in terms of wine quality such as excessive aroma losses, oxidation, and microbiological spoilage. So it is important to mention that the alcohol reduction has to be done carefully according to the manufacturers' recommendation.

Discriminative tests comparing the initial wine with samples that have 3 vol.% and 4 vol.% less alcohol showed clearer results. The panelists could differentiate more clearly and at a significant level the treated wines from the initial wine. Nevertheless there was no clear tendency in terms of preference. That is in line with other sources [39, 40].

Several comparative tests showed that the different methods for partial alcohol reduction, mentioned before, did not differ from each other when the same wines were reduced by 2 vol.% each. Even the samples that were diluted with water to have 2 vol.% less alcohol did not differ significantly from the physical methods. That is in line with other sources [2].

When the range of alcohol reduction was 4 vol.%, e.g., from initially 14.6 vol.% to 10.6 vol.%, there was a general tendency toward methods based on distillation under vacuum (vacuum rectification and spinning cone column). Here membrane processes could not deliver the same quality.

A severe alcohol reduction by distillation has the advantage that only a partial amount is treated severely. The membrane processes, in contrast, require a relatively long treatment by multiple passes of the total amount of wine through the plant to reduce the alcohol content to the same extent. If the membrane processes are to be used to produce products that are severely reduced in alcohol content, membrane plants should be in bigger size, and short-time heating could help to shorten the treatment, so that wine quality is potentially harmed less. With all tested physical methods, an aroma recovery out of the ethanol fraction could help to improve the final result in terms of quality.

The sensory effect of alcohol is very complex in terms of wine. The partial alcohol reduction of the wine changes several sensory attributes. Due to the lower alcohol content, the wines that have 3 vol.% less than the initial wine clearly show lowered sensations in terms of body and fullness. As this attribute is clearly desired, later enological interventions could aim to buffer that loss. Depending on wine style, sweetening and addition of $CO2$ or tannins could help to compensate those losses.

Bitterness and the sweetness sensation is reduced when the wines have less alcohol. The perceived acidity of the wines rises by removing alcohol. The fruitiness of the wines is reduced by the alcohol reduction contrary to the theory that wines with elevated alcohol content appear less intense in terms of fruity character. The treatment by physical methods goes along with aroma losses, and that factor is stronger than the elevated volatility of the remaining aromas due to alcohol reduction.

The theory of sweet spots in terms of alcohol has been accepted, so far, quite uncritical. With regard to wine, this term is mentioned in various publications that point out that even small differences in ranges of 0.1–0.2 vol.% can have severe influences on the taster's preference. This approach does not conform to other sources. Since an alcohol difference of less than 2 vol.% cannot be distinguished significantly, an experimental setup with alcohol steps of 0.1 or 0.2 vol.% is incomprehensible. The author's research showed that the panelist's preferences were widely spread at the respective tastings. So there was no significantly preferred spot when a set of seven samples with varying alcohol contents were tasted even though the initial and final alcohol content clearly made the wine different. It is important to note that the examiner's preferences spread evenly over the range of samples. That proves that the preferences in terms of alcohol content in wine are not uniform. Instead of small changes in terms of alcohol contents, it could be more interesting to clearly change wine style, thus creating wines that are favored by customers who prefer lighter wines [21, 31, 39, 41, 51, 57].

Author details

Matthias Schmitt* and Monika Christmann
Institute of Enology, Hochschule Geisenheim University, Geisenheim, Germany

*Address all correspondence to: matthias.schmitt@hs-gm.de

References

[1] Baker RW. Membrane Technology and Applications. West Sussex, England: Wiley; 2004

[2] Bes M. Les différentes stratégies microbiologiques et technologiques pour la production de vins à teneur réduite en alcool. Montpellier (Frankreich): Euroviti Congress; 2009

[3] Bes M, Aguera E, Athes V, Cadiere A, Cottereau P, Dequin S, et al. Les différentes stratégies microbiologiques et technologiques de production de vin à teneur réduite en alcool. Revue des Oenologes. 2010;**135**:9-11

[4] Bethge D. Destillation im Fein- und Hochvakuum. Vakuum in Forschung und Praxis. 1996;(2):84-86

[5] Bisson LF. Stuck and sluggish fermentations. American Journal of Enology and Viticulture. 1999;**50**:107-119

[6] Bocca E, Piubelli G, Stassi A, Carbognin C, Ferrarini R. Effect of dealcoholization through membrane contactor: Composition and evaluation. In: 33rd World Congress of Vine and Wine, Tiflis (Georgia). 2010

[7] Bui K, Dick R, Moulin G, Galzy P. A reverse osmosis for the production of low ethanol content wine. American Journal of Enology and Viticulture. 1986;**37**:297-300

[8] Christmann M. Neue Oenologische Verfahren. Meininger Verlag GmbH: Neustadt/Weinstraße; 2001

[9] Christmann M. Die "Spinning Cone Column"—Segen oder Fluch? Schweizer Zeitschrift für Obst- und Weinbau. 2006;**5**:6-9

[10] Clos D. Chemische und Sensorische Auswirkungen physikalischer Konzentrierungsverfahren auf Most und Wein (dissertation). Technische Universität Kaiserslautern; 2003

[11] Cotterau P. Reduction de la teneur en sucre des mouts - proédé REDUX®. Institut Français de la Vigne et du Vin; 2010, Juli 2010

[12] Curcio E, Driolo E. Membrane distillation and related operations—A review. Separation and Purification Reviews. 2005;**34**:35-86

[13] Diban N, Athes V, Bes M, Souchon I. Ethanol and aroma compounds transfer study for partial dealcoholization of wine using membrane contactor. Journal of Membrane Science. 2007;**311**:136-146

[14] Diban N, Arruti A, Barceló A, Puxeu M, Urtiaga A, Ortiz I. Membrane dealcoholization of different wine varieties reducing aroma losses. Modeling and experimental validation. Innovative Food Science and Emerging Technologies. 2013;**20**:259-269

[15] Diel B. Anreicherung deutscher Weine mittels Umkehrosmose. Vergleich zur konventionellen Saccharoseanreicherung unter analytischen, sensorischen und ökonomischen Gesichtspunkten (dissertation). Universität Gießen; 2000

[16] Ducruet J, Fast K, Noilet P. Développement d'un modèle mathématique permettant de piloter l'élimination de l'acide malique des moûts par nanofiltration. In: Eighth International Enology Symposium, Bordeaux (Frankreich). 2007

[17] Ducruet J, Fast-Merlier K, Noilet P. New application for nanofiltration: Reduction of malic acid in grape must. American Journal of Enology and Viticulture. 2010;**61**:19-25

[18] Duitschaever CL, Alba J, Buteau C, Allen B. Riesling wines made

from must concentrated by reverse osmosis. I. Experimental conditions and composition of musts and wines. American Journal of Enology and Viticulture. 1991;**61**:278-283

[19] Eriksson P. Nanofiltration extends the range of membrane filtration. Environmental Progress. 1988;7:58-62

[20] Ferrarini R, Ciman GM, Cugini L, Gostoli C, Simoni M, Camin F. Membrane contactor process to reduce ethanol in wine. In: Volatile Compounds and Stable Isotope Ratios Changes Alcohol Level Reduction in Wine. Bordeaux (Frankreich): Oenoviti International Network; 2013

[21] Fischer U. Die sensorische Bedeutung des Alkohols in Wein. Getränkeindustrie. 2010;**10**:14-17

[22] Flores JH, Heatherbell DA, McDaniel MR. Ultrafiltration of wine: Effect of ultrafiltration on white riesling and Gewürztraminer wine composition and stability. American Journal of Enology and Viticulture. 1990;**41**:207-214

[23] Flores JH, Heatherbell DA, Henderson LA, McDaniel MR. Ultrafiltration of wine: Effect of ultrafiltration on the aroma and flavor characteristics of white riesling and Gewürztraminer wines. American Journal of Enology and Viticulture. 1991;**42**:91-96

[24] Fudge AL, Ristic R, Wollan D, Wilkinson KL. Amelioration of smoke taint in wine by reverse osmosis and solid phase adsorption. Australian Journal of Grape Wine Research. 2011;**17**:41-48

[25] García-Martín N. Evaluation of several ultra- and nanofiltration membranes for sugar control in winemaking. Desalination. 2009;**245**:554-558

[26] García-Martín N, Ortega-Heras M, González-Huerta M, Prádanosa P, Hernándeza H. Sugar reduction in musts with nanofiltration membranes to obtain low alcohol-content wines. Separation and Purification Technology. 2010;**71**:158-170

[27] Gutte R. Prozessbewertung der osmotischen Destillation zur Entalkoholisierung von Wein (dissertation). Institut für Ernährungs- und Lebensmittelwissenschaften, Lehrstuhl für Lebensmitteltechnologie/–biotechnologie, Rheinische Friedrich-Wilhelms-Universität zu Bonn; 2009

[28] Hogan PA, Canning RP, Peterson PA, Johnson RA, Michaels AS. A new option: Osmotic distillation. Chemical Engineering Progress. 1998;**7**:4961

[29] Hsu JC, Heatherbell DA, Flores JH, Watson BJ. Heat-unstable proteins in grape juice and wine. II. Characterization and removal by ultrafiltration. American Journal of Enology and Viticulture. 1987;**38**:17-22

[30] Ignatowitz E. Chemietechnik (8. Auflage). Haan-Gruiten: Europa-Lehrmittel; 2007

[31] King ES, Heymann H. The effect of reduced alcohol on the sensory profiles and consumer preferences of white wine. Journal of Sensory Studies. 2014;**29**:33-42

[32] Kister HZ. Distillation design. In: New York. USA: Donnelley & Sana Company; 1992

[33] Koerner R. Returning aroma compounds removed during thermal dealcoholization of beverages. European Patent 0713911A2; 1996

[34] Kolb E, Fauth R, Frank W, Simson I, Ströhmer G. SpirituosenTechnologie (6. Auflage). Hamburg: Behr's Verlag; 2002

[35] Langrish TAG, Makarytchev SV, Fletcher DF, Prince RGH. Progress in

understanding the physical processes inside spinning cone columns. Chemical Engineering Research and Design. 2003;**81**:122-130

[36] Lawson KW, Lloyd DR. Membrane distillation. Journal of Membrane Science. 1997;**124**:1-25

[37] Liguori L, Attanasio G, Albanese D, Di Matteo M. Aglianico wine dealcoholization tests. Computer Aided Chemical Engineering. 2010;**28**:325-330

[38] Liguori L, Russo R, Albanese D, Di Matteo M. Effect of process parameters on partial dealcoholization of wine by osmotic distillation. Food and Bioprocess Technology. 2012;**6**:2514-2524

[39] Lisanti MT, Gambuti A, Genovese A, Piombino P, Moio L. Partial dealcoholization of red wines by membrane contactor technique: Effect on sensory characteristics and volatile composition. Food and Bioprocess Technology. 2012;**6**:2289-2305

[40] Massot A, Mietton-Peuchot M, Peuchot C, Milisic V. Nanofiltration and reverse osmosis in winemaking. Desalination. 2008;**231**:283-289

[41] Meillon S, Viala D, Medel M, Urbano C, Guillot G, Schlich P. Impact of partial alcohol reduction in syrah wine on perceived complexity and temporality of sensations and link with preference. Food Quality and Preference. 2010;**21**:732-740

[42] Motta S. Comparison of Two Techniques for Wine Dealcoholization, 61. Stuttgart: Deutscher Weinbaukongress; 2013

[43] Paulsen DJ, Wilson RL, Spatz DD. Osmosis and ultrafiltration applied to the processing of fruit juices, reverse osmosis and ultrafiltration. ACS Symposium Series;**281**:325-344

[44] Porter MC. Handbook of Industrial Membrane Technology. Park Ridge, USA: Noyes Publications; 1990

[45] Ribéreau-Gayon P, Dubourdieu D, Donèche B, Lonvaud A. Handbook of Enology (2. Auflage), Volume 1, the Microbiology of Wine and Vinifications. Chichester (England): John Wiley & Sons, Ltd.; 2006

[46] Röcker J, Schmitt M, Pasch L, Ebert K, Grossmann M. The use of glucose oxidase and catalase for the enzymatic reduction of the potential ethanol content in wine. Food Chemistry. 2016;**200**:660-670

[47] Schmidt O. Alkoholreduzierung mit Membranverfahren – Auf der Suche nach dem Sweet Spot. Rebe und Wein. 2010;**9**:9-13

[48] Schmidt O. Moderne Kellertechnik. Stuttgart: Ulmer Verlag; 2013

[49] Schmidtke LM, Blackman JW, Agboola SO. Production technologies for reduced alcoholic wines. Journal of Food Science. 2012;**71**:25-41

[50] Schmitt M, Christmann M. Alkoholmanagement, Ein Umdenken ist nötig. DerDeutsche Weinbau. 2011;**18**:24-26

[51] Schmitt M, Christmann M. Verminderung des Zuckergehaltes im Most als Alternative zur Alkoholreduzierung. Vol. 15. Der Deutsche Weinbau; 2012. pp. 20-22

[52] Schmitt M, Murgo M, Prieto S. Does osmotic distillation change the isotopic relation of wines? In: 37 World Congress of Vine and Wine, Mendoza (Argentina). 2014

[53] Schmitt M. Teilweise Alkoholreduzierung von Wein mittels physikalischer Verfahren: Alkohol management (dissertation). Geisenheimer Bericht 80; 2016

[54] Schmitt M, Christmann M. Verfahren zur Alkoholreduzierung im Test. Der Deutsche Weinbau. 2016;**6**:12-14

[55] Schmitt M. Gärstockungen bei Schaumweinen. Der Deutsche Weinbau. 2017;**12**:27-30

[56] Schmitt M, Broschart S, Patz CD, Rauhut D, Friedel M, Häge D. The use of yeast with lower alcohol yield for the production of sparkling base wines. In: BIO Web of Conferences, World Congress of Vine and Wine; Montevideo, Uruguay, 12. 2018

[57] Schneider V. Mikrooxidation junger Rotweine, Teil IV: Die natürliche Sauerstoffaufnahme. Das Deutsche Weinmagazin. 2001;**6**:12-16

[58] Sims CA, Bates RP, Johnson RP. Comparison of pre- and postfermentation ultrafiltration on the characteristics of sulfited and non-sulfited white wines. American Journal of Enology and Viticulture. 1990;**41**:182-185

[59] Ugarte P, Agosin E, Bordeu E, Villalobos J. Reduction of 4-ethylphenol and 4-ethylguaiacol concentration in red wines using reverse osmosis and adsorption. American Journal of Enology and Viticulture. 2005;**56**:30-36

[60] Varavuth S, Jiraratananon R, Atchariyawut S. Experimental study on dealcoholization of wine by osmotic distillation process. Separation and Purification Technology. 2009;**66**:313-321

[61] Weber D. Einfluss verschiedener Traubenmostkonzentrierungsverfahren auf die Weinqualität (dissertation). Fakultät Wissenschaftszentrum Weihenstephan für Ernährung, Landnutzung und Umwelt der Technischen Universität München; 2006

[62] Weible A, Freund M. Reduzierung des Alkohols mittels Membrantechnik. Der Deutsche Weinbau. 2011;**18**:16-18

[63] Wiederkehr M. Entfernung von Essigsäure aus Wein: Ein neues Verfahren (Teil 1). Schweizer Zeitschrift für Obst- und Weinbau. 1998;**7**:193-193

[64] Wiederkehr M. Entfernung von Essigsäure aus Wein: Ein neues Verfahren (Teil 2). Schweizer Zeitschrift für Obst- und Weinbau. 1998;**8**:393-398

[65] Wollan D. Controlling excess of alcohol in wine. Australian and New Zealand Wine Industry Journal. 2005;**20**:48-50

[66] Wollan D. The right tools for the job: Dealing with cool climate wine styles. In: Proceedings of the Sixth International Symposium for Cool Climate Viticulture and Enology; Auckland: New Zealand Society for Viticulture and Enology. 2006. pp. 159-168

[67] Wucherpfennig K. Anwendungsmöglichkeiten von Membranprozessen bei der Herstellung von Getränken, Habilitation. Fachbereich Ernährungswissenschaften der Justus Liebig-Universität Gießen; 1977

[68] Yu P, Pickering GJ. Ethanol difference thresholds in wine and the influence of mode of evaluation and wine style. American Journal of Enology and Viticulture. 2008;**59**:146-152

[69] Zürn F. Anwendungsmöglichkeiten der Umkehrosmose (Hyperfiltration) zur Qualitätsverbesserung von Most und Wein (dissertation). Justus Liebig Universität Gießen; 1979

Chapter 15

Production and Marketing of Low-Alcohol Wine

Tamara Bucher, Kristine Deroover and Creina Stockley

Abstract

Moderate wine consumption may be associated with specific health benefits and a healthy lifestyle. However, increased amounts of ethanol are cytotoxic and associated with adverse health outcomes. Alcohol reduction in wine might be an avenue to reduce alcohol related harm without forcing consumers to compromise on lifestyle and benefit from positive aspects of moderate consumption. The aim of this review is to give an overview of viticultural and pre and post fermentation methods to produce low-alcohol wine, and to summarize the current evidence on the consumer acceptance and behaviour related to low-alcohol wine. Strategies for the labelling and marketing of wines with reduced alcohol content are discussed.

Keywords: reduced-alcohol wine, wine trends, alcohol content, consumer behavior, alcohol reduction technologies

1. Alcohol and health

Alcohol consumption is associated with several social and health risks and since 2010, the WHO conducts its global strategy to reduce the harmful use of alcohol [1–4]. In a recent review, alcohol was found to be the seventh leading risk factor for premature death in 2016, contributing to 2.8 million deaths worldwide, leading to the conclusion that no amount of alcohol is safe [5]. Other sources found that alcohol can have some beneficial health effects when consumed in low-risk drinking patterns [6–10]. Research shows that there may be a beneficial cardioprotective effect of these relatively low levels of drinking for ischaemic heart disease, ischaemic stroke and diabetes mellitus, as well as death from all causes [6–10]. Low-risk drinking is also called "drinking in moderation" and is usually defined using standard units. The WHO recommends consuming a maximum of two standard drinks per day, with at least two days a week without alcohol, and never more than four drinks per consumption episode [4]. In Australia, a standard drink contains 10 g of pure alcohol [11]. Ethanol is found to be associated with favourable changes in several cardiovascular biomarkers such as higher concentration of high-density lipoprotein cholesterol and adiponectin, and lower concentration of fibrinogen and other haemostatic factors [12].

1.1 History of wine as a health food

From earliest times wine has been used as a therapeutic agent, irrespective of a lack of clinical and scientific data for a variety of ailments [13]. It has been used as a nourishment, for diuresis and hyperthermia and as an aperient, as well as an antibacterial agent for wounds. These therapies eventually became widely adopted throughout Medieval Europe until the puritanical religious movement accredited to Oliver Cromwell spread to the east coast of North America via the Pilgrim Fathers in 1620.

These puritan movements in England and the USA eventually led to the temperance movement of the nineteenth century, which condemned alcohol in all its forms. Wine only found favour again as a medicine in the last decades of the twentieth century.

1.2 Wine and health

Wine, with its grape-derived phenolic compounds, has been found to potentially have additional health benefits to other alcoholic beverages [14, 15]. Wine, when consumed in moderate amounts and when consumed together with a meal, mitigates oxidative stress and vascular endothelial damage induced by a high-fat meal [16]. Consequently, consuming red wine with meals, has been suggested to be cardioprotective and even protective against diabetes type 2 where consumers can experience better health whilst ageing as well as experience an increased lifespan [17]. This concept is now well known as the "French Paradox" and has been the subject of a considerable amount of research over the last 30 years [18]. Wine, in particular red wine, contains various phenolic compounds and their polymeric forms, which are antioxidant chemicals that interact with, and neutralize free radicals and thus prevent cell damage [19]. Phenolic compounds such as catechin, quercetin and resveratrol which are found in skins, seeds, and/or stems of the grapes are consequently in measurable concentrations in red wine have been shown individually and collectively to have antioxidant, anti-inflammatory, anti-proliferative and anti-angiogenic effects in in vitro, animal, ex vivo and limited human clinical studies, and have the potential to act as therapeutic agents in the prevention and treatment of certain chronic diseases [18]. The relative contribution of these phenolic compounds, and particularly resveratrol as a cardioprotective agent has been questioned [20], as it is yet unknown whether it is possible to absorb the necessary therapeutic amounts of resveratrol by drinking moderate amounts of wine [21]. While grape-derived resveratrol, for example, is marketed as functional ingredient and dietary supplement, it should be noted that definitive conclusions on its efficacy as a therapeutic agent are missing.

Some of the studies on the benefits of moderate wine consumption may have been limited by the possible presence of socio-economic, and other individual confounders [22]. Research on blue zones has suggested moderate wine consumption as one of the nine lifestyle behaviours found in populations worldwide that are known for their long lifespan and healthy ageing [23]. These findings suggest that moderate wine consumption may be associated with an increased longevity and a decreased risk for certain chronic diseases with for example an antioxidant, anti-inflammatory, anti-proliferative and/or anti-angiogenic basis. However, it would be incorrect to conclude that moderate wine consumption without the presence of the accompanying lifestyle behaviour determinants (such as not smoking, undertaking regular physical activity, eating a Mediterranean-style diet, not being overweight or obese, having a sense of purpose, and adequate stress management) could show the same associations. Despite these limitations and ongoing uncertainty, it may be cautiously concluded that moderate wine consumption, in addition to the positive sensory effects experienced through its aroma and taste, may have some positive health effects as well. There are, however, some potential negative health effects associated with heavier wine consumption, whether regular heavy of "binge", which are well documented. Whether the potential positive effects consistently outweigh the potential negative effects is uncertain. Therefore, it may be beneficial for the wine consumer, to consider ways to avoid or reduce the potential negative effects of their consumption.

2. Less is more

The reduction of alcoholic strength in beverages has been proposed as one strategy to reduce the harmful use of alcohol. A report by the World Cancer Research Fund in 2007 stated that a decrease in alcohol content from 14.2 to 10% would reduce the risk of breast and bowel cancer by 7% [24]. This was followed by responses from other organisations, such as the Australian National Preventative Health Taskforce. In 2008, the latter recommended the production of low-alcohol products and suggested changes to the taxation regime to encourage a shift towards the supply of lower-risk products [25, 26]. Lately, low-alcohol beverages have increased in popularity and take up a growing portion of the market. Light beers, beer that is reduced in alcohol (ethanol) content or in carbohydrate content and hence calories, have known a great success on the market worldwide, with an increased global consumption by 47.2% in volume between 2006 and 2011 [27, 28]. A first explanation for this interest in low-alcohol products could be that consumers aim to reduce their alcohol consumption because, consumption, drunkenness and intoxication in particular, may be socially unacceptable, or because they still want to be able to drive [29]. In that way, low-alcohol beverages may be perceived as a response to the alcohol-related control policies that have been adopted in many countries worldwide [27, 30, 31].

An increased health consciousness among consumers may be a second explanation for the growing interest in low-alcohol beverages [24, 32]. As health promotion efforts continue to raise awareness about the increased prevalence of dietary-related diseases, consumers may perceive low-alcohol beverages as a healthier alternative to accompany their healthy diet and lifestyle [29, 33]. In line with this, Meillon et al. [34] and Thompson and Thompson [35] found that people were motivated to drink low-alcohol beverages for calorie and weight management and perceived low-alcohol beverages as an alternative to standard alcoholic beverages.

While not at the same pace as beer, wines with a reduced alcohol content have been growing on the marketplace as well. In the UK and Germany, major supermarket chains, such as Tesco and Aldi, sell several reduced and de-alcoholised wines [36]. Recently, also the Marks and Spencer Group UK launched a new South African 5.5% wine [37], and de-alcoholised wines produced by a winery in the Hunter Valley, Australia [38].

3. What is low-alcohol wine?

Wines with reduced alcohol content are generally classified as specified in **Figure 1**. It should be noted that this classification is not explicit and varies between countries and the applicable legislations [39, 40]. For example, Standard 2.7.1 of the Australian New Zealand Food Standards Code states that an

Figure 1.
Classification of wines with reduced alcohol content [39, 40].

alcoholic beverage which contains more than 1.15% alcohol by volume must not be represented as a low alcohol beverage, while as of December 2018, the United Kingdom's Department of Health and Social Care in its Low Alcohol Descriptors Guidance states that low alcohol drinks are those of 1.2% alcohol by volume or less [41]. Legislation around the taxation of wine products varies between countries as well. Whereas some countries apply a fixed duty fee, in other countries such as the UK, the amount of duty payable depends on the strength of the wine [42]. This can make low-alcohol wine products particularly attractive in the marketplace, as they would have an important financial benefit as compared to standard strength products.

4. Production methods

The amount of sugar determines the ethanol concentration of the initial wine, such that grape berries with a lower sugar concentration that produces a lower ethanol concentration initial wine. A reduction in grape berries sugar concentration can be achieved by various viticultural techniques such as reducing the leaf area of the grapevine. The ethanol concentration in wine can also be manipulated before, during and after fermentation by particular winemaking practices, such as blending of grape juices and musts, by choosing a low ethanol-producing yeast, or post-fermentation by blending with low strength juice and the physical removal of alcohol through distillation or membrane-based technologies. Winemaking practices are subject to legislations and local and destination market regulations need to be considered. For example, this latter practice is regulated by Commission Regulation 2009 ((EC) No. 606/2009), which states that the partial removal of alcohol is allowed using physical separation techniques up to a maximum of 2% relative to the original alcohol content.

The different viticultural and winemaking techniques to reduce the ethanol concentration of wine are summarized and shown in **Table 1** and additional information can be sourced from [26] "Controlling the highs and lows of alcohol in wine" [43], "Production technologies for reduced alcoholic wines" and [44] "Production of Low-Alcohol Beverages: Current Status and Perspectives", and [45] "Microbiological strategies to produce beer and wine with reduced ethanol concentration".

Viticultural practices	**Winemaking practices**	
	Pre-fermentation	**Post-fermentation**
Reducing leaf area—the rate of sugar accumulation in berries is primarily determined by the ratio of leaf area to fruit weight (LA/FW). A relatively high LA/FW may cause the sugar concentration to reach unacceptably high levels by the time that flavour or phenolic ripeness is judged to be optimal. Therefore, a reduction of leaf area after fruit set may lead to better synchronisation of sugar and flavour/phenolic ripening, and thus a lower alcohol concentration in the resultant wine	• **Blending** Grape musts with a high sugar concentration can be blended with low strength juice (LSJ) or condensate within the constraints of wine regulations • **Enzyme additions** The enzyme glucose oxidase (GOX) from the fungus *Aspergillus niger* catalyses the conversion of glucose into gluconic acid and hydrogen peroxide. The addition of commercial preparations of the enzyme to grape juice prior fermentation has been shown to decrease the ethanol concentration in the resultant wine by 0.7% v/v compared to untreated wines. • **Fermenter design** Aeration and higher fermentation temperatures may lower alcohol concentrations in wine. Consequently, open top fermentation has been shown to yield lower alcohol concentrations after fermentation. • **Wine yeast** Commercial wine yeasts generally do not show significant variation in the amount of ethanol yielded in wine following fermentation. The yeast strain AWRI 796 has been shown, however, in some laboratory-scale trials, to yield lower ethanol concentrations than certain other commercial wine strains. For example, compared to EC1118, AWRI 796 delivered a reduction of ethanol of approximately 0.4% v/v	• **Physical removal of grape sugar or wine alcohol** Engineering options for precisely reducing sugar content of juice and alcohol concentration in wine include membrane-based systems (such as reverse osmosis and evaporative perstraction), vacuum distillation and spinning cone distillation. • **Loss of alcohol by evaporation** During barrel maturation, both water and ethanol in the wine evaporate. The ethanol concentration slowly increases in dry cellars as water evaporates faster than ethanol in this environment. Conversely, in cellars with a relative humidity over 70%, the ethanol concentration slowly decreases over time. Alcohol concentration was reported to drop by 0.2% v/v when barrels were stored for 12 months at 15°C with relative humidity over 90%

Adapted from [26, 43–45].

Table 1.
A summary of the different viticultural and winemaking techniques to potentially reduce the ethanol concentration of the final wine.

5. The low alcohol wine consumer

5.1 Consumer interest

In 2000, a review by Pickering, concluded that dealcoholized, low- and reduced-alcohol wine (DLRAW) performed well below predictions in the marketplace [39].

Ongoing limitations in sensory quality, promotional issues, and a low level of awareness of the improvements in quality based on innovations in production methods, were suggested as potential barriers for market success [39]. Additionally, a 'snobbish' attitude within both wine consumers and producers, was suggested as a possible reason for the low acceptance of low alcohol wine. Therefore, Pickering [39] described the following strategies to grow consumer interest: efforts to increase awareness of and familiarity with the products, advocacy by industry opinion leaders to improve perceived credibility and consumer acceptance, and sustained promotion and advertising campaigns to promote DLRAW. Since then, varying findings on consumer interest have been reported. An Australian survey conducted in 2010 showed 6–8% consumer interest [46] whereas another Australian survey in 2013, found an acceptance of low alcohol wine of 16% [40]. This study by Saliba et al. [40], found that consumer acceptance increases to 40% if taste were to be the same as for standard wine products. According to a study on consumer metrics in the UK, a practical desire to keep buying cheap wines, health consciousness, taste and staying in control, were the main purchase decisions for wines with a strength lower than 11% [29]. Non-availability of the products, lower quality perceptions, taste issues, lack of awareness, lack of alcohol's feel effect and absence of a lower alcohol drinking occasion, were described as the main barriers to buying reduced alcohol wine (<11%) [29].

A study by Stockley et al. [32], showed that changes in wine consumption behaviour are most influenced by health. Wine Australia stated in 2017 in 'Global Drinking Trends' [47] that an increased consciousness about the risks of excessive alcohol consumption in combination with the general wellness trend among consumers, makes them choose beverages that are perceived to be healthier alternatives, such as wine, and leads to a preference to drink less but better. Current consumer groups choose high-quality, unique, and authentic brands and flavours, and this "premiumisation" trend is also seen in the global alcoholic drinks market [47]. A non-peer-reviewed report by a marketing company from 2016 suggests increasing consumer demand for low-alcohol wine and mentions high acceptance in Germany and big growth potentials in the US and Canada [48].

5.2 Consumer profile

Research has shown that the more knowledge consumers have and the more frequently they consume wine, the less likely they are to appreciate the sensory properties of alcohol-reduced wine [34, 49]. Meillon et al. [34] found that wine professionals did not like the sensory properties of reduced alcohol wines, whereas consumer likings were less clear and masked a strong segmentation [34]. Meillon et al. [34] suggested that these findings can be explained by the theory of mere exposure [50], stating that familiarity with a product makes a consumer more likely to develop specific preferences concerning that product. Consequently, frequent wine consumers may be more likely to miss the higher alcohol level simply because that is what they are most acquainted with. Meillon et al. [49] concluded that the fewer bottles owned in the cellar, the more likely the consumers were to like the sensory properties of the reduced alcohol wine [51].

Mainly women and the younger generation (age 18–39) show interest in lower alcohol wines, according to research by Prowein which was conducted in 2012 across the US, China, Germany and the UK [52]. Similarly, an Australian study showed that females and those who drink wine with food were the consumer groups that were most likely to purchase low-alcohol wine [53]. The buyers profile in the UK was described by Bruwer et al. [29] as females, Millennials and Baby Boomers, mid to low income, who drink wine about once a week and have a medium to low level of involvement with wine [29].

5.3 Acceptability of low-alcohol wine

5.3.1 Taste

Taste is one of the most important factors in wine consumption decision making [53]. A lack of or an unfamiliar taste may be important drawbacks for low-alcohol beverages [27]. However, interestingly, experimental research has shown that, when unaware of the alcohol percentage, lay consumers were unable to discriminate between alcohol-free and alcohol-containing beer or between regular-strength and lower-strength beer [54, 55]. This is in line with research by Masson et al. [56] who found similar taste ratings for low-alcohol wine versus standard wine, however, before tasting, the subjects expected lower quality for the reduced alcohol wine. Studies have shown that low alcohol wine seems to be associated with a lower expected quality [49, 56]. Meillon et al. [49] found that a priori having tasted partially alcohol reduced wine expectations were negative, based on reasons such as; loss of tradition and authenticity of wine, worry about the quality of the wine and wine preservation, and a feeling of tempering with wine. Experimental studies with blind tasting, however, showed similar liking rates for standard wines and wines with reduced alcohol content [51, 56, 57]. It is therefore interesting to investigate the extent to which the lower expected quality based on the label or information cue affects taste ratings, i.e., are the taste ratings different when participants are aware of the reduced alcohol content versus when they are unaware of the alcohol strength. The study by Masson et al. [56], found that the expected quality for low-alcohol labelled wine was significantly lower as compared to a standard wine, however, in that same study, taste ratings of low-alcohol wines (9% alcohol) did not differ from ratings of standard wines (13% alcohol), neither under blind condition nor if participants were aware they were consuming low-alcohol wine [56]. In a recent wine study by Bucher et al., participants were randomly assigned to one of three conditions; a low-alcohol (8%) condition, a blinded low-alcohol (8%) condition, or a standard condition Sauvignon Blanc (12.5%). Participants in all three conditions reported similar results for liking of the wine and pleasantness to drink [57]. It should be noted that a further reduction in strength, i.e., <8% may have different results. A study by Meillon et al. [51], found that a reduction below or equal to −4% (resulting in a wine of 9.5%) had no significant impact on wine liking, however when the alcohol reduction reached the value of −5.5% (resulting in a wine of 7.9%), it was significantly disliked by consumers [51].

5.3.2 Price

Next to taste, price is another main determinant in wine purchase decision-making. Willingness to pay for non-standard wines with health benefits varied in previous literature. Some studies found that consumers are willing to pay more for wine made with grapes enriched in resveratrol [58]. However, others found that consumers expect lower alcohol wines to be cheaper compared to standard wine [29, 57]. As price has been described as an indicator of quality, a perceived lower quality associated with alcohol reduced wines could be a possible explanation for these findings [59]. Another explanation could be that consumers may not be aware of the additional steps, and therefore additional costs, involved in the production of low alcohol wine. Adequate information for the consumers about the process and technology involved in the production of low alcohol wines may help to establish the market potential for these wines.

5.3.3 Cultural differences in acceptance

Perception willingness-to-pay, and overall acceptance of low-alcohol wines may, however, differ between countries. A study by d'Hauteville [60] showed higher acceptance rates for the UK (27%) and Germany (20%) than for France (12% acceptance, 61% rejection). An explanation for these findings could be that the rich culture and tradition, and level of involvement with wine in France, is associated with a lower acceptance or openness towards changes in the production methods and taste of wine [48]. Another possible explanation could be that the great success of low alcohol beer in countries like the UK, made the consumers in those countries more accepting of low alcohol beverages in general, and as such, has paved the way for wine [61]. Furthermore, it has been suggested that consumers in Germany might be more willing to accept low-alcohol wine because many traditional German wine styles are naturally low in alcohol [48]. Chan et al. [62] investigated consumer preferences and perceptions on dealcoholized wine in Malaysia and found that 20% of the respondents knew about the product but only 9% consumed it. The study aimed to analyse the Malaysian consumer's attitude and how the religious regulation status affect's this. The low (9%) consumption level was explained by the finding that most respondents (90%) perceived dealcoholized wine as not halal [62]. Additionally, a study by Yoo et al. [63] showed that Koreans were more likely to choose wine based on health enhancement properties compared with Australians. Further research on the differences between countries and cultures for acceptability of wines with reduced alcohol content is needed.

6. Does low alcohol wine reduce alcohol consumption?

Offering lower alcohol wine could result in a significant decrease in total alcohol consumption. However, low alcohol labelling may not only influence product selection but also consumed amounts. Previous studies in the area of food research showed that labels on food products, such as "light" or "low-fat" could trigger an increase in consumption [64, 65]. A study by Provencher et al. [64] found that people consume more of a product when they perceived the product as healthy. Similarly, there might be a risk that people overcompensate, if they consume low alcohol wine. The evidence on this is mixed. An experimental study by Vasiljevic et al. found that the total amount of drink consumed increased as the alcohol strength on the label decreased [66]. However, two other studies showed that low alcohol labelling did not increase consumption or intended consumption. In a study by Bucher et al. [57], participants consumed equal amounts of wine, whether that be standard wine (12.5% Sauvignon Blanc) or wine with reduced alcohol content (8% Sauvignon Blanc) [57]. These results are in line with the findings of another recent experimental study that concluded that reducing wine alcohol content had neither physiological nor cognitive influence on the quantities consumed [67]. A study with 1050 wine consumers, which investigated the perceived healthiness of wine on wine consumption patterns, was in line with these findings. Saliba et al. [68] found that those perceiving wine as healthy had a higher frequency but not volume of consumption.

When a beverage with reduced alcohol content is consumed in the same quantity as a standard beverage, the total alcohol consumption is significantly lower. In the study by Bucher et al., those who drank the low-alcohol wine consumed approximately 30% less alcohol as compared to those who drank the standard wine [57]. These findings have important practical implications and suggest that reduced alcohol wine can be an effective strategy to reduce alcohol consumption and therefore decrease alcohol related risks. However, more experimental research in more natural settings is needed to investigate total alcohol consumption if consumers have the choice between a variety of different strength alcoholic wines.

7. Why marketing and labelling are important?

First, as previously mentioned by Pickering [39], and described again by Bruwer et al. [29], lack of product awareness is still an important barrier for uptake of low-alcohol wine consumption and needs to be addressed by adequate promotional activities. Second, research on the low-alcohol wine consumer in the UK found that lower alcohol on its own is not seen as a benefit [29]. Consequently, marketing strategies may need to focus on the benefits of the product rather than the literal credentials [29]. Research on consumer behaviour does suggest that consumers value the link between beverage intake and health status, and health claims may influence beverage choice [32, 69, 70]. However, even though the interest in alcohol-reduced beverages has increased, producers and marketers should carefully consider marketing strategies around alcohol-reduced wine. Experimental studies have shown that reduced alcohol claims can reduce product appeal [71] and may negatively impact expected quality [56]. A study by Masson et al. with French consumers found that the expected quality for "low-alcohol" labelled wine was significantly lower as compared to a standard wine [56]. However, in their tasting test, ratings of low-alcohol wines (9% alcohol) did not differ from ratings of standard wines (13% alcohol), neither under blind condition nor if participants were aware, they were consuming low-alcohol wine [56]. In agreement with this, the results of a recent tasting experiment with Australian consumers suggest that participants were willing to pay less for low alcohol wine [57], although quality and taste ratings were equal between the 'low alcohol wine' and the standard wine, which was labelled as 'new wine'. Therefore, a third marketing strategy may be to inform the consumers about the recent improvements in production methods and sensory properties of low-alcohol wine innovations. Furthermore, informing the consumer about the additional steps and costs involved in the production process of low-alcohol wine may have a positive effect on consumer's willingness to pay, and could be a fourth marketing strategy to consider.

A fifth marketing strategy relates to the descriptor or terminology used on the label. Terminologies for low-alcohol wine and related beverages may be tightly regulated by country specific food standards, however it might be critical to carefully consider the wording around low-alcohol wine on labels and its impact on consumer behaviour. Altered wording used on the label, e.g., light, de-alcoholised, reduced alcohol or low alcohol might evoke different consumer perceptions and reactions [71]. Vasiljevic et al. found that the terminology also has an effect on perceived strength [72]. More insight on these influences as well as on country specific differences is needed. Finally, it may be opportune to consider the way low alcohol beverages are promoted as a new product or product category. Reducing ethanol content in beverages can be an effective strategy to reduce the harmful use of alcohol [73]. However, the way the product is promoted may impact its potential positive effects. Rehm et al. described different potential mechanisms for how reduction of alcoholic strength could affect harmful use of alcohol; by replacing standard alcoholic beverages without increasing the quantity of liquid consumed; by current drinkers choosing no alcohol alternatives for part of the time and in that way reducing the average amount of alcohol consumed; or by initiating alcohol use in current abstainers [73]. Vasiljevic et al. investigated the marketing messages accompanying online selling of low/er and regular strength wine and beer products in the UK and concluded that low/er strength beverages appear to be marketed not as substitutes for higher strength products but as ones that can be consumed on additional occasions with an added implication of healthiness [74]. For reduced alcohol beverages to reduce the harmful effects of alcohol consumption, it may thus be essential to carefully consider marketing messages and product promotion.

8. Conclusions

Reduced alcohol wine may be a strategy to reduce total alcohol consumption and alcohol related harm. However, recent literature suggests that, despite good quality ratings when tasting the wines, people may still tend to perceive wines with reduced alcohol content as a lower quality product and may therefore wish to pay less for them. Consumers might falsely assume that wines with lower strength would be cheaper to produce or benefit from tax incentives (which is the case in some countries). To circumvent negative consequences of low alcohol labelling on perception, a few strategies have been suggested in the literature. These include increasing consumer knowledge related to alcohol reduction processes and increasing consumer awareness about high quality low-alcohol wines with appealing sensory properties. Media campaigns and specific awards or recognitions for lower strength wines might be helpful to promote consumer awareness of high quality low-alcohol wine products. However, research on this is warranted.

Acknowledgements

T.B. and K.D. are supported by the School of Health Sciences and the Faculty of Health and Medicine, The University of Newcastle and by Faculty of Science, School of Environmental and Life Sciences (SELS), The University of Newcastle.

C.S.S. is an adjunct Senior Lecturer in the School of Agriculture, Food and Wine at The University of Adelaide.

Conflict of interest

The authors declare no conflict of interest.

Notes/thanks/other declarations

This research received no external funding.

Author details

Tamara Bucher[1,2]*, Kristine Deroover[2] and Creina Stockley[3]

1 Faculty of Science, School of Environmental and Life Sciences (SELS), The University of Newcastle, Australia

2 Priority Research Centre for Physical Activity and Nutrition, The University of Newcastle, Callaghan, NSW, Australia

3 Stockley Health and Regulatory Solutions, Malvern, SA, Australia

*Address all correspondence to: tamara.bucher@newcastle.edu.au

References

[1] Stockwell T, Zhao J, Panwar S, Roemer A, Naimi T, Chikritzhs T. Do "moderate" drinkers have reduced mortality risk? A systematic review and meta-analysis of alcohol consumption and all-cause mortality. Journal of Studies on Alcohol and Drugs. 2016;**77**(2):185-198

[2] WHO. Global Strategy to Reduce the Harmful Use of Alcohol. Geneva, Switzerland: World Health Organization. 2010. Available from: http://www.who.int/substance_abuse/publications/global_strategy_reduce_harmful_use_alcohol/en/

[3] WHO. Global Status Report on Alcohol and Health. Geneva, Switzerland: World Health Organization. 2014. Available from: http://www.who.int/substance_abuse/publications/alcohol_2014/en/

[4] WHO. Global Status Report on Alcohol and Health. Geneva, Switzerland: World Health Organization. 2018. Available from: http://www.who.int/substance_abuse/publications/alcohol_2018/en/

[5] Griswold MG, Fullman N, Hawley C, Arian N, Zimsen SR, Tymeson HD, et al. Alcohol use and burden for 195 countries and territories, 1990-2016: A systematic analysis for the global burden of disease study 2016. The Lancet. 2018;**392**(10152):1015-1035

[6] Roerecke M, Rehm J. Alcohol consumption, drinking patterns, and ischemic heart disease: A narrative review of meta-analyses and a systematic review and meta-analysis of the impact of heavy drinking occasions on risk for moderate drinkers. BMC Medicine. 2014;**12**:182

[7] Larsson SC, Wallin A, Wolk A, Markus HS. Differing association of alcohol consumption with different stroke types: A systematic review and meta-analysis. BMC Medicine. 2016;**14**(1):178

[8] Li XH, Yu FF, Zhou YH, He J. Association between alcohol consumption and the risk of incident type 2 diabetes: A systematic review and dose-response meta-analysis. The American Journal of Clinical Nutrition. 2016;**103**(3):818-829

[9] Kunzmann AT, Coleman HG, Huang WY, Berndt SI. The association of lifetime alcohol use with mortality and cancer risk in older adults: A cohort study. PLoS Medicine. 2018;**15**(6):e1002585

[10] Saleem TM, Basha SD. Red wine: A drink to your heart. Journal of Cardiovascular Disease Research. 2010;**1**(4):171-176

[11] NHMRC. Available from: https://nhmrc.gov.au/health-advice/alcohol

[12] Brien SE, Ronksley PE, Turner BJ, Mukamal KJ, Ghali WA. Effect of alcohol consumption on biological markers associated with risk of coronary heart disease: Systematic review and meta-analysis of interventional studies. British Medical Journal. 2011;**342**:d636

[13] Stockley C. Therapeutic value of wine: A clinical and scientific perspective. In: Joshi VK, editor. Handbook of Enology: Principles, Practices and Recent Innovations. Vol. 1: Introduction to Vine and Wine. New Delhi: Asiatech Publishers, Inc; 2011. pp. 146-208

[14] Burns J, Crozier A, Lean M. Alcohol consumption and mortality: Is wine different from other alcoholic beverages? Nutrition, Metabolism, and Cardiovascular Diseases. 2001;**11**(4):249-258

[15] Snopek L, Mlcek J, Sochorova L, Baron M, Hlavacova I, Jurikova T, et al. Contribution of red wine consumption to human health protection. Molecules. 2018;**23**(7):1684

[16] Djoussé L, Ellison RC, McLennan CE, Cupples LA, Lipinska I, Tofler GH, et al. Acute effects of a high-fat meal with and without red wine on endothelial function in healthy subjects. The American Journal of Cardiology. 1999;**84**(6):660-664

[17] Boban M, Stockley C, Teissedre P-L, Restani P, Fradera U, Stein-Hammer C, et al. Drinking pattern of wine and effects on human health: Why should we drink moderately and with meals? Food & Function. 2016;**7**(7):2937-2942

[18] Catalgol B, Batirel S, Taga Y, Ozer NK. Resveratrol: French paradox revisited. Frontiers in Pharmacology. 2012;**3**:141

[19] Cancer.gov.au. Antioxidants—Fact Sheet. Available from: https://www.cancer.gov/about-cancer/causes-prevention/risk/diet/antioxidants-fact-sheet

[20] Xiang L, Xiao L, Wang Y, Li H, Huang Z, He X. Health benefits of wine: Don't expect resveratrol too much. Food Chemistry. 2014;**156**:258-263

[21] Weiskirchen S, Weiskirchen R. Resveratrol: How much wine do you have to drink to stay healthy? Advances in Nutrition. 2016;**7**(4):706-718

[22] Lindberg ML, Amsterdam EA. Alcohol, wine, and cardiovascular health. Clinical Cardiology: An International Indexed and Peer-Reviewed Journal for Advances in the Treatment of Cardiovascular Disease. 2008;**31**(8):347-351

[23] Buettner D, Skemp S. Blue zones: Lessons from the world's longest lived. American Journal of Lifestyle Medicine. 2016;**10**(5):318-321

[24] North West Public Health Observatory. Can Promotion of Lower Alcohol Products Help Reduce Alcohol Consumption? 2012. Available from: http://www.cph.org.uk/wp-content/uploads/2012/08/can-promotion-of-lower-alcohol-products-help-reduce-consumption--a-rapid-review.pdf

[25] Australian National Preventive Health Taskforce. Preventing Alcohol-Related Harm in Australia: A Window of Opportunity. Technical Report No 3. Commonwealth of Australia. 2008

[26] Stockley C, Varela C, Coulter A, Dry P, Francis I, Muhlack R, et al. Controlling the Highs and the Lows of Alcohol in Wine. New York, USA: Nova Science Publishers; 2012

[27] Chrysochou P. Drink to get drunk or stay healthy? Exploring consumers' perceptions, motives and preferences for light beer. Food Quality and Preference. 2014;**31**:156-163

[28] Euromonitor. Available from: https://www.euromonitor.com

[29] Bruwer J, Jiranek V, Halstead L, Saliba A. Lower alcohol wines in the UK market: Some baseline consumer behaviour metrics. British Food Journal. 2014;**116**(7):1143-1161

[30] Howley M, Young N. Low-alcohol wines: The consumer's choice? International Journal of Wine Marketing. 1992;**4**(3):45-56

[31] Schaefer JM. On the potential health effects of consuming "non-alcoholic" or "de-alcoholized" beverages. Alcohol. 1987;**4**(2):87-95

[32] Stockley CS, Taylor AW, Montgomerie A, Dal Grande E. Changes in wine consumption are influenced most by health 58; results from a

population survey of South Australians in 2013. International Journal of Wine Research. 2016;**55**:13-22

[33] WHO. Diet, Nutrition and The Prevention of Chronic Diseases: Report of a Joint WHO/FAO Expert Consultation. Geneva, Switzerland: World Health Organization. 2003. Available from: http://www.who.int/nutrition/publications/obesity/WHO_TRS_916/en/

[34] Meillon S, Dugas V, Urbano C, Schlich P. Preference and acceptability of partially dealcoholized white and red wines by consumers and professionals. The American Journal of Enology and Viticulture. 2010;**61**(1):42-52

[35] Thompson NJ, Thompson KE. Reasoned action theory: An application to alcohol-free beer. Journal of Marketing Practice: Applied Marketing Science. 1996;**2**(2):35-48

[36] The Drinks Business. Tesco Launches Low-Alcohol Wine Range to Keep Up With Demand. 2017

[37] Independent. This is a Breakthrough—Marks & Spencer is Selling a Wine With Half the Normal Calorie-Count. 2017

[38] The Drinks Business. McGuigan Launches De-alcoholised Wine Range at M&S. 2018

[39] Pickering GJ. Low-and reduced-alcohol wine: A review. Journal of Wine Research. 2000;**11**(2):129-144

[40] Saliba AJ, Ovington LA, Moran CC. Consumer demand for low-alcohol wine in an Australian sample. International Journal of Wine Research. 2013;**5**(5):1-8

[41] Department of Health and Social Care. Low Alcohol Descriptors Guidance. 2018

[42] Gov.UK. Tax on Shopping Alcohol and Tobacco. Available from: https://www.gov.uk/tax-on-shopping/alcohol-tobacco

[43] Schmidtke LM, Blackman JW, Agboola SO. Production technologies for reduced alcoholic wines. Journal of Food Science. 2012;**77**(1):R25-R41

[44] Liguori L, Russo P, Albanese D, Di Matteo M. Production of low-alcohol beverages: Current status and perspectives. In: Food Processing for Increased Quality and Consumption. San Diego, USA: Elsevier; 2018. pp. 347-382

[45] Varela J, Varela C. Microbiological strategies to produce beer and wine with reduced ethanol concentration. Current Opinion in Biotechnology. 2019;**56**:88-96

[46] Mueller S, Lockshin L, Louviere J. Alcohol in moderation: Market potential for low alcohol wine before and after excise tax. In: 6th AWBR International Conference; 2011

[47] Wine Australia. Global Drinking Trends. 2017

[48] Wine Intelligence. All News, France, Lower Alcohol—Going Down? 2016

[49] Meillon S, Urbano C, Guillot G, Schlich P. Acceptability of partially dealcoholized wines–measuring the impact of sensory and information cues on overall liking in real-life settings. Food Quality and Preference. 2010;**21**(7):763-773

[50] Zajonc RB. Mere exposure: A gateway to the subliminal. Current Directions in Psychological Science. 2001;**10**(6):224-228

[51] Meillon S, Viala D, Medel M, Urbano C, Guillot G, Schlich P. Impact of partial alcohol reduction in Syrah wine on perceived complexity and temporality

of sensations and link with preference. Food Quality and Preference. 2010;**21**(7):732-740

[52] Decanter. Consumers Across Three Continents Prefer Lower Alcohol Wines: Prowein. 2012

[53] Saliba A, Ovington L, Moran CC, Bruwer J. Consumer Attitudes to Low Alcohol Wine: An Australian Sample. Wine Titles Pty Limited. 2013

[54] Segal DS, Stockwell T. Low alcohol alternatives: A promising strategy for reducing alcohol related harm. International Journal of Drug Policy. 2009;**20**(2):183-187

[55] King ES, Heymann H. The effect of reduced alcohol on the sensory profiles and consumer preferences of white wine. Journal of Sensory Studies. 2014;**29**(1):33-42

[56] Masson J, Aurier P, D'hauteville F. Effects of non-sensory cues on perceived quality: The case of low-alcohol wine. International Journal of Wine Business Research. 2008;**20**(3):215-229

[57] Bucher T, Wilczynska M, Dohle S. Consumer perception and behavior related to low alcohol wine: Do people overcompensate? In: ISBNPA Conference; 3-6 June 2018; Hong Kong. 2018

[58] Barreiro-Hurlé J, Colombo S, Cantos-Villar E. Is there a market for functional wines? Consumer preferences and willingness to pay for resveratrol-enriched red wine. Food Quality and Preference. 2008;**19**(4):360-371

[59] Leavitt HJ. A note on some experimental findings about the meanings of price. The Journal of Business. 1954;**27**(3):205-210

[60] d'Hauteville F. Consumer acceptance of low alcohol wines. International Journal of Wine Marketing. 1994;**6**(1):35-48

[61] Euromonitor. Country Sector Briefing: Beer—United Kingdom. 2010. Available from: www.euromonitor.com

[62] Chan SM, Adzahan NM, Ab Karim MS, Karim R, Lasekan O, Regenstein JM. Consumer preferences and perceptions on dealcoholised wine. Journal of Food Products Marketing. 2012;**18**(1):65-77

[63] Yoo YJ, Saliba AJ, MacDonald JB, Prenzler PD, Ryan D. A cross-cultural study of wine consumers with respect to health benefits of wine. Food Quality and Preference. 2013;**28**(2):531-538

[64] Provencher V, Polivy J, Herman CP. Perceived healthiness of food. If it's healthy, you can eat more! Appetite. 2009;**52**(2):340-344

[65] Wansink B, Chandon P. Can "low-fat" nutrition labels lead to obesity? Journal of Marketing Research. 2006;**43**(4):605-617

[66] Vasiljevic M, Couturier D-L, Frings D, Moss AC, Albery IP, Marteau TM. Impact of lower strength alcohol labelling on consumption: A randomized controlled trial. Health Psychology. 2018;**37**(7):658-667

[67] Masson J, Aurier P. Modifying wine alcohol content: Sensory and non-sensory impacts on quantities consumed. International Journal of Entrepreneurship and Small Business. 2017;**32**(1-2):102-117

[68] Saliba AJ, Moran CC. The influence of perceived healthiness on wine consumption patterns. Food Quality and Preference. 2010;**21**(7):692-696

[69] Samoggia A. Wine and health: Faraway concepts? British Food Journal. 2016;**118**(4):946-960

[70] Higgins LM, Llanos E. A healthy indulgence? Wine consumers and the health benefits of wine. Wine Economics and Policy. 2015;**4**(1):3-11

[71] Vasiljevic M, Couturier D-L, Marteau TM. Impact on product appeal of labeling wine and beer with (a) lower strength alcohol verbal descriptors and (b) percent alcohol by volume (% ABV): An experimental study. Psychology of Addictive Behaviors: Journal of the Society of Psychologists in Addictive Behaviors. 2018;**32**(7):779-791

[72] Vasiljevic M, Couturier DL, Marteau TM. Impact of low alcohol verbal descriptors on perceived strength: An experimental study. British Journal of Health Psychology. 2018;**23**(1):38-67

[73] Rehm J, Lachenmeier DW, Llopis EJ, Imtiaz S, Anderson P. Evidence of reducing ethanol content in beverages to reduce harmful use of alcohol. The Lancet Gastroenterology and Hepatology. 2016;**1**(1):78-83

[74] Vasiljevic M, Coulter L, Petticrew M, Marteau TM. Marketing messages accompanying online selling of low/er and regular strength wine and beer products in the UK: A content analysis. BMC Public Health. 2018;**18**(1):147